COMPUTER-AIDED NETWORK DESIGN

COMPUTER-AIDED NETWORK DESIGN

Revised Edition

Donald A. Calahan
Professor of Electrical Engineering
University of Michigan

New York St. Louis San Francisco Düsseldorf Johannesburg Kuala Lumpur London
Mexico Montreal New Delhi Panama Rio de Janeiro Singapore Sydney Toronto

Library of Congress Cataloging in Publication Data

Calahan, Donald Albert, 1935-
Computer-Aided Network Design.

Includes bibliographical references.
1. Electronic data processing—Electric networks. I. Title.
TK454.2.C35 621.319'2'02854 78-39800
ISBN 07-009601-5

COMPUTER-AIDED NETWORK DESIGN

3 4 5 6 7 8 9 0 KPKP 7 9 8 7 6 5 4

This book was set by Scripta Technica, Inc., and printed and bound by Kingsport Press, Inc. Cover design by Nicholas Krenitsky. The editors were Charles R. Wade and David Damstra. Thomas J. LoPinto supervised production.

to Martha
for her wonderful patience

Contents

Preface

The spectrum of interests represented in the field termed "Computer-Aided Network Design" is characterized by the following two rather extreme viewpoints:

(1) the design viewpoint, where the motivation is to achieve a prescribed circuit function and computer assistance is sought only when intuitive design methods fail to yield an acceptable design;
(2) the algorithmic viewpoint, where equation formulation and numerical solution methods are studied for their potential application to broad classes of simulation and design problems.

With the power and user-orientation of problem-oriented languages now being developed for circuit analysis, most design needs can be accommodated without serious concern for the computing algorithms involved. Indeed, if we consider that an increasing share of sophisticated circuit design is being performed by integrated circuit manufacturers, the need for formal instruction in circuit analysis and design algorithms to satisfy the design viewpoint (1) above is open to serious question.

As a source of model problems for learning digital-computer-based problem solving, however, circuit design offers the following advantages.

(1) Elementary circuit analysis - on which most of this text is based - is a familiar or easily learned discipline, not requiring special

mathematics which can mask principles of equation formulation and numerical solution.

(2) Most important numerical methods can be related to practical circuit design problems. These include solution of simultaneous linear and nonlinear algebraic equations, solution of ordinary differential equations, function minimization, and approximation techniques. The major exception is the numerical solution of partial differential equations, which is at present a strongly problem-dependent process.

(3) The electrical network is an excellent analog of many physical, biological, and natural (ecological) systems. The "through" and "across" variables - current and voltage - appear in easily identifiable forms in structures, continuum mechanics, resource and transportation distribution systems, and the communication system of the human body, to name a few. The equation formulation procedures for such systems is remarkably similar.

(4) The study of the digital simultation and automatic design of large static and dynamic systems, such as networks, is a natural complement to the systems theoretic design of small, approximate models of these same systems. The dimensionality and ill-conditioning problems which accompany analysis of larger systems are foreign to these traditional disciplines.

(5) Most of the solution methods used in network simulation are based on general numerical analysis methods. However, efficient implementation of these methods for problem solving often requires data handling techniques which are not usually taught in numerical analysis courses. Also, formal numerical analysis procedures can often be explained in terms of network models, thus exploiting the insight and natural interests of the reader.

From an instructional viewpoint, a major attraction of computer-aided network design is the ease with which the student can combine elementary formulation and solution methods to produce his own analysis and design programs. These programs can be made quite user-oriented, similar to the popular problem-oriented languages which the student may have used in circuit design courses.

To exploit this motivational opportunity, the material of this text is divided into two parts:

(1) Part I, Chapters 2-7, which
  (a) introduce certain basic numerical methods such as solution of simultaneous linear equations;
  (b) interpret other numerical processes - such as linearization and discretization - in terms of network models;
  (c) examine certain forms of ill-conditioning and their solution;
  (d) present complete computer programs which solve specific model problems;
  (e) present major portions of programs intended to solve general classes of network problems;
  (f) introduce the student to the rudiments of automatic design.

(2) Part II, Chapters 8-12, which
  (a) generalize and extend key algorithms of Part I using relatively more sophisticated mathematical analysis,
  (b) examine techniques of data handling which considerably enhance solution speed.

This division may be considered a natural one for undergraduate/graduate instruction. Alternatively, it provides an opportunity for the undergraduate interested in an application other than electronic circuit design to advance selectively in one of the topics of Part II.

Acknowledgments

The author is indebted jointly to the Air Force Office of Scientific Research (Applied Mathematics Division) for their continuing research support in the subject area of this text, and to the IBM Thomas J. Watson Research Center (Yorktown Heights, New York) for the opportunity to work with the staff of their Mathematical Sciences group. In particular, Drs. Robert Brayton, Fred Gustavson, Gary Hachtel, and Ralph Willoughby have been most challenging and cooperative in developing concepts and algorithms represented in this text. Locally, the assistance of Mr. Thomas Grapes has been of continuing value in the area of sparse matrix methods. Finally, the most important man-machine interface was provided by the secretaries of the Systems Engineering Laboratory. Many thanks to them.

Donald A. Calahan

1

Network Design by Computer

1.1 INTRODUCTION

As the size and sophistication of physical man-made systems has increased, a corresponding need has developed for thorough study and optimal design of these systems prior to construction. The design of electrical circuits is one of the best examples of this escalation, where 20 - 50 component discrete circuits have been replaced by 300 - 3000 component integrated circuits. Further, these circuits are usually produced in large quantities, so that significant effort can be expended in design of a prototype circuit.

A model for a modern computer-aided design (CAD) process is shown in Figure 1.1. Any block or partition of this flow chart can represent an investment in time from minutes to years, depending on the application. Network design fits this model well; moreover, for certain network design problems the model-design-build (B-A-B) loop can be compactly represented on a table-top. For example, the power of the computer can be brought into the laboratory, as indicated by the juxtaposition of graphic terminal and oscilloscope waveform displays in Figure 1.2. Any discrepancy between measured, computed, and ideal responses can be immediately detected and either the computer model or the physical circuit adjusted accordingly.

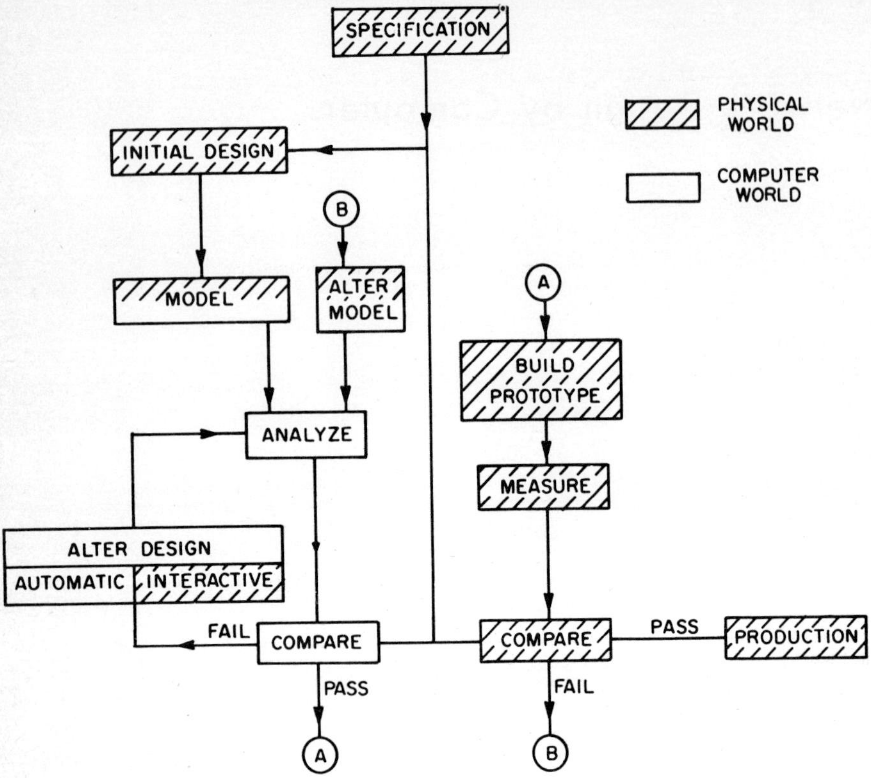
SPECIFICATION
PHYSICAL WORLD
COMPUTER WORLD
INITIAL DESIGN
B
MODEL
ALTER MODEL
A
BUILD PROTOTYPE
ANALYZE
MEASURE
ALTER DESIGN
AUTOMATIC
INTERACTIVE
FAIL
COMPARE
COMPARE
PASS
PRODUCTION
PASS
FAIL
A
B

FIG. 1.1 Computer-Aided Design Model

computek

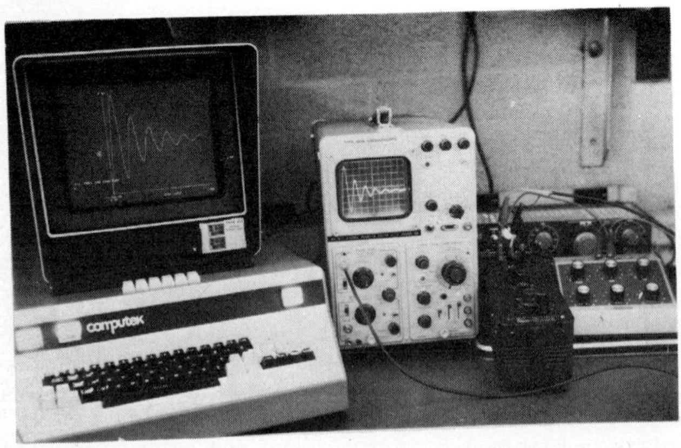

FIG. 1.2 Graphics in the Laboratory

1.2 NETWORK DESIGN

Specializing our interest to network design, another view of the design process is given by Table 1.1, with an emphasis on the computer-aided steps. Our interest will be strictly in Part II, the algorithmic aspects, having to do with problem formulation and numerical solution.

TABLE 1.1 Circuit Design Considerations

I. Circuit Considerations	
Design Specification	DC Transient Frequency Sensitivity
Model	Linear, nonlinear Large signal, small signal DC, dynamic Lumped, distributed Fixed, time varying
II. Algorithmic Considerations	
Equation Formulation	Response analysis (mesh, node, etc.) Sensitivity analysis
Numerical Solution	Matrix methods Iteration methods Integration methods Approximation methods
Automatic Design	Error functions Weighting functions Constraints
III. CAD Implementation	
Program Preparation	New special purpose program Existing general purpose program
Design Mode	Cut-and-try Automatic
Interaction	Batch Teletypewriter Graphical

There are at least two approaches that we can take to the topic. First, we can emphasize "textbook" design procedures, with demonstrations of how the computer can shorten certain computations as a fast slide rule. This approach invariably fails to disclose the full potential of the computer as a design tool, for the following reasons.

(1) The formulation of problems for digital computer solutions is often different than for hand computation. Hand formulation also limits the size of problems that can be considered.

(2) Certain numerical problems can arise with analysis of even small networks and must be expected in analysis of larger networks. Understanding of the origin of these problems and the means of avoiding them requires detailed mathematical analysis, beyond that possible in even an advanced electronics course.

In contrast, we will adopt the ultimate goal of developing general, efficient, and flexible algorithms for design of a variety of classes of networks. We will find that this goal can be achieved without becoming entangled in a nightmare of programming details and that, in fact, a number of elementary network concepts can be easily extended to formulate and solve what may seem to be difficult design problems.

To illustrate the design potential offered by this attention to generality and efficiency, consider the so-called "automatic design" problem illustrated in Figure 1. 3 [1. 1].*

The logic chain displayed in Figure 1. 3c is composed of eight logic circuits. The complexity of each is shown in Figure 1. 3b and the nonlinear dynamic transistor model complexity indicated in Figure 1. 3a. The total problem complexity is given in Table 1. 2.

TABLE 1. 2 Problem Complexity

Nodes	180
Branches	460
Nonlinear elements	112
Differential equations	192
Total equations	1300

The design problem is stated as follows. Given an input $V_{IN}(t)$ and an initial set of element values, adjust R_1, R_2, and R_3 (assumed the same for each circuit) to achieve a desired rise time t_{RISE} and delay time t_{DELAY} in the output $V_{OUT}(t)$ as shown in Figure 1. 4.

This design problem is made amenable to automatic computer design using the methods of Chapters XI and XII by requiring that the output $V_{OUT}(t)$ match a specified output $\hat{V}_{OUT}(t)$; i. e., the quantity

$$P = \int_0^{t_f} (\hat{V}_{OUT}(t) - V_{OUT}(t))^2 \, dt$$

is to be minimized so that P = 0 signifies a precise match.

The computer algorithm adjusts the three resistance values in a series of 36 steps, realizing the final $V_{OUT}(t)$ shown in Figure 1. 4. The value of P in this case is decreased (and the match correspondingly improved) by a factor of 200. The computing time is one hour on IBM 360/91, one of the fastest of modern computers.

1. 3 RESPONSE AND NETWORK CLASSIFICATION

Most network specifications are given in terms of one of these types of responses:

(1) dc (direct current) response; this is the response of a purely resistive network to constant sources such as batteries;

(2) frequency or ac response; this is the response of a network of (linear) resistances, capacitances, etc., to a sinusoidal signal input;

(3) transient response; this is the time domain response of any network.

*Number in bracket points to a reference at the end of each chapter.

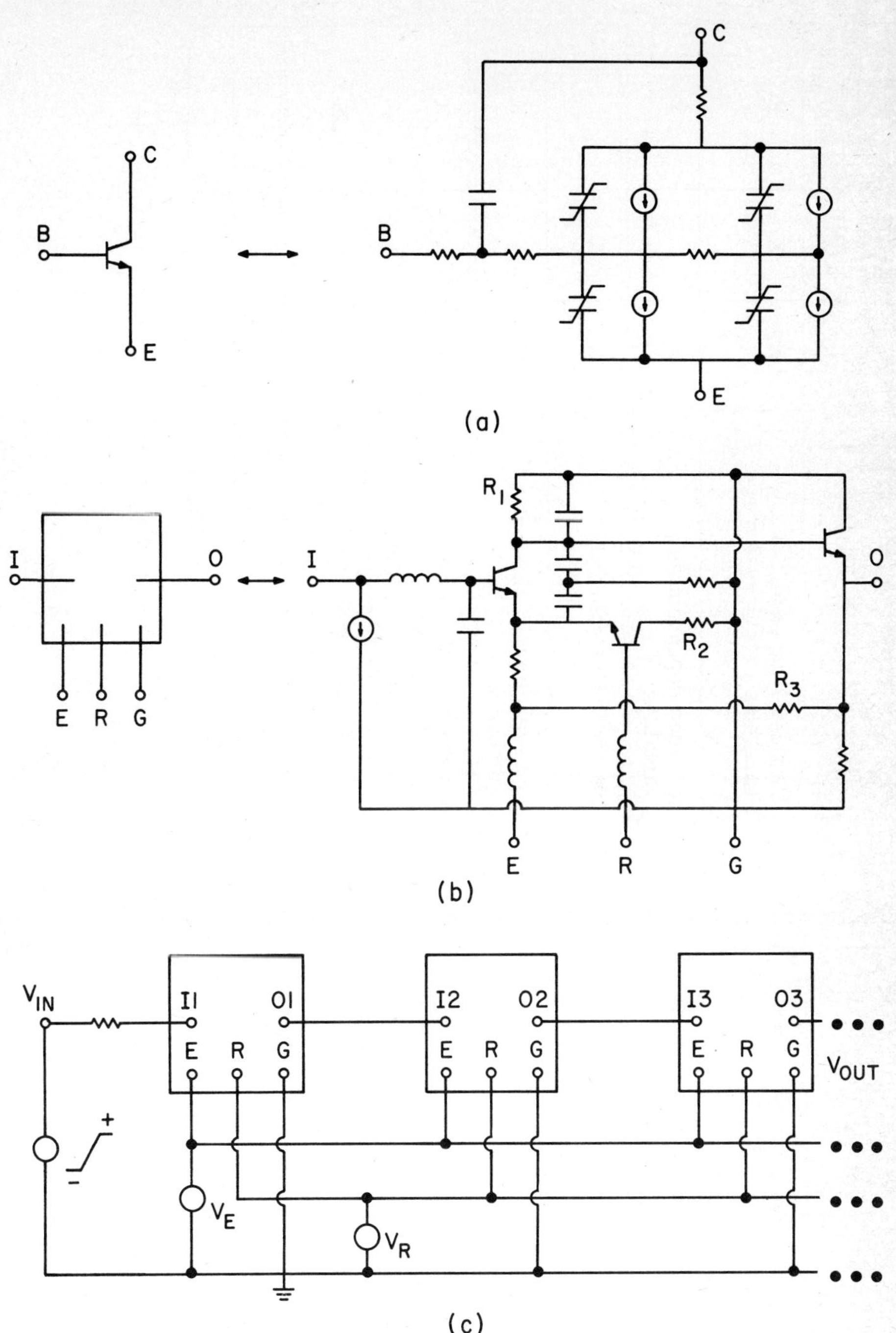

FIG. 1.3 Current Switch-Emitter Follower Circuit Logic Chain
(a) two-dimensional transistor model;
(b) current switch-emitter follower logic stage;
(c) logic chain

TABLE 1.3 Network and Response Classification

Network Type	Property	Component Identification	Types of Response		
			DC	AC	Transient
Linear (Linearized) (Small signal)	Response proportional to excitation	Element values independent of circuit variables (v(t), i(t))	X	X	X
Nonlinear	Response not proportional to excitation	Element values dependent on circuit variables	X		X
Resistive (Memoryless)	No energy storage	Resistors, resistive controlled sources	X	X	
Dynamic (Memory)	Energy storage	Inductors, capacitors, resistors, controlled sources	X	X	X
Fixed (Time-invariant)	RCL fixed network cannot produce energy	Element values constant, or if values vary, they must depend on circuit variables (and so are nonlinear)	X	X	X
Time-varying	RCL time-varying network can produce energy	Element values vary with time, but are independent of excitation			X
Lumped	Network described by algebraic or differential equations				
Distributed	Network described by partial differential equations				

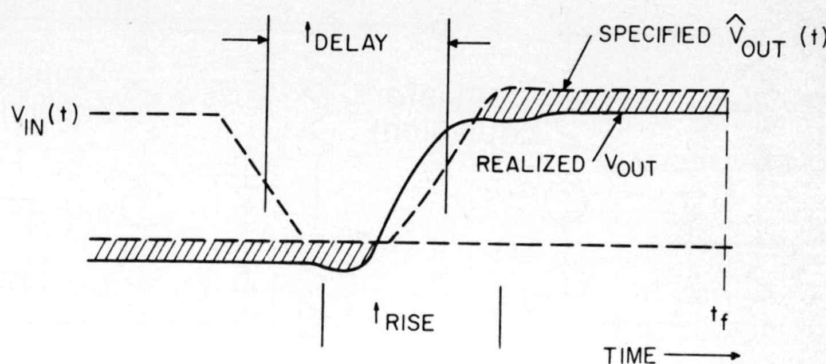

FIG. 1.4 Input-Output Behavior of Logic Chain of Figure 1.3

Networks may similarly be classified, according to the types of elements they contain. These classifications are shown in Table 1.3, together with associated types of responses. These formal distinctions can be interpreted in terms of common circuit analysis problems with the assistance of Table 1.4 and Figure 1.5

1.4 CONCLUSION

The serious reader will be able to formulate and numerically solve the equations for any of the networks of Figure 1.5 by the end of Chapter IV, even though only one of them (Figure 1.5(a)) has a symbolic solution. Aside from this direct goal, however, we might recall from the preface the more important purpose of our study: the approaches to computer solution we will learn turn out to be of significance in a variety of simulation and engineering design problems.

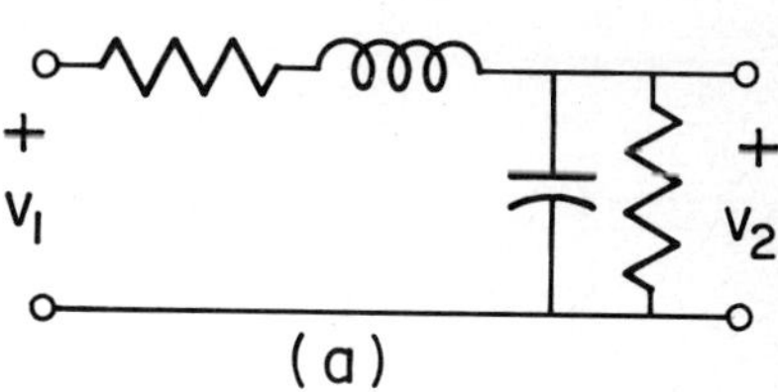

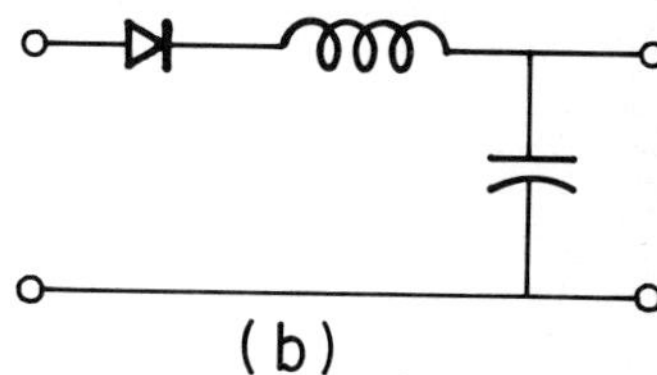

FIG. 1.5 Examples of Common Networks (see Table 1.4)

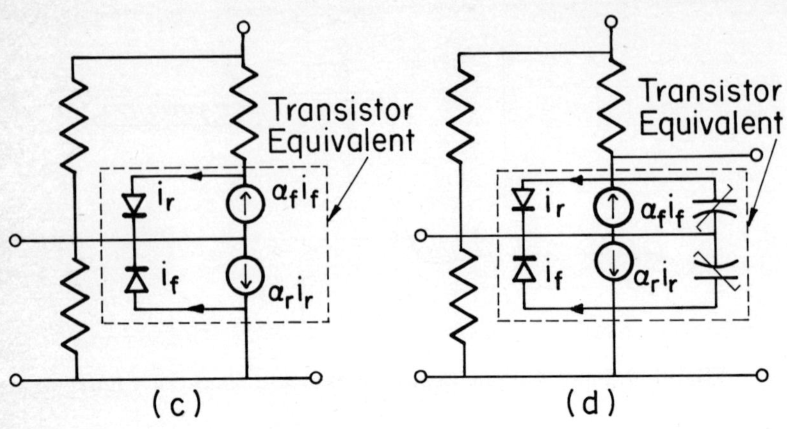

FIG. 1.5 Examples of Common Networks (see Table 1.4) (Continued)

TABLE 1.4 See Figure 1.5

			Type of Analysis (typical)		
Part	Description	Network Type	DC	AC	Trans.
a	LC Filter	Linear, dynamic, lumped, fixed		X	X
b	Rectifier and filter	Nonlinear, dynamic, lumped, fixed			X
c	Model of biased transistor	Nonlinear, resistive, lumped, fixed	X		
d	Switching model of biased transistor	Nonlinear, dynamic, lumped, fixed	X		X

REFERENCES

1.1 Hatchel, G.D., R.K. Brayton, and F.G. Gustavson, "The Sparse Tableau Approach to Network Analysis and Design," Trans. IEEE, Vol. CT-18, No. 1, pp. 101-113; January, 1971.

2

Analysis of Linear Networks

2.1 INTRODUCTION

The need to analyze a linear network is a recurring requirement in computer-aided network analysis. Not only are a majority of the network problems to be solved posed as linear problems; nonlinear resistive and dynamic networks are usually solved by analysis of a sequence of "linearized" networks.

The analysis of such networks can commonly be viewed as a two-stage process: (1) equation formulation and (2) numerical solution. The numerical solution of linear network equations will be of special interest to us. Besides presenting constructive algorithms for equation solution, we shall uncover a number of interesting numerical problems that occur in analysis of practical linear networks. Some relate to efficiency and some to accuracy; some challenge traditional approaches to network analysis, and others exploit these same procedures. In any case, the methodology will be highly suggestive of procedures to be adopted for solution of related problems in later chapters.

2.2 NODE ANALYSIS

Any network analysis problem that we present in this text can be analyzed using node analysis, regardless of whether the network is linear, nonlinear, resistive, dynamic, etc. This perhaps surprising fact should be ample motivation for examining this traditional method of analysis with renewed vigor!

Node equations are formulated using Kirchhoff's Current Law (KCL). As an example, consider the linear resistive network of Figure 2.1. Using the basic branch relationship i = Gv, a summation of currents flowing away from each node yields the following equations.

node 1 $$G_1v_1 + G_2(v_1 - v_2) + G_6(v_1 - v_3) = i_s$$

node 2 $$G_2(v_2 - v_1) + G_3v_2 + G_4(v_2 - v_3) = 0$$

node 3 $$G_6(v_3 - v_1) + G_4(v_3 - v_2) + G_5v_3 = 0$$

where the summation at the arbitrarily chosen "ground" node 4 is not written, but

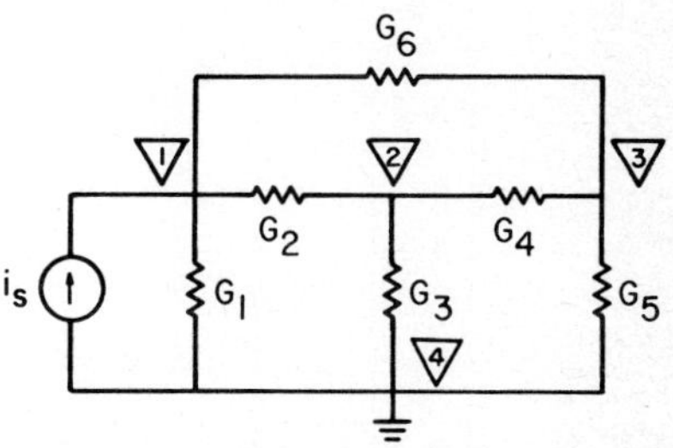

FIG. 2.1 Network Analyzed on Node Basis

all voltages are measured with respect to this ground node. In matrix form, these equations can be rearranged as

$$\begin{bmatrix} G_1 + G_2 + G_6 & -G_2 & -G_6 \\ -G_2 & G_2 + G_3 + G_4 & -G_4 \\ -G_6 & -G_4 & G_4 + G_5 + G_6 \end{bmatrix} \begin{bmatrix} v_1 \\ v_2 \\ v_3 \end{bmatrix} = \begin{bmatrix} i_s \\ 0 \\ 0 \end{bmatrix} \quad (2.1)$$

or more succinctly as

$$\underset{\sim}{G}_N \underline{v} = \underline{i} \quad (2.2)$$

where $\underset{\sim}{G}_N$ is termed the node conductance matrix.

In general, entries of $\underset{\sim}{G}_N$ (g_{ij}) have one of the two following forms:

(1) g_{ii}, a diagonal term, is the sum of all the conductances connected to node i;

(2) g_{ij}, $i \neq j$, an off-diagonal term, is the negative sum of all conductances connected between nodes i and j.

The analytical or symbolic solution of the node equations usually is of the following form. Let $\underset{\sim}{G}_N$, $\underline{v}$, and $\underline{i}$ have the general form

$$\underset{\sim}{G}_N = \begin{bmatrix} g_{11} & g_{12} & \cdots & g_{1n} \\ g_{21} & g_{22} & \cdots & g_{2n} \\ \vdots & \vdots & & \vdots \\ g_{n1} & g_{n2} & \cdots & g_{nn} \end{bmatrix}, \quad \underline{v} = \begin{bmatrix} v_1 \\ v_2 \\ \vdots \\ v_n \end{bmatrix}, \quad \underline{i} = \begin{bmatrix} i_1 \\ i_2 \\ \vdots \\ i_n \end{bmatrix} \tag{2.3}$$

Then the solution for $\underline{v}$ in (2. 2) can be written

$$\underline{v} = \begin{bmatrix} v_1 \\ v_2 \\ \vdots \\ v_n \end{bmatrix} = \begin{bmatrix} \frac{\Delta_{11}}{\Delta} & \frac{\Delta_{21}}{\Delta} & \cdots & \frac{\Delta_{n1}}{\Delta} \\ \frac{\Delta_{12}}{\Delta} & \frac{\Delta_{22}}{\Delta} & \cdots & \frac{\Delta_{n2}}{\Delta} \\ \vdots & \vdots & & \vdots \\ \frac{\Delta_{1n}}{\Delta} & \frac{\Delta_{2n}}{\Delta} & \cdots & \frac{\Delta_{nn}}{\Delta} \end{bmatrix} \begin{bmatrix} i_1 \\ i_2 \\ \vdots \\ i_n \end{bmatrix} \tag{2.4}$$

Here Δ is the determinant of $\underset{\sim}{G}_N$ and Δ_{ij} is the i-j cofactor given by $(-1)^{i+j}M_{ij}$, where M_{ij} is the determinant obtained by eliminating the i-th row and the j-th column of $\underset{\sim}{G}_N$, For example, the voltage v_3 of Figure 2. 1 is determined from

$$v_3 = \frac{\Delta_{13}}{\Delta} i_s + \frac{\Delta_{23}}{\Delta}(0) + \frac{\Delta_{33}}{\Delta}(0) = \frac{\Delta_{13}}{\Delta} i_s \tag{2.5}$$

where

$$\Delta_{13} = G_2G_4 + G_6(G_2 + G_3 + G_4)$$

$$\Delta = (G_1 + G_2 + G_6)[(G_2 + G_3 + G_4)(G_4 + G_5 + G_6) - G_4^2]$$
$$+ G_2[-G_2(G_4 + G_5 + G_6) - G_4G_4] - G_6[G_2G_4 + G_6(G_2 + G_3 + G_4)]$$

where Laplace's expansion down the first column has been used to find Δ.

2.3 NUMERICAL SOLUTION OF SIMULTANEOUS LINEAR EQUATIONS

2.3.1 Introduction

Determination of network voltages and currents by evaluation of determinants and cofactors is never performed numerically. A simple calculation shows why: evaluation of the determinant of a full matrix (i.e., one with nonzero values in every (i, j) position) using the Laplaçe expansion requires $(n-1)n!$ multiplications! We will shortly show that a solution for $\underline{v}$ - which is our goal - can be obtained in approximately $n^3/3$ operations.

2.3.2 The Gauss Elimination Procedure

<u>The Algorithm</u>. Consider the problem of solving sets of simultaneous linear equations expressed in the equivalent forms

$$\begin{aligned} a_{11}x_1 + a_{12}x_2 + \cdots a_{1n}x_n &= b_1 \\ a_{21}x_1 + a_{22}x_2 + \cdots a_{2n}x_n &= b_2 \\ &\vdots \\ a_{n1}x_1 + a_{n2}x_2 + \cdots a_{nn}x_n &= b_n \end{aligned}$$

or

$$\begin{bmatrix} a_{11} & a_{12} \cdots & a_{1n} \\ a_{21} & a_{22} \cdots & a_{2n} \\ \vdots & \vdots & \vdots \\ a_{n1} & a_{n2} & a_{nn} \end{bmatrix} \begin{bmatrix} x_1 \\ x_2 \\ \vdots \\ x_n \end{bmatrix} = \begin{bmatrix} b_1 \\ b_2 \\ \vdots \\ b_n \end{bmatrix}$$

or

$$\underset{\sim}{A}\,\underline{x} = \underline{b}$$

where the a_{ij} and b_i are given and the x_i are to be found. For example, the equation

$$\begin{bmatrix} 4 & -2 & -1 \\ -2 & 17/2 & -2 \\ -1 & -2 & 11 \end{bmatrix} \begin{bmatrix} x_1 \\ x_2 \\ x_3 \end{bmatrix} = \begin{bmatrix} 2 \\ 0 \\ -2 \end{bmatrix}$$

has the $\underline{x}$ solution

$$\underline{x} = \frac{1}{85} \begin{bmatrix} 44 \\ 8 \\ -10 \end{bmatrix}$$

In this section we will discuss a simple yet often effective method of finding $\underline{x}$ from $\underset{\sim}{A}$ and $\underline{b}$, namely the Gauss elimination procedure.

The Gauss procedure begins by attempting to eliminate x_1 from all the equations we can. If we divide the first equation by a_{11}, we obtain

$$x_1 + (a_{12}/a_{11})x_2 + \cdots + (a_{1n}/a_{11})x_n = b_1/a_{11} \tag{2.6}$$

This equation is now used as a reference in the following procedure:

(1) Multiply this reference equation by a_{21} and subtract the result from the second equation, term by term:

$$\begin{array}{l} a_{21}x_1 + a_{22}x_2 + \ldots + a_{2n}x_n = b_2 \\ \underline{-a_{21}x_1 - a_{21}(a_{12}/a_{11})x_2 + \ldots - a_{21}(a_{1n}/a_{11})x = -a_{21}b_1/a_{11}} \\ (0)\,x_1 + (a_{22} - a_{21}a_{12}/a_{11})\,x_2 + \ldots + (a_{2n} - a_{21}a_{1n}/a_{11})\,x_n = \\ \qquad b_2 - a_{21}b_1/a_{11} \end{array}$$

Note that x_1 has purposefully vanished in the new equation.

(2) In general, to eliminate the first unknown from the i-th equation, multiply the reference equation by a_{i1} and subtract the result from the i-th equation.

The process is more readily followed in matrix form. After all the a_{i1} $(i \neq 1)$ have been reduced to zero, the matrix equation has the form

$$\left[\begin{array}{c|ccc} 1 & a_{12}/a_{11} & \cdots & a_{1n}/a_{n} \\ \hline 0 & a_{22} - a_{21}a_{12}/a_{11} & \cdots & a_{2n} - a_{21}a_{1n}/a_{11} \\ \cdot & \cdot & \cdots & \cdot \\ \cdot & \cdot & \cdots & \cdot \\ 0 & a_{n2} - a_{n1}a_{12}/a_{11} & \cdots & a_{nn} - a_{n1}a_{1n}/a_{11} \end{array}\right] \left[\begin{array}{c} x_1 \\ \hline x_2 \\ \cdot \\ \cdot \\ x_n \end{array}\right] = \left[\begin{array}{c} b_1/a_{11} \\ \hline b_2 - a_{21}b_1/a_{11} \\ \cdot \\ \cdot \\ b_n - a_{n1}b_1/a_{n} \end{array}\right] \tag{2.7}$$

The above example would be of the form

$$\left[\begin{array}{c|cc} 1 & -1/2 & -1/4 \\ \hline 0 & 15/2 & -5/2 \\ 0 & -5/2 & 43/4 \end{array}\right] \left[\begin{array}{c} x_1 \\ \hline x_2 \\ x_3 \end{array}\right] = \left[\begin{array}{c} 1/2 \\ \hline 1 \\ -3/2 \end{array}\right] \tag{2.8}$$

at this stage of solution.

If we examine (2.7) or (2.8), we observe that the lower right partition shown involves n-1 equations in n-1 unknowns. We can therefore repeat the same algorithm to eliminate x_2, where the reference equation is now the second one. For the example of (2.8), the second equation is divided by

15/2, multiplied by -5/2, and then subtracted from the third equation to yield

$$\begin{bmatrix} 1 & -1/2 & -1/4 \\ 0 & 1 & -1/3 \\ 0 & 0 & \frac{119}{12} \end{bmatrix} \begin{bmatrix} x_1 \\ x_2 \\ x_3 \end{bmatrix} = \begin{bmatrix} 1/2 \\ 2/15 \\ -7/6 \end{bmatrix} \tag{2.9}$$

This process of eliminating matrix entries below the diagonal continues until the matrix is in upper triangular form as in (2.9). The final equation is then divided by the diagonal element,

$$\begin{bmatrix} 1 & -1/2 & -1/4 \\ 0 & 1 & -1/3 \\ 0 & 0 & 1 \end{bmatrix} \begin{bmatrix} x_1 \\ x_2 \\ x_3 \end{bmatrix} = \begin{bmatrix} 1/2 \\ 2/15 \\ -2/17 \end{bmatrix} \tag{2.10}$$

leaving the matrix in the general form

$$\begin{bmatrix} 1 & a'_{12} & \cdots & a'_{1n} \\ 0 & 1 & & a'_{2n} \\ \vdots & \ddots & & \\ \vdots & & 1 & a'_{n-1,n} \\ 0 & 0 & \cdots & 1 \end{bmatrix} \begin{bmatrix} x_1 \\ x_2 \\ \vdots \\ x_{n-1} \\ x_n \end{bmatrix} = \begin{bmatrix} b'_1 \\ b'_2 \\ \vdots \\ b'_{n-1} \\ b'_n \end{bmatrix} \tag{2.11}$$

This completes the forward reduction of the matrix.

We can immediately solve for x_n in (2.11), since the last equation requires $x_n = b'_n$. The next to last equation has the form

$$x_{n-1} = b'_{n-1} - a_{n-1,n} x_n$$

from which x_{n-1} can be determined. This back substitution continues up the $\underline{x}$ vector, solving successively for x_n, x_{n-1}, x_{n-2}, ... x_1 by the formula

$$x_r = b'_r - \sum_{j=r+1}^{n} a_{rj} x_j \qquad r = n-1, n-2, .. 1 \tag{2.12}$$

Pivoting. A set of n equations in n unknowns may not have a unique solution for several reasons. There may be no solution because the equations are inconsistent, or there may be an infinite number of solutions, viz,

$$\begin{bmatrix} 1 & 2 \\ 2 & 4 \end{bmatrix} \begin{bmatrix} x_1 \\ x_2 \end{bmatrix} = \begin{bmatrix} 1 \\ 3 \end{bmatrix} \qquad \text{no solution}$$

$$= \begin{bmatrix} 2 \\ 4 \end{bmatrix} \qquad \text{no unique solution}$$

Both of these possibilities must be reflected in the forward reduction procedure, since the back substitution cannot fail to yield a unique solution once (2.11) has been formed (i.e., (2.12) cannot fail).

The forward reduction will in fact fail only if the r-th reference equation has a zero-valued leading coefficient (a_{rr}), for then the equation cannot be divided by a_{rr} as required. For example, suppose (2.8) had been of the form

$$\begin{bmatrix} 1 & -1/2 & -1/4 \\ 0 & 0 & -5/2 \\ 0 & -5/2 & 43/4 \end{bmatrix} \begin{bmatrix} x_1 \\ x_2 \\ x_3 \end{bmatrix} = \begin{bmatrix} 1/2 \\ 1 \\ -3/2 \end{bmatrix}$$

The failure of the algorithm in this case is only apparent, however. We can exchange the second and third equations (row pivoting) to form

$$\begin{bmatrix} 1 & -1/2 & -1/4 \\ 0 & -5/2 & 43/4 \\ 0 & 0 & -5/2 \end{bmatrix} \begin{bmatrix} x_1 \\ x_2 \\ x_3 \end{bmatrix} = \begin{bmatrix} 1/2 \\ -3/2 \\ 1 \end{bmatrix}$$

and the reduction can continue. Or, the order of the second and third variables can be interchanged, yielding

$$\begin{bmatrix} 1 & -1/4 & -1/2 \\ 0 & -5/2 & 0 \\ 0 & 43/4 & -5/2 \end{bmatrix} \begin{bmatrix} x_1 \\ x_3 \\ x_2 \end{bmatrix} = \begin{bmatrix} 1/2 \\ 1 \\ -3/2 \end{bmatrix}$$

This is termed column pivoting. Obviously, as we proceed through the reduction of a large matrix, we may adopt a mixed strategy of row and column pivoting (double pivoting). The only unresolvable contingency is the case when all entries in the unreduced lower right hand partition of the matrix are zero. For example, (2.9) could be of the form

$$\begin{bmatrix} 1 & -1/2 & 0 \\ 0 & 1 & b \\ 0 & 0 & 0 \end{bmatrix} \begin{bmatrix} x_1 \\ x_2 \\ x_3 \end{bmatrix} = \begin{bmatrix} 1/2 \\ 2/15 \\ c \end{bmatrix}$$

If $c \neq 0$, no solution exists since the third equation cannot be satisfied. If $c = 0$ and $b = 0$, the solution for all the x_i's is not unique. What does $b = c = 0$ imply?

The Computational Effort. Although a truly realistic appraisal of the CPU time required to carry out an algorithm such as Gauss reduction include details such as time to load and store words, it is common to regard the number of elementary arithmetic operations alone as the measure of the computational effort. In this section, we examine this question for full matrices.

When the matrix is full, the number of multiplications and divisions required for generation of a unit diagonal or a zero off-diagonal is readily calculated (Table 2.1a), for every $a_{ij} (i \geq j)$ involved in the forward reduction. Unfortunately, the table is not easily summed to arrive at a total operations count. However, it can be shown that the operation count must be of the form*

$$N_g(n) = a_r n^r + a_{r-1} n^{r-1} \dots + a_0 \tag{2.13}$$

where $r \leq 3$. By matching this equation to the readily summed operation count for several low order cases (Table 2.1b), we can calculate the dependence to be

$$N_g(n) = \frac{n^3}{3} + \frac{n^2}{2} + \frac{n}{6} \tag{2.14}$$

By similar reasoning, the back substitution step can be shown to require $n(n-1)/2$ operations. The total count for calculation of $\underline{x}$ is then

$$N_g(n) = \frac{n^3}{3} + n^2 - \frac{n}{3} \tag{2.15}$$

TABLE 2.1 Developing an Operation Count

$$\begin{bmatrix} n & & & & \\ n & n-1 & & & \\ n & n-1 & & & \\ \cdot & \cdot & & & \\ \cdot & \cdot & & 2 & \\ n & n-1 & \cdot\ \cdot & 2 & 1 \end{bmatrix}$$

(a)

	Operation Count	
n	Forward Reduction	Back Substitution
1	1	0
2	5	1
3	14	3
4	30	6

(b)

*A careful study of Gauss reduction would reveal that a difference equation in n could be written for the operation count. The solution of this difference equation can be shown to have the form of (2.13). The bound on r represents the case if n were to appear in <u>every</u> position of the tableau of Table 2.1a.

The fastest of present-day serial computers could be expected to require .5μ sec for each operation and related loading and storage instructions. A 1000 x 1000 full matrix reduction would thus require 5-6 minutes! This prodigious amount of CPU time should not be surprising, if we recall that each nonzero matrix position commonly represents at least one network branch, requiring a total of 10^6 branches to fill a matrix of such size. As we shall see later, both the size n <u>and</u> the sparsity of a matrix are important in estimating computation time.

2.3.3 Solution by LU Factorization

The Algorithm. Although the operations involved in Gaussian elimination are present in most other matrix solution methods, the Gauss ordering of these operations is often too restrictive, and the discarding of sometimes useful intermediate data is computationally costly if repeated solutions are necessary as parts of $\underset{\sim}{A}$ or $\underline{b}$ change. A more flexible solution in many cases is matrix factorization (also called Crout reduction or triangularization).

Consider the possibility of representing the example

$$\begin{bmatrix} 4 & -2 & -1 \\ -2 & 17/2 & -2 \\ -1 & -2 & 11 \end{bmatrix} \begin{bmatrix} x_1 \\ x_2 \\ x_3 \end{bmatrix} = \begin{bmatrix} 2 \\ 0 \\ -2 \end{bmatrix} \tag{2.16}$$

in the form

$$\underbrace{\begin{bmatrix} 4 & 0 & 0 \\ -2 & 15/2 & 0 \\ -1 & -5/2 & 119/12 \end{bmatrix}}_{\underset{\sim}{L}} \overbrace{\underbrace{\begin{bmatrix} 1 & -1/2 & -1/4 \\ 0 & 1 & -1/3 \\ 0 & 0 & 1 \end{bmatrix}}_{\underset{\sim}{U}} \underbrace{\begin{bmatrix} x_1 \\ x_2 \\ x_3 \end{bmatrix}}_{\underline{x}}}^{\underline{y}} = \underbrace{\begin{bmatrix} 2 \\ 0 \\ -2 \end{bmatrix}}_{\underline{b}} \tag{2.17}$$

It is easy to verify by multiplying $\underset{\sim}{L}$ and $\underset{\sim}{U}$ that this factorization of $\underset{\sim}{A}$ into the lower triangular matrix $\underset{\sim}{L}$ and an upper triangular matrix $\underset{\sim}{U}$ is valid for this example. With $\underset{\sim}{A}$ in this form, we could solve (2.16) with a forward and a back substitution as follows.

(1) Solve $\underset{\sim}{L}\underline{y} = \underline{b}$ for $\underline{y}$, viz,

$$\begin{bmatrix} 4 & 0 & 0 \\ -2 & 15/2 & 0 \\ -1 & -5/2 & 119/12 \end{bmatrix} \begin{bmatrix} y_1 \\ y_2 \\ y_3 \end{bmatrix} = \begin{bmatrix} 2 \\ 0 \\ -2 \end{bmatrix}$$

yielding after a <u>forward substitution</u>

$$\underline{y} = \begin{bmatrix} 1/2 \\ 2/15 \\ -2/17 \end{bmatrix}$$

which is identical to the right hand side after the forward Gauss elimination (Eq. (2.10)).

(2) Solve $\underset{\sim}{U}\underline{x} = \underline{y}$ for $\underline{x}$, viz,

$$\begin{bmatrix} 1 & -1/2 & -1/4 \\ 0 & 1 & -1/3 \\ 0 & 0 & 1 \end{bmatrix} \begin{bmatrix} x_1 \\ x_2 \\ x_3 \end{bmatrix} = \begin{bmatrix} 1/2 \\ 2/15 \\ -2/17 \end{bmatrix}$$

yielding again

$$\underline{x} = \begin{bmatrix} 44/85 \\ 8/85 \\ -2/17 \end{bmatrix}$$

One advantage of this procedure is immediately evident: if $\underline{b}$ changes, $\underset{\sim}{A}$ does not have to be re-factored and two substitution steps yield a new solution.

The calculation of $\underset{\sim}{L}$ and $\underset{\sim}{U}$ proceeds in a remarkably straight - forward manner. To show this, we consider a general 4 x 4 matrix which is to be factored. We write

$$\begin{bmatrix} a_{11} & a_{12} & a_{13} & a_{14} \\ a_{21} & a_{22} & a_{23} & a_{24} \\ a_{31} & a_{32} & a_{33} & a_{34} \\ a_{41} & a_{42} & a_{43} & a_{44} \end{bmatrix} = \begin{bmatrix} \ell_{11} & 0 & 0 & 0 \\ \ell_{21} & \ell_{22} & 0 & 0 \\ \ell_{31} & \ell_{32} & \ell_{33} & 0 \\ \ell_{41} & \ell_{42} & \ell_{43} & \ell_{44} \end{bmatrix} \begin{bmatrix} 1 & u_{12} & u_{13} & u_{14} \\ 0 & 1 & u_{23} & u_{24} \\ 0 & 0 & 1 & u_{34} \\ 0 & 0 & 0 & 1 \end{bmatrix} \tag{2.18}$$

$$= \begin{bmatrix} \ell_{11} & \ell_{11}u_{12} & \ell_{11}u_{13} & \ell_{11}u_{14} \\ \ell_{21} & \ell_{21}u_{12}+\ell_{22} & \ell_{21}u_{13}+\ell_{22}u_{23} & \ell_{21}u_{14}+\ell_{22}u_{24} \\ \ell_{31} & \ell_{31}u_{12}+\ell_{32} & \ell_{31}u_{13}+\ell_{32}u_{23}+\ell_{33} & \ell_{31}u_{14}+\ell_{32}u_{24}+\ell_{33}u_{34} \\ \ell_{41} & \ell_{41}u_{12}+\ell_{42} & \ell_{41}u_{13}+\ell_{42}u_{23}+\ell_{43} & \ell_{41}u_{14}+\ell_{42}u_{24}+\ell_{43}u_{34}+\ell_{44} \end{bmatrix} \tag{2.19}$$

We can immediately determine the first column of $\underset{\sim}{L}$,

$$\ell_{i1} = a_{i1} \qquad i = 1, \ldots 4 \tag{2.20}$$

Proceeding to the second column of $\underset{\sim}{L}$, we equate

$$\begin{aligned} \ell_{11} \; u_{12} &= a_{12} \\ \ell_{21} \; u_{12} + \ell_{22} &= a_{22} \\ \ell_{31} \; u_{12} + \ell_{32} &= a_{32} \\ \ell_{41} \; u_{12} + \ell_{42} &= a_{42} \end{aligned} \tag{2.21}$$

Solving for u_{12} in (2.21), we can easily show that

$$\begin{aligned} u_{12} &= a_{12}/\ell_{11} \\ \ell_{22} &= a_{22} - \ell_{21} u_{12} \\ \ell_{32} &= a_{32} - \ell_{31} u_{12} \\ \ell_{42} &= a_{42} - \ell_{41} u_{12} \end{aligned}$$

Moving to the next column, we can similarly find

$$\begin{aligned} u_{13} &= a_{13}/\ell_{11} \\ u_{23} &= (a_{23} - \ell_{21} u_{13}) / \ell_{22} \\ \ell_{33} &= a_{33} - \ell_{31} u_{13} - \ell_{32} u_{23} \\ \ell_{43} &= a_{43} - \ell_{41} u_{13} - \ell_{42} u_{23} \end{aligned}$$

In general, we calculate, at the j-th column

$$u_{ij} = (a_{ij} - \sum_{k=1}^{i-1} \ell_{ik} u_{kj}) / \ell_{ii} \qquad i < j \tag{2.22}$$

$$\ell_{ij} = a_{ij} - \sum_{k=1}^{j-1} \ell_{ik} u_{kj} \qquad i \geq j \tag{2.23}$$

The computer implementation of the algorithm proceeds directly from (2.22-2.23). However, the following points should be observed.

(1) The procedure is halted by a zero-valued ℓ_{ii}, since (2.22) requires division by this term. Some form of row-column interchange is then necessary to locate a nonzero ℓ_{ij}. This does not imply a physical relocation of data in memory, but only that row-column pointers to the pivot position be rearranged (see Ref. [2.1] for details).

(2) Once u_{ij} or ℓ_{ij} is calculated, a_{ij} is never required thereafter. Thus, we can simply replace a_{ij} with u_{ij} or ℓ_{ij}.

Example 2.1. The first step in the factorization of $\underset{\sim}{A}$ of (2.16) is the calculation of the first column of $\underset{\sim}{L}$ and the first row of $\underset{\sim}{U}$. We immediately replace the entries of $\underset{\sim}{A}$ as follows:

$$\begin{bmatrix} \ell_{11} & u_{12} & u_{13} \\ \ell_{21} & a_{22} & a_{23} \\ \ell_{31} & a_{32} & a_{33} \end{bmatrix} = \begin{bmatrix} a_{11} & a_{12}/\ell_{11} & a_{13}/\ell_{11} \\ a_{21} & a_{22} & a_{23} \\ a_{31} & a_{32} & a_{33} \end{bmatrix}$$

$$= \begin{bmatrix} 4 & -1/2 & -1/4 \\ -2 & 17/2 & -2 \\ -1 & -2 & 11 \end{bmatrix}$$

The second and third rows are similarly reduced by the following steps.

$$\begin{bmatrix} \ell_{11} & u_{12} & u_{13} \\ \ell_{21} & \ell_{22} & u_{23} \\ \ell_{31} & \ell_{32} & a_{33} \end{bmatrix} = \begin{bmatrix} 4 & -1/2 & -1/4 \\ -2 & a_{22} - u_{12}\ell_{21} & (a_{23} - u_{13}\ell_{21})/\ell_{22} \\ -1 & a_{32} - u_{12}\ell_{31} & a_{33} \end{bmatrix}$$

$$= \begin{bmatrix} 4 & -1/2 & -1/4 \\ -2 & 15/2 & -1/3 \\ -1 & -5/2 & 11 \end{bmatrix}$$

$$\begin{bmatrix} \ell_{11} & u_{12} & u_{13} \\ \ell_{21} & \ell_{22} & u_{23} \\ \ell_{31} & \ell_{32} & \ell_{33} \end{bmatrix} = \begin{bmatrix} 4 & -1/2 & -1/4 \\ -2 & 15/2 & -1/3 \\ -1 & -5/2 & a_{33} - u_{13}\ell_{31} - u_{23}\ell_{32} \end{bmatrix}$$

$$= \begin{bmatrix} 4 & -1/2 & -1/4 \\ -2 & 15/2 & -1/3 \\ -1 & -5/2 & 119/12 \end{bmatrix}$$

Computational Cost of Factorization. We can use the same technique for determining the operation count now as we used for evaluating Gauss reduction. That is, we assume that the dependence of operation count N_f on matrix size n is of the form

$$N_f(n) = a_3 n^3 + a_2 n^2 + a_1 n + a_0$$

and solve for the coefficients a_i necessary to yield the operation counts for the low order cases calculated in Table 2.2b. The result is that the number of multiplications-divisions grows as

$$N_f(n) = \frac{n^3}{3} - \frac{n}{3}$$

for a full matrix. If the effort in two substitution steps is added (n^2), the total number of operations required to solve a set of simultaneous equations is

$$N_f(n) = \frac{n^3}{3} + n^2 - \frac{n}{3}$$

which, of course, is identical to the cost of Gauss reduction. Note that if $\underline{b}$ changes, only n^2 operations must be performed to obtain a new solution.

TABLE 2.2 Operation Count for Factorization

	position	column 1	2	3	4	. .	n
row	1	0	1	1	1	1 ·	1
	2	0	1	2	2	2 ·	2
	3	0	1	2	3	3 ·	3
	4	0	1	2	3	4 ·	4
	.	.	.	.	.	. .	.
	n	0	1	2	3	4 ·	n-1

(a) Op count to calculate u_{ij} or ℓ_{ij}

n	no. of operations
1	0
2	2
3	8
4	20

(b) Low order case

2.3.4 A Linear Network Analysis Example

To illustrate the use of LU factorization and to provide a model for a general circuit analysis program to be discussed in Section 2.7, we consider the analysis of the amplifier of Figure 2.2. Specifically, we want to calculate the gain v_o/v_s of the small signal model of Figure 2.2(b).

The solution program (RNET) of Appendix Table 2.1 is divided into four parts:

(1) data input phase (subroutine READ),
(2) node equation formulation phase (subroutine NODEQ),
(3) node equation solution phase using LU factorization (subroutine FCTR and BKSB),
(4) data output phase (subroutine OUTPUT).

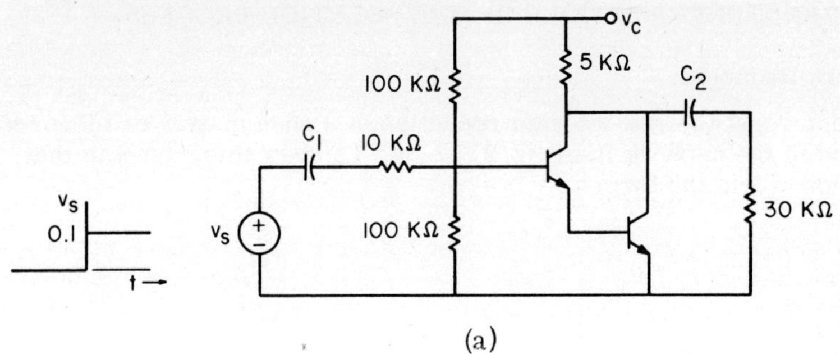

(a)

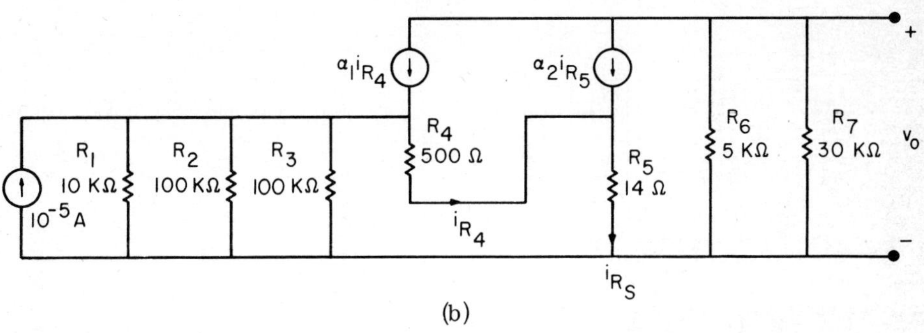

(b)

FIG. 2.2 Amplifier Analysis Problem

The voltage data corresponding to this network is shown in Table 2.3(b). The voltage gain is observed to be

$$\mu = v_o/v_s = -130.6$$

TABLE 2.3 Results of Computer Analysis of Amplifier of Figure 2.2

```
.98          .98         10000.      100000.     100000.     500.
14.          5000.       30000.      .00001
 3  1  2  0
```

(a) Input Data

```
ALF1 =   0.9800D 00 ALF2 =   0.9800D 00 R1 =   0.1000D 05
R2 =   0.1000D 06 R3 =   0.1000D 06 R4 =   0.5000D 03
R5 =   0.1400D 02 R6 =   0.5000D 04 R7 =   0.3000D 05
IS =   0.1000D-04
OUTPUT NODES =  2   0   OUTPUT VOLTAGE =     -13.0609756
```

(b) Output Data

2.4 NETWORK INTERPRETATIONS OF THE SOLUTION PROCESS

2.4.1 Network Reduction

We can interpret the row-column reduction of a node matrix to $\underset{\sim}{L}\underset{\sim}{U}$ form as a reduction of the network itself [2.2] [2.3]. To show this, assume that $\underset{\sim}{G}_N$ is partitioned into the form

$$\underset{\sim}{G}_N = \begin{bmatrix} \ell_{11} & \ell_{11}\underline{u}_1^T \\ \underline{\ell}_1 & \underset{\sim}{g}_{22} \end{bmatrix}$$

where $\underline{\ell}_1$ and $\underline{u}_1$ are column vectors representing the first column of $\underset{\sim}{L}$ and the first row of $\underset{\sim}{U}$ respectively. We similarly partition $\underline{v}$ and $\underline{i}$ so that we can write

$$\begin{bmatrix} \ell_{11} & \ell_{11}\underline{u}_1^T \\ \underline{\ell}_1 & \underset{\sim}{g}_{22} \end{bmatrix} \begin{bmatrix} v_1 \\ \underline{v}_2 \end{bmatrix} = \begin{bmatrix} i_1 \\ \underline{i}_2 \end{bmatrix}$$

Solving the first equation for v_1 yields

$$v_1 = i_1/\ell_{11} - \underline{u}_1^T \underline{v}_2$$

The second equation can now be written as

$$[\underset{\sim}{g}_{22} - \underline{\ell}_1\underline{u}_1^T]\, \underline{v}_2 = \underline{i}_2 - (i_1/\ell_{11})\,\underline{\ell}_1$$

where v_1 has been eliminated. This equation represents the node equations of a network with one fewer nodes.

If the original network contained only two-terminal resistances, we can obtain the element values of the new network by expanding the node matrix in terms of the element values as follows. Let an n-node network be described by

$$\underset{\sim}{G}_N = \begin{bmatrix} \sum_{j=1}^{n} G_{1j} & -G_{12} & -G_{13} & \cdot & -G_{1,n-1} \\ -G_{21} & \sum_{j=1}^{n} G_{2j} & -G_{23} & \cdot\ \cdot & -G_{2,n-1} \\ -G_{31} & -G_{32} & \sum_{j=1}^{n} G_{3j} & \cdot\ \cdot\ \cdot & -G_{3,n-1} \\ \cdot & \cdot & \cdot & & \cdot \\ \cdot & \cdot & \cdot & & \cdot \\ -G_{n-1,1} & -G_{n-1,2} & -G_{n-1,3} & \cdot\ \cdot & \sum_{j=1}^{n} G_{n-1,j} \end{bmatrix}$$

where conductance $G_{ji} = G_{ij}$ is connected between nodes i and j. The reduced matrix equation analogous to g_{22} has the form

$$\begin{bmatrix} \sum_{j=1}^{n} G_{2j} - G_{12} G_{21}/\sum_{j=1}^{n} G_{1j} & -G_{23} - G_{13} G_{21}/\sum_{j=1}^{n} G_{1j} \cdots \\ -G_{32} - G_{12} G_{31}/\sum_{j=1}^{n} G_{1j} & \sum_{j=1}^{n} G_{3j} - G_{13} G_{31}/\sum_{j=1}^{n} G_{1j} \cdots \\ \vdots & \vdots \end{bmatrix} \tag{2.24}$$

The network described by this node equation then has a conductance of G'_{ij} between nodes i and j given by the negative of the off-diagonal terms of (2.24), namely

$$G'_{ij} = G_{ij} + G_{ki} G_{jk}/\sum_{j=1}^{n} G_{kj} \tag{2.25}$$

for $k = 1$. It is left as an exercise to show that this formula applies to the conductance from node to ground $(i = j)$ as well.

Example 2.2. The matrix $\underset{\sim}{A}$ of (2.16) is the node conductance matrix of the network of Figure 2.3a. Successive application of the above element value transformations yields the sequence of networks shown in Figure 2.3 b-c, simulating the solution steps of Example 2.1.

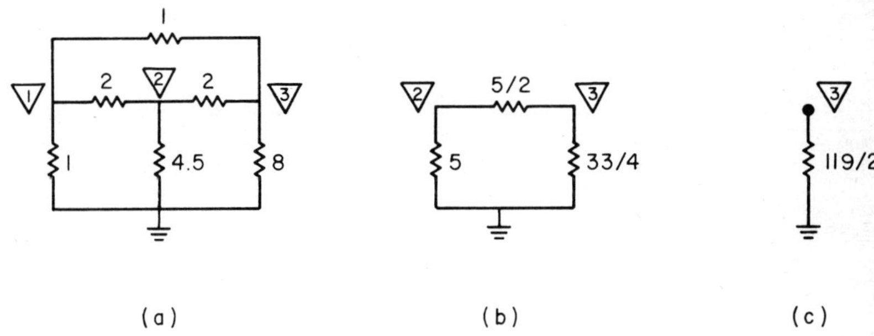

FIG. 2.3 Network Reduction Illustration (values in mhos)

Another interpretation of the network reduction viewpoint is given in Problem 2.7.

2.4.2 Superposition and Sparsity

We next examine from a numerical point of view the familiar principle of superposition. To facilitate a node-analysis-based description of this concept, let all the excitations be current sources contained in a vector $\underline{i}$, viz,

$$\underset{\sim}{G}_N \underline{v} = \underline{i} \tag{2.26}$$

Superposition requires that we calculate the explicit dependence of each node voltage response on each current excitation. This is equivalent to inverting $\underset{\sim}{G}_N$, i.e.,

$$\underline{v} = \underset{\sim}{G}_N^{-1} \underline{i} \tag{2.27}$$

where the inverse of $\underset{\sim}{G}_N$ is a matrix of transfer impedances. Now consider the calculation of $\underline{v}$ from $\underline{i}$: if $\underset{\sim}{G}_N$ is a full matrix then usually $\underset{\sim}{G}_N^{-1}$ will be full as well; therefore to calculate $\underline{v}$ if $\underline{i}$ changes requires n^2 multiplications, which is identical in effort to the two substitution steps involved in repeated solutions from the $\underset{\sim}{L}\underset{\sim}{U}$ factored form of $\underset{\sim}{G}_N$.

The factored form of the matrix has one outstanding advantage over the explicit inverse, however. To illustrate, consider the ladder network of Figure 2.4a. The node matrix is

$$\underset{\sim}{G}_N = \begin{bmatrix} G_1 + G_2 & -G_2 & 0 & \\ -G_2 & G_2 + G_3 + G_4 & -G_4 & 0 \\ 0 & -G_4 & G_4 + G_5 + G_6 & -G_6 \\ 0 & 0 & -G_6 & G_6 + G_7 \end{bmatrix}$$

which is known as a tridiagonal matrix. The structure of this matrix may be represented by indicating the nonzero positions with an 'x'

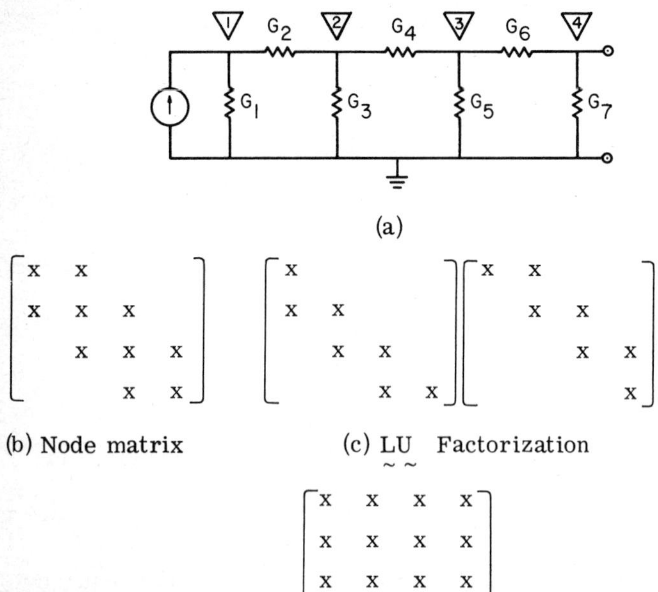

(a)

(b) Node matrix

(c) $\underset{\sim}{L}\underset{\sim}{U}$ Factorization

(d) Inverse

FIG. 2.4 Example for Comparison of Factorization and Inverse Calculation as displayed in Figure 2.4b. Now consider the structure of L and U on the one hand and G_N^{-1} on the other (Figure 2.4c-d). L and U together have the same structure as G_N; however, G_N^{-1} has a nonzero value in every position, since $\Delta_{ij} \neq 0$ (check this by eliminating any row and column and evaluating the remaining cofactor). Therefore, a solution based on (2.27) will require n^2 multiplications even though G_N is sparse, whereas two substitution steps will require only 3n multiplications and divisions.

We conclude from this example that the inverse matrix - together with the associated network concept of superposition - can be a numerically inefficient solution device. Factorization is a preferred procedure provided sparsity can be exploited (Chapter X).

2.4.3 Pivoting and Sparsity

We have already shown that matrix factorization can be superior to more familiar network solution methods by illustrating the relative sparsity of the matrices it produces. But the sparsity of L and U can also depend on the ordering of the equations and the solution variables.

To illustrate, consider a matrix of the form of Table 2.4a. Following the procedure of Example 2.1 where the entries of L and U

TABLE 2.4 Comparison of Sparsity of Factored Matrices

$$\begin{bmatrix} a_{11} & a_{12} & a_{13} & a_{14} \\ a_{21} & a_{22} & 0 & 0 \\ a_{31} & 0 & a_{33} & 0 \\ a_{41} & 0 & 0 & a_{44} \end{bmatrix} \Longrightarrow \begin{bmatrix} \ell_{11} & u_{12} & u_{13} & u_{14} \\ \ell_{21} & a_{22} & 0 & 0 \\ \ell_{31} & 0 & a_{33} & 0 \\ \ell_{41} & 0 & 0 & a_{44} \end{bmatrix}$$

(a) (b)

$$\begin{bmatrix} a_{44} & 0 & 0 & a_{41} \\ 0 & a_{22} & 0 & a_{21} \\ 0 & 0 & a_{33} & a_{31} \\ a_{14} & a_{12} & a_{13} & a_{11} \end{bmatrix} \Longrightarrow \begin{bmatrix} \ell_{11} & 0 & 0 & u_{14} \\ 0 & a_{22} & 0 & a_{21} \\ 0 & 0 & a_{33} & a_{31} \\ \ell_{41} & a_{12} & a_{13} & a_{11} \end{bmatrix}$$

(c) (d)

overlay the entries of A, the matrix has the form of Table 2.4b after calculation of the first column of L and the first row of U. Even without making further calculations, we observe that all future ℓ_{ij}'s and u_{ij}'s will be nonzero since each will contain a term of the form $\ell_{i1}u_{1j}$ (see Eqs. (2.22 to 23)). An example of this propagation is given in Table 2.4b. Thus, L and U will have a number of non-zero-valued positions which were zero-valued in A. These are known as fill positions or simply fills. In contrast to this alarming situation, consider the result of interchanging rows and columns 1 and 4 as displayed in Table 2.4c. The first column of L and row of U each have only one non-zero-valued position; further, these are strategically situated to cause no later fills, as indicated in Table 2.4d. In fact, inspection shows that no fills are created in the entire LU factorization of the matrix of Table 2.4c.

The algorithm for reducing the solution effort by re-ordering is known as optimal ordering. Alternatively, we can view this process as one of selecting pivot positions without physically altering the matrix. The orderings of Table 2.4a and 2.4c could thus be indicated pictorially as

$$\begin{bmatrix} \boxed{a_{11}}^{1} & a_{12} & a_{13} & a_{14} \\ a_{21} & \boxed{a_{22}}^{2} & 0 & 0 \\ a_{31} & 0 & \boxed{a_{33}}^{3} & 0 \\ a_{41} & 0 & 0 & \boxed{a_{44}}^{4} \end{bmatrix} \qquad \begin{bmatrix} \boxed{a_{11}}^{4} & a_{12} & a_{13} & a_{14} \\ a_{21} & \boxed{a_{22}}^{2} & 0 & 0 \\ a_{31} & 0 & \boxed{a_{33}}^{3} & 0 \\ a_{41} & 0 & 0 & \boxed{a_{44}}^{1} \end{bmatrix}$$

or more succinctly as

$$(1,1),\ (2,2),\ (3,3),\ (4,4) \text{ and } (4,4),\ (2,2),\ (3,3),\ (1,1)$$

respectively. Indeed, there is no reason to be restricted to diagonal pivoting. We could as easily have chosen the pivot order

$$(3,3),\ (4,1),\ (2,2),\ (1,4)$$

The only inviolable rules are that (1) a zero-valued position cannot be chosen for a pivot, and (2) two pivots cannot be selected from the same row or column.

The organization of an ordering algorithm is discussed in Chapter X. In practice, these algorithms can reduce the solution time by an order of magnitude or more for large networks.

Example 2.3. The node matrix of a network (see Figure 2.6) has nonzero positions indicated by X's in Table 2.5a. After an LU factorization without re-ordering of equations and variables, fills occur in the positions indicated by O's. After re-ordering of the diagonal positions, the number of fills has been drastically reduced from 140 to 10. If this fill reduction is properly exploited, the solution time can be reduced by approximately a factor of four.

2.5 FREQUENCY-DOMAIN ANALYSIS

2.5.1 Admittance Branch Models

It is well known that if we assume a sinusoidal steady state behavior in each branch of the form

$$i_k(t) = |I_k| \cos(\omega t + \phi_{i_k}) = \text{Re}[I_k e^{j\omega t}] \qquad (2.28)$$

$$v_k(t) = |V_k| \cos(\omega t + \phi_{v_k}) = \text{Re}[V_k e^{j\omega t}] \qquad (2.29)$$

where $I_k = |I_k| e^{j\phi_{i_k}}$, $V_k = |V_k| e^{j\phi_{v_k}}$, then

TABLE 2.5 Effects of Ordering on Fills

```
XX XXX
XXXOOO
 XXXOO
XOXXOOXX   X XXX
XOOOXXOO   O OOO
XOOOXXXO X O XXOXX
   XOXXXXOXO OOOOO
   XOOXXOOOO OOOOO
      XOXXOO OOOOO
     XOOXXOO OOOOO
      XOOOXO OOOOO
   XOOOOOOOXXOOOOOO
           XXXOOOO
   XOXOOOOOOXXXOXO
   XOXOOOOOOOXXXOOX
   XOOOOOOOOOOXXOOO
     XOOOOOOOXOOXOO
     XOOOOOOOOOOOXX
              XOOXX
```

(a)

```
X      X       X
 X        X  X
  X       X
   X         X  X
    X         XX
     XXX
     XXO     X
X    XOX     X X
        XXX
        XXO    X
 XX     XOX  X X
           XXX
           XXOX
 X X  XX  XXOXXOX
    X       XXXXX
X   X  X XX  OXXXX
   X         XXXXOX
               XOXX
                XXX
```

(b)

(1) $I_k = (G)V_k$, for a conductance (2.30)

(2) $I_k = (j\omega C)V_k$, for a capacitance (2.31)

(3) $I_k = (1/j\omega L)V_k$, for an inductance. (2.32)

These admittances $Y(j\omega)$ which relate $I_k(j\omega)$ to $V_k(j\omega)$ may then be treated as complex-valued conductances. This implies that a network model with the same structure as the original but with these admittances substituted for the original elements can be solved to yield I_k and V_k directly, from which i_k and v_k of (2.28-2.29) are determined.

Example 2.4. The sinusoidal excitation of the network of Figure 2.5a

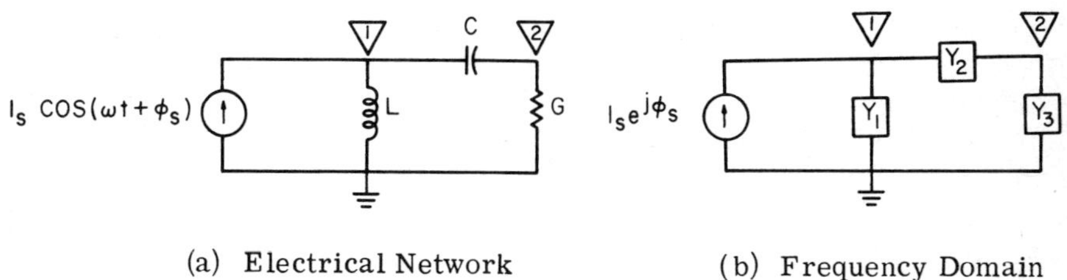

(a) Electrical Network

(b) Frequency Domain Model

FIG. 2.5 Representation of Frequency Domain Calculations

will yield the steady state branch currents and voltage of the form of (2.28-2.29). These responses are found by solving the network of Figure 2.5b, which is described by the node matrix

$$\begin{bmatrix} Y_1 + Y_2 & -Y_2 \\ -Y_2 & Y_2 + Y_3 \end{bmatrix} = \begin{bmatrix} 1/j\omega L + j\omega C & -j\omega C \\ -j\omega C & G + j\omega C \end{bmatrix} \tag{2.33}$$

so that we need to solve

$$\begin{bmatrix} Y_1 + Y_2 & -Y_2 \\ -Y_2 & Y_2 + Y_3 \end{bmatrix} \begin{bmatrix} V_1 \\ V_2 \end{bmatrix} = \begin{bmatrix} |I_s| e^{j\phi_s} \\ 0 \end{bmatrix} \tag{2.34}$$

2.5.2 Numerical Solution

Introduction

For any value of ω (ω_i), complex-valued admittances of the form of Eqs. (2.30-2.32) can be calculated and the node matrix $\underset{\sim}{Y}_N$ constructed. The complex-valued node equations can then be solved by the methods of Sections 2.3.2, 2.3.3, and 2.4.1, i.e., by

(1) Gauss elimination
(2) LU factorization
(3) network reduction.

Stepping ω over a frequency interval ($\omega_{min} \le \omega_i \le \omega_{max}$) and re-solution of the node equations for each ω_i then yields complex-valued node voltages as functions of frequency - the so-called frequency response. Branch voltages and currents are then routinely found from Eqs. (2.30-2.32).

Unfortunately, this solution process can become quite ill-conditioned.* To illustrate one cause of numerical problems, consider the solution of the network of Figure 2.5a as ω approaches $1/\sqrt{LC}$. An increasingly severe subtraction occurs in the (1, 1) position of $\underset{\sim}{Y}_N$ of (2.33). This type of unavoidable subtraction does not occur in resistive network analysis when all conductances are positive-valued. Further, this subtraction can make the entire solution process far more sensitive to other types of avoidable numerical problems. In the following sections, we study two such sources of avoidable ill-conditioning.

Cancellation

Although we indicated in Section 2.4.1 that solution of the node matrix is related to node-by-node reduction of the network, these are not numerically identical techniques. In fact, it is easily shown that network reduction can be a far better conditioned solution method.

Consider again the node matrix of (2.33),

$$\underset{\sim}{Y}_N = \begin{bmatrix} Y_1 + Y_2 & -Y_2 \\ -Y_2 & Y_2 + Y_3 \end{bmatrix}$$

*"Ill-conditioned" is a term used to describe a calculation which yields an exceptionally inaccurate result or which requires excessive computational effort.

Either the forward reduction step of Gauss elimination or LU factorization results in a final calculation of the symbolic form

$$\text{①}\; Y_2 + Y_3 - \frac{Y_2^2}{Y_1 + Y_2} = \text{②}\; \frac{Y_1Y_2 + Y_2^2 + Y_1Y_3 + Y_2Y_3 - Y_2^2}{Y_1 + Y_2}$$

$$= \text{③}\; \frac{Y_1Y_2 + Y_1Y_3 + Y_2Y_3}{Y_1 + Y_2} \qquad (2.35)$$

Unfortunately, the symbolic cancellation $Y_2^2 - Y_2^2$ cannot be distinguished numerically. If $|Y_2| >> |Y_1|$ and $|Y_2| >> |Y_3|$, then the evaluation of ① will differ significantly from the evaluation of ③. This type of cancellation is therefore most serious when nodes are connected by large admittances or are otherwise closely coupled.

To demonstrate the effect this cancellation can have on round-off error, consider expression ①. There are two principal opportunities for round-off errors to occur: (1) in the $Y_2 + Y_1$ addition, and (2) in the combined $Y_2^2/(Y_1 + Y_2)$ calculation. To study the propagation of these errors, we write the calculation of ① as

$$(Y_2 + Y_1)\,(1 + \epsilon_1) - \frac{Y_2^2\,(1 + \epsilon_2)}{Y_1 + Y_2} \qquad (2.36)$$

where ϵ_i is the round-off error ($|\epsilon_i| \le 10^{-7}$ in IBM 370 single precision Fortran). The relative error is then

$$E_{rel} = \frac{\left\{(Y_2 + Y_1)\,(1 + \epsilon_1) - \dfrac{Y_2^2\,(1 + \epsilon_2)}{Y_1 + Y_2}\right\} - \left\{Y_1 + Y_2 - \dfrac{Y_2^2}{Y_1 + Y_2}\right\}}{\dfrac{Y_1Y_2 + Y_1Y_3 + Y_2Y_3}{Y_1 + Y_2}}$$

For the assumed relationships between the Y's,

$$E_{rel} \approx (\epsilon_1 - \epsilon_2)\,Y_2/(Y_1 + Y_3) \qquad (2.37)$$

Thus, the round-off error with which the calculation is performed is magnified by the ratio $Y_2/(Y_1 + Y_3)$. For example, if $\omega = 1$ and $C >> G$ in the network of Figure 2.5a and if $|\epsilon_i| \le 10^{-7}$, then

$$|E_{rel}| \le 2 \times 10^{-7}\,(C/G)$$

The network reduction method of Section 2.4.1 does not suffer from this cancellation, since the network admittance Y'_{ij} would be obtained without symbolic subtraction from (2.25) as

$$Y'_{ij} = Y_{ij} + Y_{ki}Y_{jk} / \sum_{j=1}^{n} Y_{kj}$$

A large error in one step of the solution process does not of course always propagate to the final solution variables with the same severity. To gain an appreciation for the practical significance of this effect, we consider the following example [2.4].

Example 2.5a. The normalized crystal filter of Figure 2.6 presents a particularly difficult computer analysis problem. First, the network has a 2% bandwidth ($\pm$.01 rad/sec at a center frequency of 1 rad/sec). This requires nearly complete cancellation of large positive and negative reactive terms over the passband in the manner of Example 2.4 near $\omega = 1$. This cancellation is unavoidable. However, very small resistances are also shown to represent the loss of each crystal. The corresponding admittances (conductances) are therefore large, so that significant cancellation can be expected if Gauss elimination or LU factorization is used for solution.

The results of analysis by factorization and by network reduction are shown in Table 2.6b-c for radian frequencies in the passband.* These single precision calculations, when compared with the double precision calculations of Table 2.6a, show a complete loss of accuracy using factorization for most frequencies. In contrast, only 1 - 2 significant figures are lost when network reduction is performed.

The back substitution step of matrix solution methods is equally prone to cancellation. This can occur when the input and the output do not share the ground node (i.e., the node from which all other node voltages are referenced).

Example 2.5b. In the calculations of Table 2.6b-c, node (1) is chosen as the ground node; these analyses are duplicated in Table 2.6d-e with node (12) chosen as the ground node. Even when all cancellation is avoided by use of the network reduction algorithm, all significance in the output voltage is again lost for the latter ground selection (Table 2.6d).

The mechanism for the loss of accuracy is relatively clear. When a large admittance is across output nodes N_{0_1} and N_{0_2}, the node voltages V_{0_1} and V_{0_2} defined with respect to a third node will be nearly equal, i.e., $|(V_{0_1} - V_{0_2})/V_{0_1}| << 1$. Again, a significant cancellation will occur since V_{0_1} and V_{0_2} and not their difference are solution variables. If either node is the ground node, however, no subtraction is necessary to calculate the output voltage. A similar argument applies when the input is connected across a large admittance and neither input node is the ground node. This latter situation occurs in Example 2.5b.

Pivoting and Round-off Error. In Section 2.3.2, row and/or column pivoting was shown to be a necessity when a zero-valued diagonal entry is encountered. This can occur in reduction of $\underset{\sim}{Y_N}$ at discrete values of ω. For

*The node equations were optimally ordered as discussed in Section 2.4.3.

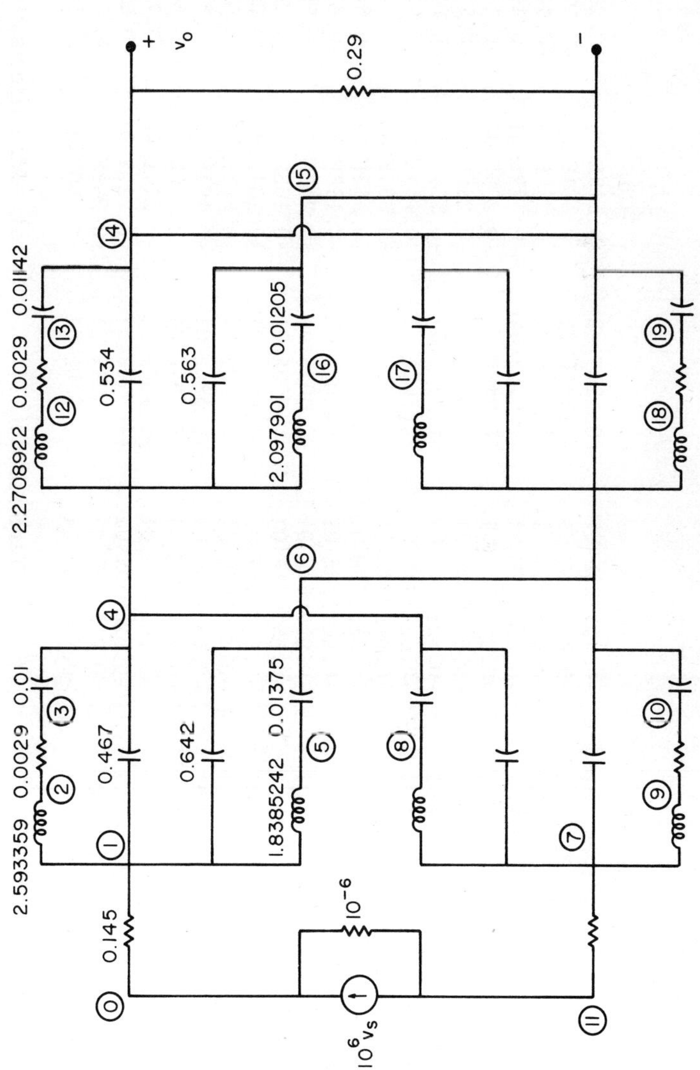

FIG. 2.6 Crystal Filter Example

TABLE 2.6 Accuracy Comparisons of Node Formulations
(b-e are single precision Fortran calculations)

Freq. (htz)	Voltage Gain (V_o/V_s) (db) (a)	(b)	(c)	(d)	(e)
.990	-0.954707D 01	-0.954729E 01	-0.120E 02	0.545E 01	-0.200E 02
.991	-0.670710D 01	-0.670694E 01	-0.716E 01	0.521E 01	-0.171E 02
.992	-0.665932D 01	-0.665996E 01	-0.667E 01	0.233E 01	-0.103E 02
.993	-0.690077D 01	-0.690092E 01	-0.725E 01	-0.372E 01	-0.105E 02
.994	-0.677751D 01	-0.677753E 01	-0.746E 01	-0.200E 01	-0.107E 02
.995	-0.647863D 01	-0.647866E 01	-0.677E 01	-0.285E 01	-0.718E 01
.996	-0.626918D 01	-0.626914E 01	-0.617E 01	-0.361E 01	-0.929E 01
.997	-0.627406D 01	-0.627394E 01	-0.607E 01	-0.481E 01	-0.951E 01
.998	-0.642985D 01	-0.642983E 01	-0.627E 01	-0.667E 01	-0.103E 02
.999	-0.658778D 01	-0.658759E 01	-0.649E 01	-0.945E 01	-0.881E 01
1.000	-0.663133D 01	-0.663126E 01	-0.603E 01	-0.138E 02	-0.856E 01
1.001	-0.652278D 01	-0.652264E 01	-0.601E 01	-0.216E 02	-0.797E 01
1.002	-0.631037D 01	-0.631031E 01	-0.690E 01	-0.168E 02	-0.760E 01
1.003	-0.612115D 01	-0.612114E 01	-0.655E 01	-0.104E 02	-0.660E 01
1.004	-0.610659D 01	-0.610655E 01	-0.662E 01	-0.681E 01	-0.654E 01
1.005	-0.631401D 01	-0.631400E 01	-0.684E 01	-0.471E 01	-0.634E 01
1.006	-0.659045D 01	-0.659041E 01	-0.694E 01	-0.350E 01	-0.642E 01
1.007	-0.664833D 01	-0.664831E 01	-0.700E 01	-0.271E 01	-0.774E 01
1.008	-0.623969D 01	-0.623970E 01	-0.674E 01	-0.212E 01	-0.726E 01
1.009	-0.625464D 01	-0.625460E 01	-0.634E 01	-0.249E 01	-0.716E 01
1.010	-0.927594D 01	-0.927596E 01	-0.939E 01	-0.650E 01	-0.968E 01

(a) Double Precision Cal.

(b) Network Reduction
⓪ Node Ground

(c) LU Factorization
⓪ Node Ground

(d) Network Reduction
⑫ Node Ground

(e) LU Factorization
⑫ Node Ground

example, for $\omega = 1/\sqrt{LC}$, the (1, 1) entry of Y_N of Eq. (2.33) becomes zero-valued.

Even when the prospective pivot position is not zero-valued, its adoption may still cause avoidable round-off errors if it is small in relation to other entries of the matrix. To show this, consider the reduction of the (i, j) entry to zero in the Gauss elimination procedure. The new entries a'_{ik} in the i-th row are $a_{ik} - (a_{ij}/a_{jj})a_{jk}$. There are two opportunities for round-off error to occur: (1) in the multiplication-division operation, and (2) in the subtraction (or addition, depending on the sign). These errors can be represented by

$$a'_{ik} = [a_{ik} - (a_{ij}/a_{jj})(a_{jk})(1 + \epsilon_1)](1 + \epsilon_2) \qquad (2.38)$$

The total error is then readily found as

$$\begin{aligned} \epsilon_{ik} &= a^1_{ik} - [a_{ik} - (a_{ij}/a_{jj})a_{jk}] \\ &= \frac{a'_{ik}\epsilon_2}{1 + \epsilon_2} - (a_{ij}/a_{jj})a_{jk}\,\epsilon_1 \end{aligned} \qquad (2.39)$$

We are concerned with making ϵ_{ik} as small as possible. Since the Gauss procedure reduces an entire column at a time, the only term common to all calculations in a column reduction is the quantity a_{jj}. It is typical, therefore, in the reduction of the j-th column, to interchange rows so that the largest entry appears in the (j, j) position. This is of course row pivoting. The columns can also be interchanged, so that the largest entry of the entire remaining matrix is located in the (j, j) position. We have termed this double pivoting.

Pivoting to reduce round-off error is a general stratagem, and is not peculiar to network problems. Indeed, its significance in the network problems is not well understood or even easily documented. It is far more common to select pivots on the basis of optimal ordering (Section 2.4.3) rather than reduction of round-off error, especially with the easy availability of double precision arithmetic on most computers.

2.5.3 Network Scaling

The element values present in common electrical circuits are such that a numerical solution can result in overflows or underflows. It becomes essential in these cases to scale element values so that the entries in the matrix are near unity in magnitude, or, if this is impossible, at least centered about unity. It is also common to think of element values in units other than ohms, farads, and henries. For example, at high frequencies, we often consider inductances in millihenries, resistance in kilohms, and capacitance in picofarads.

There are two factors which can be chosen arbitrarily in scaling a network: the frequency scaling factor a and the impedance (admittance) scaling factor b. To illustrate, consider the division of all the admittances of Figure 2.7 by b; so that the total admittance I/V is also divided by b.

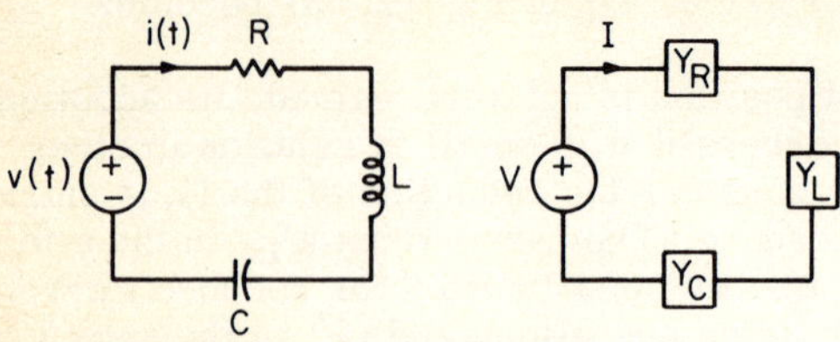

FIG. 2.7 A Network to be Scaled

Then designating the new admittances by primes we require

$$Y_R' = \frac{1}{R'} = \frac{Y_R}{b} = \frac{1}{bR}$$

$$Y_L' = \frac{1}{j\omega L'} = \frac{Y_L}{b} = \frac{1}{j\omega(bL)}$$

$$Y_C' = j\omega C' = \frac{Y_C}{b} = j\omega\left(\frac{C}{b}\right)$$

We can satisfy these equations by writing simply

$$R' = bR \qquad L' = bL \qquad C' = \frac{C}{b}$$

To scale the frequency variable ω by a, we define a new frequency variable $\omega'' = \omega/a$. Then each admittance $Y''(j\omega'')$ of the frequency scaled network must equal $Y(j\omega)$ of the unscaled network, viz,

$$Y_L'' = \frac{1}{j\omega''L''} = \frac{1}{j\frac{\omega}{a}L''} = \frac{1}{j\omega\left(\frac{L''}{a}\right)} = Y_L = \frac{1}{j\omega L}$$

$$Y_C'' = j\omega''C'' = j\,\frac{\omega}{a}\,C'' = j\omega\left(\frac{C''}{a}\right) = Y_C = j\omega C$$

Combining frequency and impedance scaling, we then have

$$\begin{aligned} R'' &= R' = bR \\ L'' &= aL' = abL \\ C'' &= aC' = \frac{a}{b}C \end{aligned} \qquad (2.40)$$

To exemplify use of these formulas, suppose

$$R = 5000\ \Omega \qquad L = 10^{-9}\ \text{h}, \qquad C = 2 \times 10^{-12}\ \text{f}$$

We desire the computer model to have the same response when $\omega'' = 1$ rad/sec as the original circuit when $\omega = 10^9$ rad/sec. We determine <u>a</u> from

$$\omega'' = 1 = \frac{\omega}{a} = \frac{10^9}{a}$$

$$a = 10^9$$

Since no scaling of the admittance levels is specified, we set b = 1. Then the computer model would have the values

$$R'' = R = 5000\ \Omega$$

$$L'' = aL = 1h$$

$$C'' = aC = 2 \times 10^{-3}\ f$$

The scaled elements can be made to have a value close to unity by choosing b = .001; then the above values would become

$$R'' = 5\ \Omega$$

$$L'' = .001\ h$$

$$C'' = 2\ f$$

The circuit admittance I''/V'' is now equal to (I/V)/1000 when

$$\omega = 10^9 \quad \text{or} \quad \omega'' = 1$$

2.6 HIGH FREQUENCY CIRCUIT ANALYSIS

2.6.1 Introduction

An increasing percentage of frequency domain analysis problems involve two-port or multiport network elements in addition to the branch (one-port) elements already studied. In general, an n-port element is described by an n x n matrix relating a total of n currents and n voltages defined at pairs of nodes assumed to be accessible for measurement or calculation purposes. For example, four pairs of currents and voltages define the four-port of Figure 2.8.

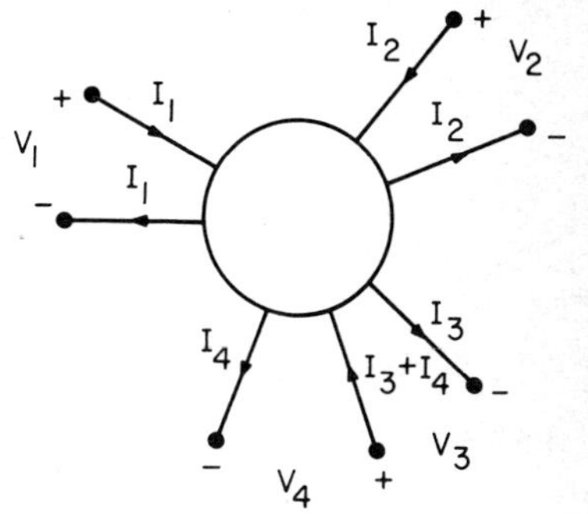

FIG. 2.8 A Four-Port Network

An example of this matrix description is the so-called $\underset{\sim}{Y}$ or short circuit matrix. Here, the port currents are assigned to one vector $\underline{I}$ and port voltages to another vector $\underline{V}$. $\underset{\sim}{Y}$ is then defined by

$$\underset{\sim}{Y}\,\underline{V} = \underline{I}$$

or in detail by

$$\begin{bmatrix} y_{11}(j\omega) & y_{12}(j\omega) & \cdot\ \cdot & y_{1n}(j\omega) \\ y_{21}(j\omega) & y_{22}(j\omega) & \cdot\ \cdot & y_{2n}(j\omega) \\ \cdot & \cdot & \cdot & \\ \cdot & \cdot & \cdot & \\ y_{n1}(j\omega) & y_{n2}(j\omega) & \cdot\ \cdot & y_{nn}(j\omega) \end{bmatrix} \begin{bmatrix} V_1(j\omega) \\ V_2(j\omega) \\ \cdot \\ \cdot \\ V_n(j\omega) \end{bmatrix} = \begin{bmatrix} I_1(j\omega) \\ I_2(j\omega) \\ \cdot \\ \cdot \\ I_n(j\omega) \end{bmatrix} \tag{2.41}$$

In the limiting case $n = 1$, only $y_{11}(j\omega)$ is defined and the one-port becomes a branch element.

Examples of network elements requiring port descriptions are given below.

(a) At the small wavelengths associated with high frequencies, certain network elements which are lumped at lower frequencies appear distributed and must be modeled accordingly. The transmission line (Figure 2.9) is a frequent component in such models. It is usually used as a two-port, i.e., with an input and output. The $\underset{\sim}{Y}$ matrix, determined from the defining

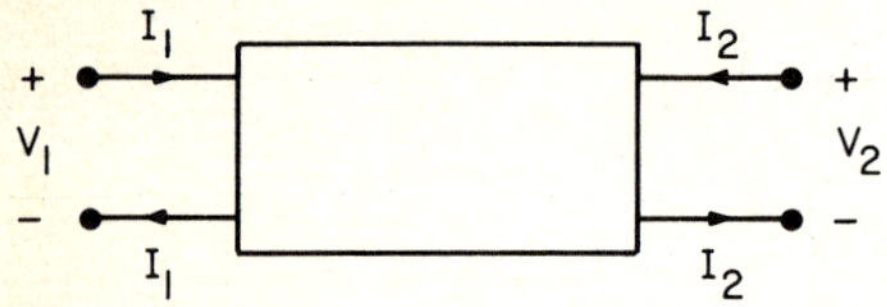

FIG. 2.9 The Transmission Line as a Two-Port Network

partial differential equations, is given by

$$\underset{\sim}{Y} = \begin{bmatrix} Y_o \coth \beta & -Y_o/\sinh \beta \\ -Y_o/\sinh \beta & Y_o \coth \beta \end{bmatrix} \tag{2.42}$$

where

$Y_o = \gamma/(r + j\omega\ell)$, the characteristic admittance $(=1/Z_o)$

$\gamma = \sqrt{(r + j\omega\ell)(g + j\omega c)}$

$d =$ length of line

$\beta = d\gamma$

$r =$ resistance/unit length

$\ell =$ inductance/unit length

g = conductance/unit length

c = capacitance/unit length

Two frequently occurring special lines are

(1) the lossless line ($r = g = 0$, $\gamma = j\omega\sqrt{\epsilon}/3 \times 10^8$, $Z_0 = R_0 = \sqrt{\ell/c}$ where

ϵ = relative dielectic constant

$$\coth\beta = -j\cot(d\omega\sqrt{\epsilon}/3 \times 10^8)$$

$$\sinh\beta = j\sin(d\omega\sqrt{\epsilon}/3 \times 10^8)$$

(2) the RC line ($\ell = g = 0$, $\beta = \omega rce^{j\pi/4}$, $Y_0 = \sqrt{\frac{r}{\omega c}e^{j\pi/4}}$) where

$$\coth\beta = \frac{\cosh a \cos a + j \sinh a \sin a}{\sinh a \cos a + j \cosh a \sin a}$$

$$\sinh\beta = \sinh a \cos a + j \cosh a \sin a$$

$$a = d\omega rc/\sqrt{2}$$

(3) At high frequencies, devices such as transistors are often difficult to describe with lumped models based on the equations of physics. Indeed, if the model is intended only to represent the device in a frequency domain network analysis, then it suffices to determine by measurement the port matrix at each frequency where a network analysis is desired (we will shortly show this). This latter model, a collection of complex-valued port matrices, is a precise representation of the device independent of the physical principles on which the device depends for operation (Figure 2. 10).

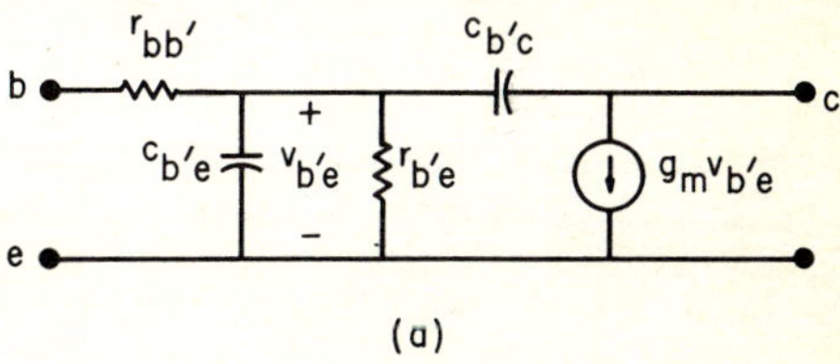

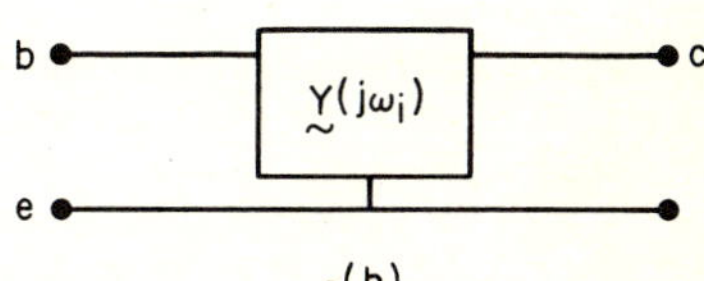

FIG. 2. 10 Lumped and Port Description of Transistor

2. 6. 2 Microwave Circuit Analysis [2. 5]

In general, a microwave circuit model will contain both branch and port models. For example, a one-stage microwave amplifier has the form of Figure 2. 11a. The thick lines represent interconnections that are modeled by lossless transmission lines, whereas source and load resistors are validly modeled as lumped elements. The lines a and d are quarter wave length lines* at some frequency in the range of interest, and so appear as

*A lossless quarter wavelength line terminated with R_L has a look-in impedance of R_0^2/R_L at the other end.

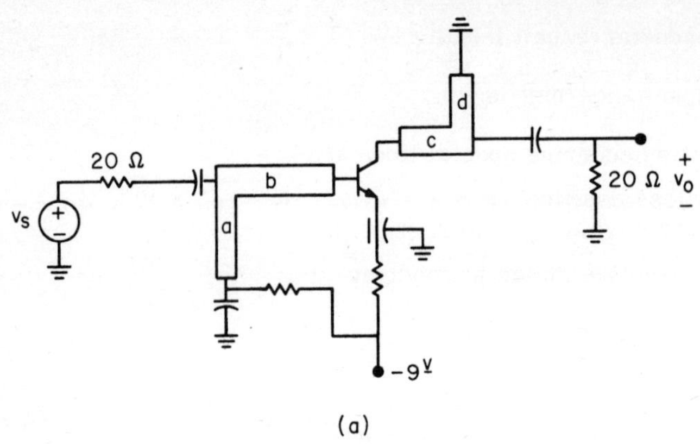

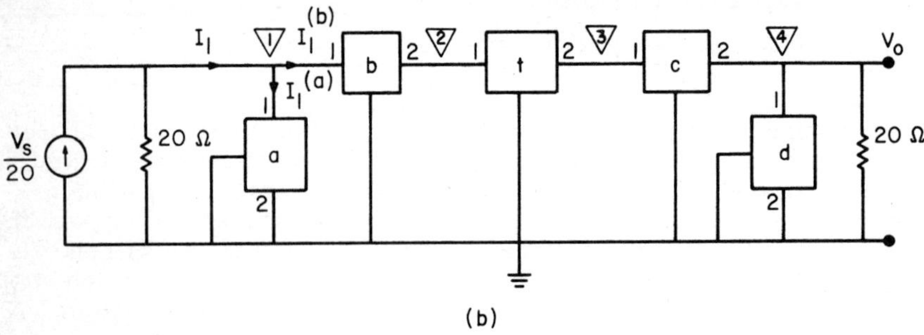

FIG. 2.11 Microwave Amplifier and Its Small Signal Model

open circuits as that frequency. Lines b and c have the purpose of matching the transistor to source and load resistances.**

Such circuits may be analyzed in the frequency domain by methods closely akin to traditional node and mesh analysis. For example, consider the amplifier of Figure 2.11a. The small signal model is shown in Figure 2.11b. Let us assume that the two-port Y matrix of the transistor is given. We can calculate the Y matrix of a lossless transmission line given the characteristic impedance R_0, the line length d, and the dielectric constant ϵ. Let the entries of Y be denoted $y_{ij}^{(k)}$, where k denotes the transmission line or the transistor. We can then write a set of node equations for the circuit as follows. At node ∇1, KCL gives.

$$\frac{V_s}{20} = \frac{V_1}{20} + I_1 \tag{2.43}$$

But I_1 is the sum of two currents flowing into transmission line two-ports a and b. Each of these currents may in turn be related to transmission line port voltages as follows:

**A lossless line will match two resistances R_1 and R_2 for maximum power transfer if $R_0 = \sqrt{R_1 T_2}$.

$$I_1^{(a)} = y_{11}^{(a)} V_1^{(a)} + y_{12}^{(a)} V_2^{(a)}$$

$$= y_{11}^{(a)} V_1$$

since $V_2^{(a)} = 0$; further

$$I_1^{(b)} = y_{11}^{(b)} V_1^{(b)} + y_{12}^{(b)} V_2^{(b)}$$

$$= y_{11}^{(b)} V_1 + y_{12}^{(b)} V_2$$

Equation (2.43) then becomes

$$\frac{V_s}{20} = \left(\frac{1}{20} + y_{11}^{(a)} + y_{11}^{(b)}\right) V_1 + y_{12}^{(b)} V_2$$

Similarly, at node $\nabla\!\!\!\!2$

$$0 = y_{21}^{(b)} V_1 + (y_{22}^{(b)} + y_{11}^{(t)}) V_2 + y_{12}^{(t)} V_3$$

This formulation process is continued for all nodes. The $\tilde{Y}_N$ matrix of the network is evaluated at each frequency ω_i by (1) calculating the $\tilde{Y}$ matrix for each transmission line, (2) measuring the transitor $\tilde{Y}^{(t)}$ matrix at frequency ω_i, and (3) calculating the branch admittances.

Example 2.6. For the network model of Figure 2.11, assume the transmission line and transistor data of Table 2.7 (the transistor data is given in terms of scattering matrix measurements described in Summary 2.1). A computer program to evaluate the voltage transfer function V_o/V_s is given in Appendix Table 2.2; however, the node equations are only partially programmed. The completed program yields the voltage gains of Table 2.8 at the four frequencies for which transistor measurements are given.

TABLE 2.7 Data for Network of Figure 2.11

Line	R_o	length
a	20 ohms	4.66 mm
b	20 ohms	5.55 mm
c	20 ohms	5.78 mm
d	20 ohms	4.66 mm

$\epsilon = 9.$

(a) Transmission Line Data

TABLE 2. 7 (Continued)

Frequency (ghtz)	Scattering matrix s_{11}	(magnitude, phase in degrees) s_{12}	s_{21}	s_{22}
1. 5	(. 5120, 193.)	(. 0750, 56. 0)	(2. 24, 72. 0)	(. 5350, 340.)
2. 0	(. 5750, 170.)	(. 0760, 56. 0)	(1. 10, 67. 0)	(. 530, 329.)
2. 5	(. 6120, 170.)	(. 09000, 51.)	(1. 40, 52. 5)	(. 528, 322.)
3. 0	(. 650, 168.)	(. 0900, 66.)	(1. 15, 47. 3)	(. 528, 332.)

$$Z_T = \begin{bmatrix} 20. & 0 \\ 0 & 20. \end{bmatrix}$$

(b) Measured Transistor Data

TABLE 2. 8 Response of Network of Figure 2. 11

FREQUENCY HTZ	GAIN DB	PHASE DEG
0.15000D 10	-4.630	94.819
0.20000D 10	-8.435	53.257
0.25000D 10	-5.623	-1.440
0.30000D 10	-6.566	-35.770

The scattering matrix $\underset{\sim}{S}$ is a port description which describes power transfer between ports just as the $\underset{\sim}{Y}$ matrix describes current and voltage transfer relationships. However, to describe power transfer, the ports must be terminated in impedances which absorb and reflect power. For a two port with terminating impedance as shown, let the scattering matrix and its terminating im-

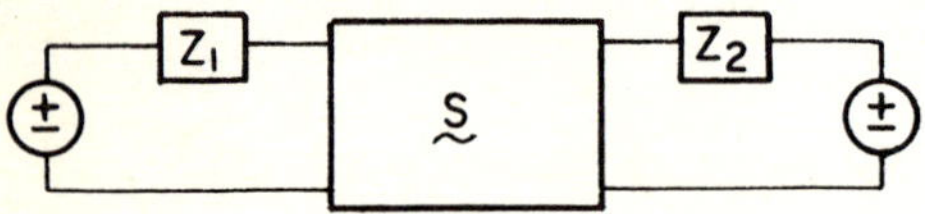

pedances be denoted by

$$\underset{\sim}{S} = \begin{bmatrix} s_{11} & s_{12} \\ s_{21} & s_{22} \end{bmatrix} \qquad \underset{\sim}{Z}_T = \begin{bmatrix} Z_1 & 0 \\ 0 & Z_2 \end{bmatrix}$$

The relationships between the $\underset{\sim}{Y}$ and $\underset{\sim}{S}$ matrices are

$$\underset{\sim}{Y} = \underset{\sim}{Z}_T^{-1/2} (\underset{\sim}{I} + \underset{\sim}{S})^{-1} (\underset{\sim}{I} - \underset{\sim}{S}) \underset{\sim}{Z}_T^{-1/2}$$

The scattering matrix is easier to measure at high frequencies, where it is difficult to achieve the exact short circuit necessary to determine $\underset{\sim}{Y}$ directly [2. 5].

Summary 2. 1 The Scattering Matrix

2.6.3 Network "Peeling"

One of the most intriguing and highly useful computer applications in high frequency circuit design is the process of network "peeling." The concept is illustrated in Figure 2.12. Measurements yielding the high frequency $\underset{\sim}{Y}$ matrix of a transistor are recorded.

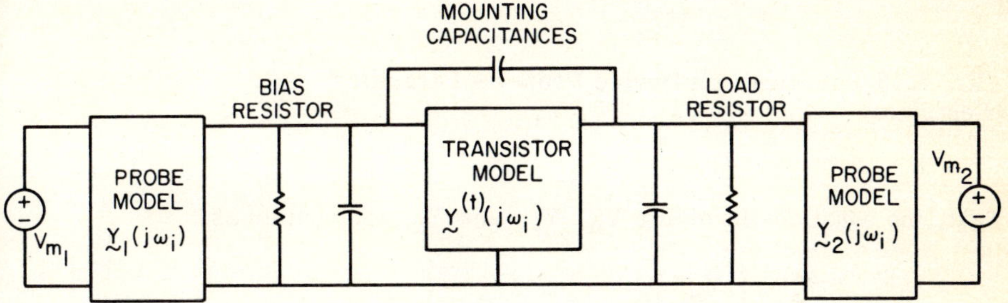

FIG. 2.12 Measurement of $\underset{\sim}{Y}$ Matrix of Transistor

Readings are subject to attenuation and phase shift errors introduced by (1) the measuring probe and associated measurement electronics, and (2) mounting capacitances (and possibly inductance) as well as biasing resistances, which are external to the transistor. However, all of these parasitics can usually be modeled and measured in lumped or port form. Thus, the following problem is posed: given a measured $\underset{\sim}{Y}$ matrix, "peel" away known parasitics to determine the true transistor $\underset{\sim}{Y}^{(t)}$ matrix.

This problem is generalized to the model of Figure 2.13a, where a device description $\underset{\sim}{Y}_d$ is to be found in terms of a parasitic description $\underset{\sim}{Y}_p$ and a measured port matrix $\underset{\sim}{Y}_m$. As a first step in solution, the network is formally partitioned as in Figure 2.13b, where voltage-current notation at adjacent ports defines the conditions for connection.

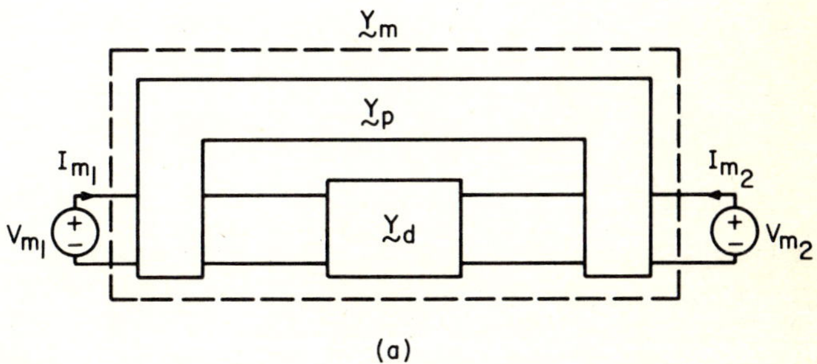

FIG. 2.13 Partitioning a Device From Its Parasitics

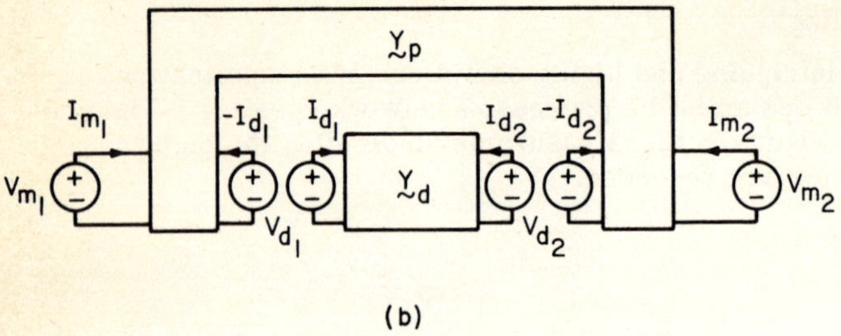

FIG. 2.13 Partitioning a Device From Its Parasitics (Continued)

Next, the equations involving $\underset{\sim}{Y}_d$, $\underset{\sim}{Y}_p$, and $\underset{\sim}{Y}_m$ are written as

$$\begin{bmatrix} I_{d_1} \\ I_{d_2} \end{bmatrix} = \underset{\sim}{Y}_d \begin{bmatrix} V_{d_1} \\ V_{d_2} \end{bmatrix} \tag{2.44}$$

$$\begin{bmatrix} I_{m_1} \\ I_{m_2} \end{bmatrix} = \underset{\sim}{Y}_m \begin{bmatrix} V_{m_1} \\ V_{m_2} \end{bmatrix} \tag{2.45}$$

$$\begin{bmatrix} -I_{d_1} \\ -I_{d_2} \\ \hline I_{m_1} \\ I_{m_2} \end{bmatrix} = \underset{\sim}{Y}_p \begin{bmatrix} V_{d_1} \\ V_{d_2} \\ \hline V_{m_1} \\ V_{m_2} \end{bmatrix} \tag{2.46}$$

$$= \left[\begin{array}{c|c} \underset{\sim}{Y}_1 & \underset{\sim}{Y}_2 \\ \hline \underset{\sim}{Y}_3 & \underset{\sim}{Y}_4 \end{array}\right] \begin{bmatrix} V_{d_1} \\ V_{d_2} \\ \hline V_{m_1} \\ V_{m_2} \end{bmatrix} \tag{2.47}$$

where the parasitic matrix $\underset{\sim}{Y}_p$ has been partitioned for later convenience. Substitution of (2.45) into (2.47) yields

$$\begin{bmatrix} -I_{d_1} \\ -I_{d_2} \\ \hline 0 \\ 0 \end{bmatrix} = \left[\begin{array}{c|c} \underset{\sim}{Y}_1 & \underset{\sim}{Y}_2 \\ \hline \underset{\sim}{Y}_3 & \underset{\sim}{Y}_4 - \underset{\sim}{Y}_m \end{array}\right] \begin{bmatrix} V_{d_1} \\ V_{d_2} \\ \hline V_{m_1} \\ V_{m_2} \end{bmatrix}$$

Elimination of V_{m_1} and V_{m_2} then gives

$$\begin{bmatrix} -I_{d_1} \\ -I_{d_2} \end{bmatrix} = [\, \underset{\sim}{Y}_1 - \underset{\sim}{Y}_2 [\, \underset{\sim}{Y}_4 - \underset{\sim}{Y}_m]^{-1} \underset{\sim}{Y}_3] \begin{bmatrix} V_{d_1} \\ V_{d_2} \end{bmatrix}$$

so that

$$\underset{\sim}{Y}_d = -\underset{\sim}{Y}_1 + \underset{\sim}{Y}_2 [\, \underset{\sim}{Y}_4 - \underset{\sim}{Y}_m]^{-1} \underset{\sim}{Y}_3 \tag{2.48}$$

Equation (2.48) therefore represents the process of "peeling" away circuit parasitics.

Example 2.7. The $\underset{\sim}{Y}_d$ matrix of the transistor of Figure 2.14 is to be determined at a scaled frequency f = .0003 from the $\underset{\sim}{Y}_p$ and $\underset{\sim}{Y}_m$ matrices, both of which are calculated at f = .0003 from the given parasitics (see Table 2.9). This $\underset{\sim}{Y}_d$ matrix will then be compared with the precise $\underset{\sim}{Y}$ matrix of the pi transistor model shown, which is

$$\underset{\sim}{Y} = \begin{bmatrix} .00101 & -.00001 \\ .09999 & .00011 \end{bmatrix} \tag{2.49}$$

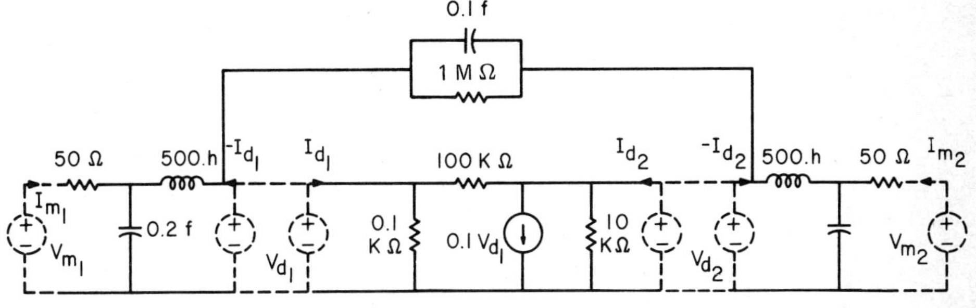

FIG. 2.14 Circuit to Be Peeled

TABLE 2.9 Calculated Port Matrices for Example 2.7
(Matrix entries are given in (magnitude, degrees))

$$\underset{\sim}{Y}_m = \begin{bmatrix} (1.75626 \times 10^{-3},\ 49.810) & (1.77500 \times 10^{-4},\ 260.979) \\ (9.40021 \times 10^{-2},\ -5.849) & (1.45575 \times 10^{-3},\ 78.449) \end{bmatrix}$$

$$\underset{\sim}{Y}_p = \left[\begin{array}{cc|cc} (2.00161 \times 10^{-2},\ .539) & (1.886319 \times 10^{-4},\ 267.535) & (2.00071 \times 10^{-2},\ 178.919) & (3.55562 \times 10^{-6},\ -2.465) \\ (1.886319 \times 10^{-4},\ 267.535) & (2.00161 \times 10^{-2},\ .539) & (3.55562 \times 10^{-6},\ -2.465) & (2.00071 \times 10^{-2},\ 178.919) \\ \hline (2.00071 \times 10^{-2},\ 178.919) & (3.55562 \times 10^{-6},\ -2.465) & (1.99964 \times 10^{-2},\ -1.080) & (6.702186 \times 10^{-8},\ 87.535) \\ (3.55562 \times 10^{-6},\ -2.465) & (2.00071 \times 10^{-2},\ 178.919) & (6.702186 \times 10^{-8},\ 87.535) & (1.99964 \times 10^{-2},\ -1.080) \end{array}\right]$$

TABLE 2.10 $\underset{\sim}{Y}_d$ Matrices for $\underset{\sim}{Y}_p$ and $\underset{\sim}{Y}_m$ of Different Accuracies

Y_d			
	0.001011		-0.000010
+J	0.000004	-J	0.000001
	0.099905		0.000111
+J	0.000094	+J	0.000005

(a) Full Accuracy

Y_d			
	0.001031		-0.000011
-J	0.000107	-J	0.000002
	0.100077		0.000126
+J	0.000065	-J	0.000107

(b) Magnitude to three significant figures phase to nearest degree

Y_d			
	0.001015		-0.000018
+J	0.000591	-J	0.000002
	0.100175		0.000109
-J	0.000037	+J	0.000592

(c) Magnitude to two significant figures phase rounded to 0°, 90°, 180°, 270°

First, the matrices $\underset{\sim}{Y}_i$ of (2.48) are identified as shown by the partitions in Table 2.9. Next, (2.48) is implemented in program form in Appendix Table 2.3. This program is written to allow reading and printing in rectangular or polar form. The calculated $\underset{\sim}{Y}_d$ matrix is shown in Table 2.10a, to be compared with the $\underset{\sim}{Y}$ of (2.49).

This peeling technique is by nature a subtractive process. This hints that care must be observed in its formulation and numerical solution to avoid ill-conditioning in the solution. This observation is substantiated by examining $\underset{\sim}{Y}_m$ and $\underset{\sim}{Y}_p$ in Table 2.8 and Y_d in Table 2.9a. Whereas the phase angles of $\underset{\sim}{Y}_m$ indicate comparable real and imaginary parts, the large components of $\underset{\sim}{Y}_p$ are all nearly real valued. Thus, (2.48) results in the subtraction of nearly real-valued data from complex-valued data to yield the nearly real valued $\underset{\sim}{Y}_d$ of Table 2.9a. A premium is therefore placed on accurate calculation or measurement of $\underset{\sim}{Y}_p$. For example, if $\underset{\sim}{Y}_p$ and $\underset{\sim}{Y}_m$ were measured, the accuracy of $\underset{\sim}{Y}_d$ of Table 2.10b-c would be more representative of "peeling" results. One way of avoiding this problem is indicated in Problem 2.6.

2.7 S - PLANE ANALYSIS

2.7.1 Review

The node analysis of small signal models of electronic circuits using the Laplace Transform variable s is certainly one of the most common uses of circuit analysis. If implemented improperly (noniteratively), this procedure has computational liabilities that limit its utility to circuits in the ten-to-fifteen node category. We now take a critical view of this calculation, with constructive algorithms saved for Chapter X.

To review, s-plane analysis requires a calculation of a network function in the Laplace Transform variable s representing the response-excitation behavior of a linear network. Restricting our interest to the rational functions of s which represent lumped networks, this function usually has one of two forms.

Coefficient form:

$$T(s) = \frac{\sum_{i=0}^{n} a_i s^i}{\sum_{j=0}^{m} b_j s^j}$$

where $a_n \neq 0$ and $b_m \neq 0$;

Factored form:

$$T(s) = \frac{K \prod_{i=1}^{n} (s - z_i)}{\prod_{j=1}^{m} (s - p_j)}$$

Whichever form is available, the frequency response $O(j\omega)$ of the response is simply

$$O(j\omega) = T(j\omega)\, I(j\omega)$$

where $I(j\omega)$ is the frequency domain representation of the input.

2.7.2 The Accuracy Problem for Narrow-band Networks

There are two sources of numerical difficulties in generating $T(s)$ using conventional techniques such as node analysis. The first is related to the cancellation already noted in Section 2.5.2 and is considered in Problem 2.8. We will discuss the second, which occurs in the analysis of narrow-band networks.

In certain types of network structures, the coefficients of the network function $T(s)$ can be assembled without cancellation and common factors by using special reduction techniques. Ladder networks are one important class. Except for round-off errors, then, the coefficients can be guaranteed to be accurate to the same number of significant figures as carried in the calculation. Even so, problems still exist in computing a frequency response, especially in the case of narrow-band networks.* Consider, for example, the network of Figure 2.15. The driving point admittance is

$$Y(s) = \frac{N(s)}{D(s)} = \frac{500\,(s)\,(s^2 + 0.02s + 1)}{250,000.\,s^4 + 1000.\,s^3 + 500,002.\,s^2 + 1000.\,s + 250,000.} \tag{2.50}$$

If each coefficient is assumed to have an infinite number of zeros beyond the decimal point, then

---

*Polynomials of high order can be ill-conditioned even with widely spaced roots; see [2.1], p. 378.

$$Y(s) = \frac{500\,(s)\,(s^2 + 0.02s + 1)}{250,000.\,(s + 0.0010 \pm j0.9990)\,(s + 0.0010 \pm j1.0010)}$$

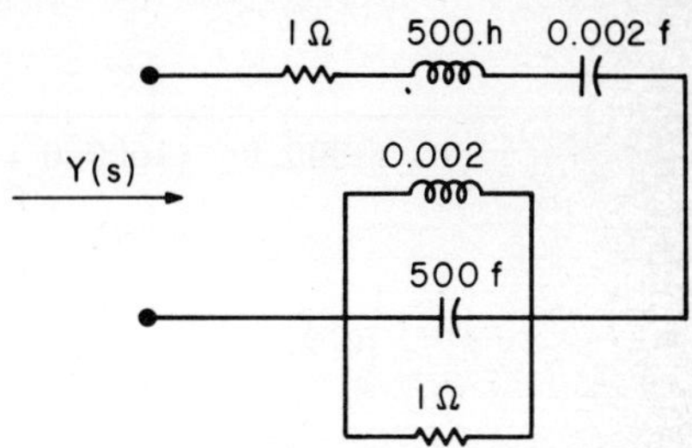

FIG. 2.15 Narrow-band Network

If we are given the factored form of $Y(s)$, the frequency response may be found to normal accuracy using five significant figures in the calculations. However, suppose the same five significant figures had been used in calculation of the poles from (2.50); then the coefficient of s^2 would be rounded to 5×10^5 and the resultant polynomial would appear to have a root at $s = \pm j\,1$! We may conclude that, associated with a given polynomial coefficient accuracy, there is a region in the s-plane where the roots may lie. Obviously, the frequency response $Y(j\omega)$ is then highly dependent on the number of significant figures carried in the coefficients.

The number of significant figures required can be estimated if the relative bandwidth and the degree of the polynomial can be determined a priori. For example, if the roots of $D(s)$ are required to five significant figures, then $D(s_i)$ is permitted to be zero for

$$-0.00105 \leq \text{Re } s_i \leq -0.000995$$

$$0.099895 \leq |\text{Im } s_i| \leq 0.99905$$

$$1.0095 \leq |\text{Im } s_i| \leq 1.00105$$

Choosing a number slightly beyond this range, say $s_i = -0.00106 + j1.00106$, we find the true value of $D(s_i)$ is

$$|D(s_i)| = |250,000\,(-0.00006 + j2.00006)\,(-0.00006 + j0.00206)$$

$$(-0.00006 + j2.00206)\,(-0.00006 + j0.00006)|$$

$$\approx |10^6\,(-0.00006 + j0.00206)\,(-0.00006 + j0.00006)|$$

$$\approx 10^{-1}$$

But $|D(s_i)|$ may also be evaluated from the coefficients as given in (1.22); therefore, if the coefficients are not known to at least one decimal place, the remainder would be zero rather than 10^{-1}, and the required root resolution could not be achieved. Note that the manner of extracting the roots does not change this inherent accuracy restriction.

The accuracy problem may usually be avoided in narrow-band network analysis by separately analyzing at a single frequency the subnetworks causing multiple resonances and then combining the results. Thus, retaining the same five significant figures in our calculation, the total admittance Y(j1) of the network of Figure 2.15 is

$$Y(j1) = \frac{1}{(1 + j1000.0 - j1000.0 \pm j0.1) + \dfrac{1}{1 + j1000.0 - j1000.0 \pm j0.1}}$$

$$= \frac{1}{2. \pm j0.2}$$

which now has one significant figure. We may thus visualize the descrepancy in accuracy required for single frequency and for s-plane analysis as arising from the loss of identification between resonances and the elements producing the resonances.

2.8 A GENERAL LINEAR RESISTIVE CIRCUIT ANALYSIS PROGRAM (RCAP)

We conclude this chapter by presenting a circuit analysis program which can analyze a variety of circuits, using node analysis as an equation formulation procedure. Although the algorithm is complicated relative to the RNET program of Appendix Table 2.1, it is easily extendable to handle analysis of nonlinear and transient analysis problems. Therefore, an investment in studying the program code is justifiable.

An electrical network is described by two general types of information:

(a) topological or structural, given by branch node numbers, and
(b) constitutive, as given by the branch element type (e.g., a resistance) and value (e.g., 2 Ω).

In the RNET program of Appendix Table 2.1 only the element value information is entered as input data with the type and topological information included in the NODEQ subroutine. To make the equation formulation subroutine NODEQ applicable to any network, all three types of element information must be read as input data. It is common to describe each element on a single card in a form similar to

```
  R1      01     02     5.                          (2.51)
  type       nodes(m, n)    value(c)
```

IBNCH (1, NEL)	Element name
IBNCH (2, NEL)	Type of contro;
IBNCH (3, NEL)	Controlling branch
IBNCH (4, NEL)	Node number (positive reference)
IBNCH (5, NEL)	Node number (negative reference)
IBNCH (6, NEL)	Code number denoting type of element (assigned by NCODE subroutine)

(a) Description of IBNCH Array in Subroutine INPUT

FIG. 2.16 Explanation of Parts of RCAP

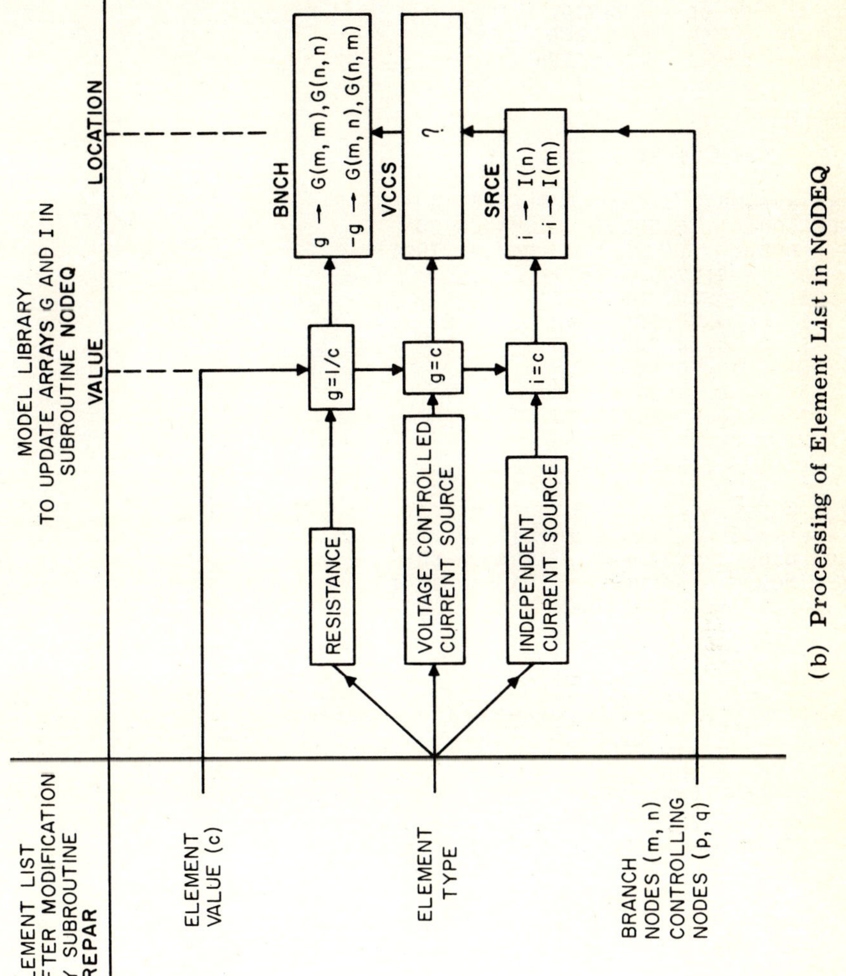

(b) Processing of Element List in NODEQ

FIG. 2.16 Explaration of Parts of RCAP

TABLE 2.11 Amplifier Analysis Using RCAP

```
SMALL SIGNAL AMPLIFIER ANALYSIS
I1          00 01      .00001,.000005
R1          01 00      100000.
R2          01 00      100000.
R3          01 00      10000.,20000.
R4          01 03      500.
R5          03 00      14.
R6          02 00      5000.
R7          02 00   V  30000.
I3   VR4    02 01      .00196
I2   VR5    02 03      .07
AMEN
```

(a) Computer Input

```
**********SMALL SIGNAL AMPLIFIER ANALYSIS
BRANCH NODE     ELEMENT      BRANCH CON-
NAME    NOS.    VALUE        NAME   TROL
I1     0  1  0.1000000D-04
R1     1  0  0.1000000D 06
R2     1  0  0.1000000D 06
R3     1  0  0.1000000D 05
R4     1  3  0.5000000D 03
R5     3  0  0.1400000D 02
R6     2  0  0.5000000D 04
R7     2  0  0.3000000D 05
I3     2  1  0.1960000D-02 R4      V
I2     2  3  0.7000000D-01 R5      V
OUTPUT NODES =  2  0  OUTPUT VOLTAGE =    -13.0609756 -
**********SMALL SIGNAL AMPLIFIER ANALYSIS
BRANCH NODE     ELEMENT      BRANCH CON-
NAME    NOS.    VALUE        NAME   TROL
I1     0  1  0.5000000D-05
R1     1  0  0.1000000D 06
R2     1  0  0.1000000D 06
R3     1  0  0.2000000D 05
R4     1  3  0.5000000D 03
R5     3  0  0.1400000D 02
R6     2  0  0.5000000D 04
R7     2  0  0.3000000D 05
I3     2  1  0.1960000D-02 R4      V
I2     2  3  0.7000000D-01 R5      V
OUTPUT NODES =  2  0  OUTPUT VOLTAGE =    -10.2980769
**********SMALL SIGNAL AMPLIFIER ANALYSIS
```

(b) Computer Output

A general linear Resistive Circuit Analysis Program (RCAP) accepting data of the form of (2.51) is listed in Appendix Table 2.4. A detailed description of the input data format for this program is given in Appendix Table 2.5.

The subroutines called by the main program are similar in function to those in Appendix Table 2.1. The key to construction of the node equations from the elements specified as in (2.51) is a model library. From knowledge of the element type and value the library generates conductances and current sources that are inserted in appropriate locations of G and I. This process is summarized in Figure 2.16 and implemented in subroutine NODEQ in Appendix Table 2.4.

Sample input and output data to solve the amplifier of Figure 2.2 are given in Tables 2.11a and 2.11b for different values of source resistance. It should be noted that the model library is restricted to only one type of controlled source - the voltage-controlled current source. As observed in Table 2.11a, the input data must be similarly restricted.

Problems

2.1 The LU factorization and back substitution program of Appendix Table 2.1 is incomplete, certain subscripts being absent in the forward and back substitution steps. Edit the program to obtain the answer in Table 2.3.

2.2 A representation of a two-stage amplifier is shown in Figure P2.1. The output voltage v_o resulting from a unit input $i_s = 1$ is to be found. The program of Appendix Table 2.1 can be used except for the node equation formulation subroutine NODEQ and the changes required by Problem 2.1.

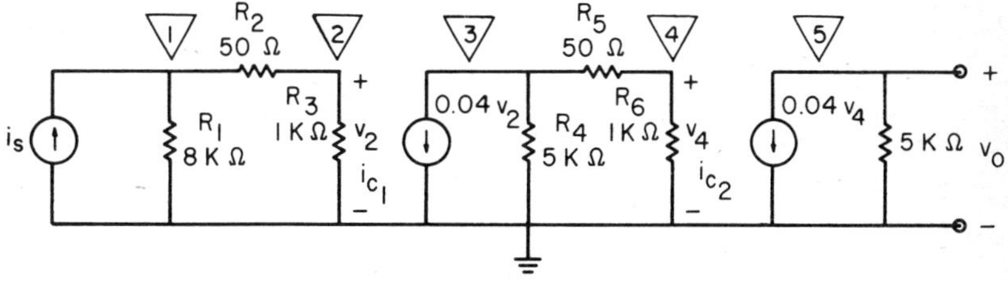

FIG. P2.1 Two Stage Amplifier Model

2.3 Edit the RCAP program of Appendix Table 2.4 in subroutine VCCS so that a voltage controlled current source can be accommodated. Then analyze the amplifier model of Figure P2.1.

In performing this edit, the reader will have to trace the components of arrays IBNCH and ELVAL through the READ and PREPAR subroutines. The subroutine VCCS to be written is of the same general form as the BNCH subroutine used to update $\tilde{G}_n$ when a resistance is processed.

2.4 Convert the RCAP program to a general frequency domain analysis program (FCAP). The frequency plotting information can be read similarly to the transient plotting information in TCAP (see Appendix Table 2.5).

2.5 The parasitics associated with the external leads, can enclosure, and bonding leads of a mounted transistor are shown in Figure P2.2. The measured Y parameters are shown in Figure P2.3. Calculate the $\tilde{Y}_d$ matrix of the transistor at f = 200, 400, 600, and 800 MHz.

First it is suggested that the problem be frequency scaled by 10^5 and impedance scaled by 10^4. The $\tilde{Y}_m$ matrix of the parasitic network can be calculated using FCAP in Problem 2.4.

2.6 In developing the network peeling algorithm, we arbitrarily choose the parasitic elements to be defined by short circuit matrix $\tilde{Y}_p$. To obtain a better conditioned peeling algorithm, we note that in practice the parasitic network does not result in gross differences between $\tilde{Y}_d$ and $\tilde{Y}_m$. That is,

$$\begin{bmatrix} I_{m_1} \\ I_{m_2} \end{bmatrix} \approx \begin{bmatrix} -I_{d_1} \\ -I_{d_2} \end{bmatrix} \qquad \begin{bmatrix} V_{d_1} \\ V_{d_2} \end{bmatrix} \approx \begin{bmatrix} V_{m_1} \\ V_{m_2} \end{bmatrix}$$

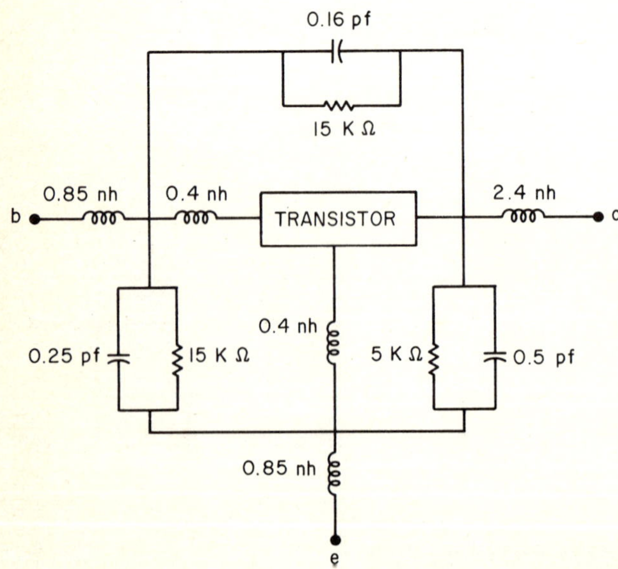

FIG. P2.2

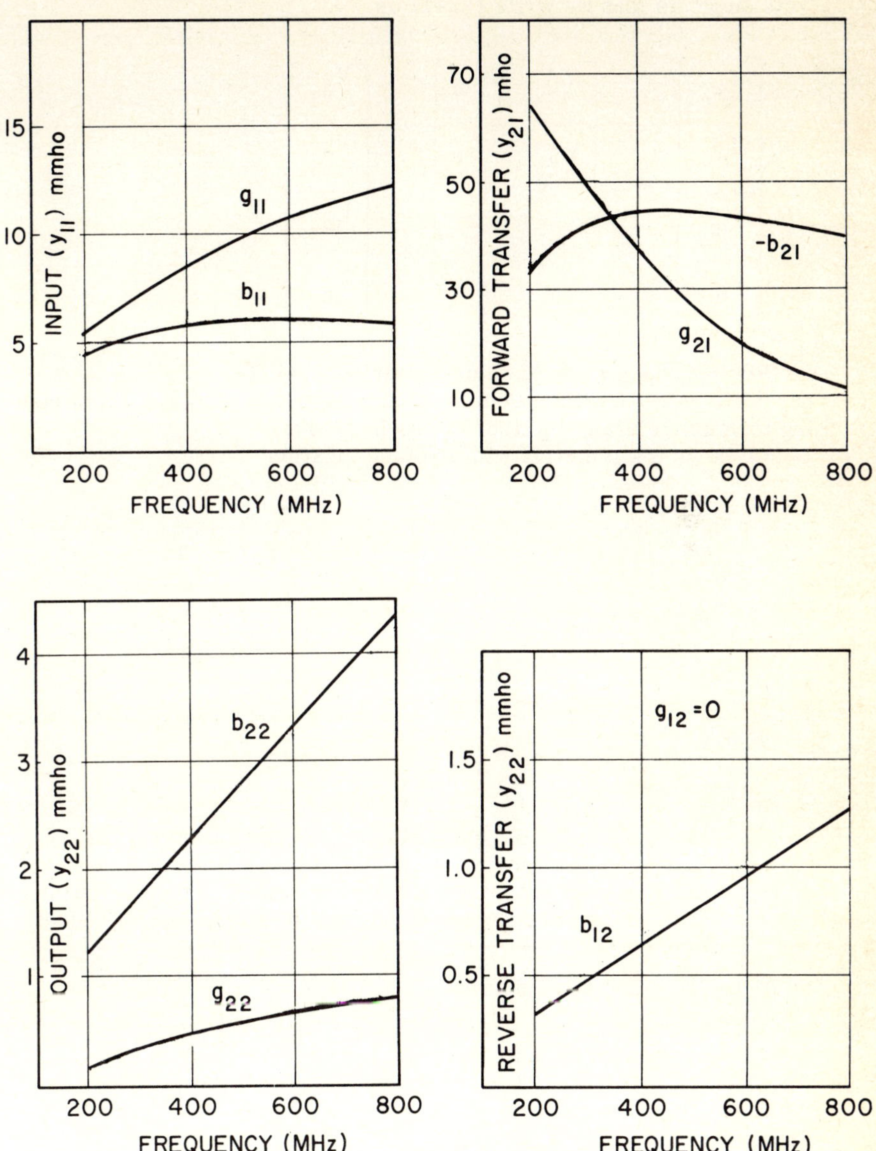

FIG. P2.3 Common-emitter $\underset{\sim}{Y}_m$ Matrix Parameters

This suggests that we write (2.47) as

$$\begin{bmatrix} -I_{d_1} \\ -I_{d_2} \\ \hline V_{m_1} \\ V_{m_2} \end{bmatrix} = \begin{bmatrix} \underset{\sim}{Y}_1 & \underset{\sim}{\alpha}_2 \\ \underset{\sim}{\mu}_3 & \underset{\sim}{Z}_4 \end{bmatrix} \begin{bmatrix} V_{d_1} \\ V_{d_2} \\ \hline I_{m_1} \\ I_{m_2} \end{bmatrix}$$

so that $\underset{\sim}{\alpha}_2$ and $\underset{\sim}{\mu}_3$ would be unit matrices and $\underset{\sim}{Y}_1 = \underset{\sim}{Z}_4 = \underset{\sim}{0}$ if the parasitic networks were simply connecting wires. Derive a new peeling formula relating $\underset{\sim}{Y}_d$ and $\underset{\sim}{Y}_m$ as in (2.48).

2.7 Let $\underset{\sim}{A}$ be an n x n matrix with the factorization

$$\underset{\sim}{A} = \underset{\sim}{L}\underset{\sim}{U}$$

and let $\underline{e}^T = [1\ 1\ 1\ .\ .\ 1]$.

(a) Show that the matrix

$$\underset{\sim}{A}_{zs} = \begin{bmatrix} \underset{\sim}{A} & -\underset{\sim}{A}^T \underline{e} \\ -\underline{e}^T \underset{\sim}{A} & \underline{e}^T \underset{\sim}{A}\, \underline{e} \end{bmatrix}$$

is zero sum, i.e., each row and column sums to zero.

(b) Find the LU factorization of $\underset{\sim}{A}_{zs}$ in terms of $\underset{\sim}{L}$, $\underset{\sim}{U}$, and $\underline{e}$.

(c) What special property does the factorization of $\underset{\sim}{A}_{zs}$ possess?

(d) Discuss how the answer to (c) can be used to avoid cancellation.

See Ref. [2.6] for more details.

2.8 Let the subtraction in Eq. (2.36) cause a round-off error denoted by $1 + \epsilon_3$. Is ϵ_3 magnified by $Y_2/(Y_1 + Y_3)$?

2.9 The network of Figure P2.4 is to be analyzed by node analysis. Determine V_2/I_1 in terms of s and note the complete cancellation of certain powers of s.

2.10 Let it be assumed the complete solution (all x_i variables) is required of a set of linear simultaneous equations. Consider the possibility of reducing an entire column to zero at a time, i.e., both above and below the diagonal term (this is termed Gauss-Jordan reduction). Compare the computational efficiency of this procedure with Gauss elimination.

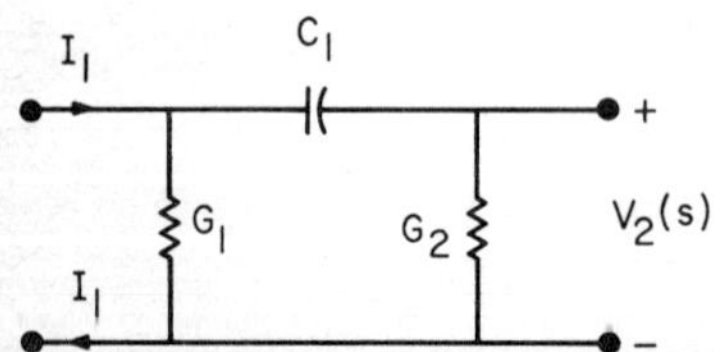

FIG. P2.4

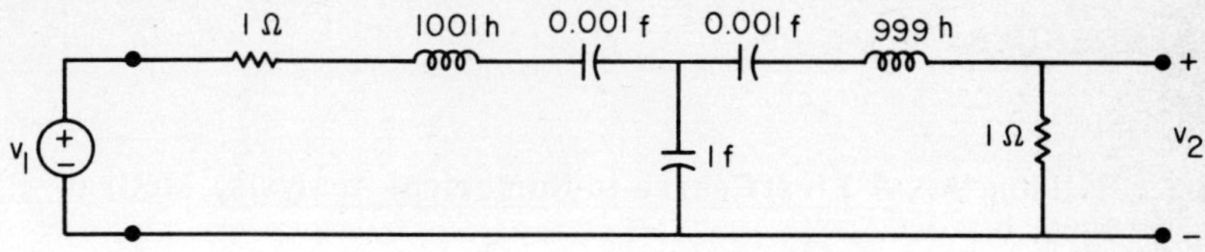

FIG. P2.5

2.11 Calculate the voltage transfer function of the narrow-band filter of Figure P2.5, in terms of s. Now test the denominator for right half plane roots. How many significant figures are required in the coefficients to establish stability? (Requires a calculator or computer.)

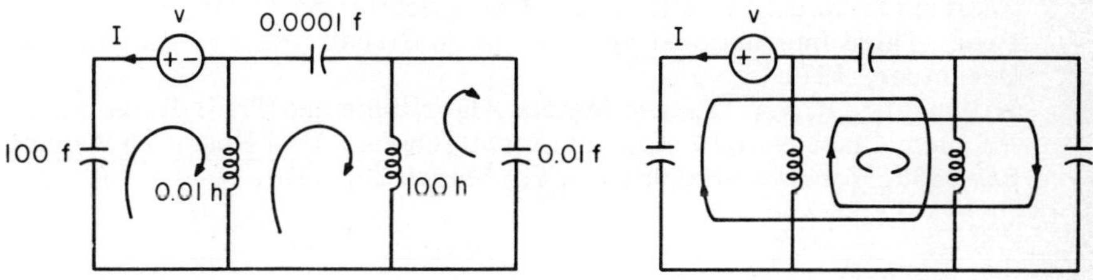

FIG. P2.6

2.12 To illustrate the effect of poor formulation on accuracy, write the equations associated with the two sets of loop currents in Figure P2.6. Solve for I/V at $\omega = 1.005$, by Gauss elimination. (Requires a calculator or computer.)

2.13 How many operations are required to solve a matrix equation given below for an unsymmetric matrix? Compare with a full matrix in each case. An x represents a nonzero entry.

$$\begin{bmatrix} x & x & x & & & & & & \\ x & x & x & & & & & & \\ x & x & x & & & & & x & \\ & & & x & x & x & & & \\ & & & x & x & x & & & \\ & & & x & x & x & & & x \\ & & & & & & x & x & x \\ & & x & & & & x & x & x \\ & & & & & x & x & x & x \end{bmatrix} \begin{bmatrix} x_1 \\ \cdot \\ \cdot \\ \cdot \\ \cdot \\ x_9 \end{bmatrix} = \begin{bmatrix} b_1 \\ \cdot \\ \cdot \\ \cdot \\ \cdot \\ b_9 \end{bmatrix}$$

2.14 Study the computational cost of factoring a symmetric matrix into the form

$$\underset{\sim}{A} = \underset{\sim}{L}\,\underset{\sim}{L}^T$$

It may be shown that, even if $\underset{\sim}{A}$ is real and symmetric, $\underset{\sim}{L}$ will be complex valued if $\underset{\sim}{A}$ is not positive definite [2.1]. See Problem 10.9 for more general factorization of symmetric matrix.

REFERENCES

2.1 Ralston, A., _A First Course in Numerical Analysis,_ McGraw-Hill; 1965.

2.2 Bingham, J. A. C., "A Method of Avoiding Loss of Accuracy in Nodal Analysis," _Proc. IEEE,_ vol. 55, no. 3, pp. 409-410; March, 1967.

2.3 Calahan, D. A., _Computer-Aided Network Design,_ Preliminary Edition, McGraw-Hill; 1968.

2.4 Calahan, D. A., and T. E. Grapes, "Cancellation Avoidance in Matrix-Based Network Analysis," _Proc. Third International Symposium on Circuit Theory,_ pp. 109-110; December, 1970.

2.5 Monaco, V. A., and P. Tiberio, "A Computer Program for Circuit Analysis from d.c. to Microwave Using Scattering Parameters," _Proc. Third International Symposium on Circuit Theory,_ pp. 119-120; December, 1970.

2.6 Willoughby, R. A., "Sparse Matrix Algorithms and Their Relation to Problem Classes and Computer Architecture," _IBM Research Report_ RC-2833, Yorktown Heights, N. Y.; March 30, 1970.

3

Nonlinear DC Circuit Analysis

3.1 INTRODUCTION

3.1.1 Types of Resistive Nonlinearities

DC (direct current) analysis refers to the analysis of network models containing no dynamic elements, or alternatively, to the steady state analysis of networks with constant excitations.

In contrast to linear, fixed resistive networks which possess unique solutions, nonlinear resistive networks may have many dc solutions. A common computer circuit exhibiting such behavior is the flip-flop, where two stable states exist. The state existing at any time is determined by the past history of the excitations. This observation serves warning that new considerations are added when nonlinear elements are allowed. Although the network equations can be formulated by methods remarkably similar to those used for linear networks, new methods of numerical solution must be learned beyond the matrix reduction techniques considered in the last chapter.

The analysis of nonlinear networks is greatly simplified by restricting the class of allowed nonlinearities. We will initially consider only those two-terminal elements that are described by v-i equations either of the form

$$v = g(i)$$

or of the form

$$i = g(v)$$

where g is a continuous single-valued function of the independent variable. (Note that the inverse g^{-1} may or may not exist.) Examples included in this class are the nonideal diodes described by the diode equation* $i_d = I_s (e^{\lambda v_d} - 1)$ and, more generally, by two-terminal memoryless elements having piecewise continuous v-i characteristics as shown in Figure 3.1a. Note that the continuity and single-valued requirements, inserted to avoid certain numerical problems, depend on the choice of independent variable; viz, the v-i curve of Figure 3.1b is continuous in i but not in v, whereas the characteristic of Figure 3.1c is not single-valued in i. Cases not to be considered are nonlinear multiterminal elements, whether represented explicitly as the nonlinear transitor beta of Figure 3.2a or implicitly from a set of device characteristics. Our methods of formulating the equations for two-terminal elements will suggest means of handling such cases.

3.1.2 Scaling

As we noted in Chapter II, an important aspect of digital computation is proper problem scaling. Since resistive circuits are not frequency-dependent and since the concept of impedance is not directly applicable to nonlinear elements such as the above, we introduce two generalizations of impedance scaling, namely, current and voltage scaling.

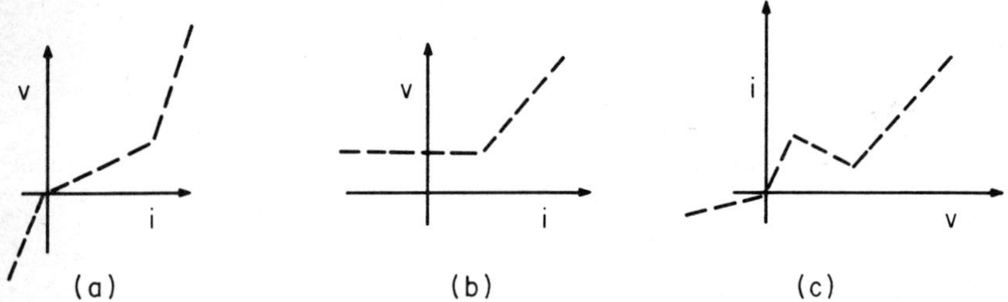

FIG. 3.1 Allowable Two-terminal Characteristics

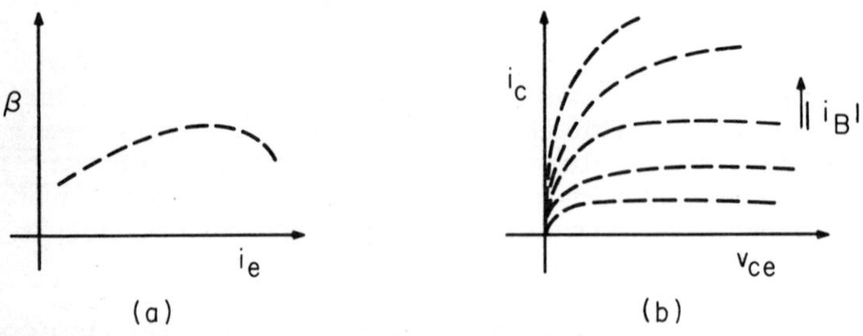

FIG. 3.2 Disallowed Characteristics

---

*$\lambda = q/kT$, where q is the charge on an electron, k is the Boltzman constant, and T is the temperature in degrees (Kelvin).

To exemplify, consider a network containing a diode and a resistor described by

$$i_d = I_s(e^{\lambda v_d} - 1), \quad i_R = v_R/R$$

We could replace i by bi, yielding

$$i_d = \frac{I_s}{b}(e^{\lambda v_d} - 1), \quad i_R = v_R/(bR)$$

or we could replace v by v/b as

$$i_d = I_s(e^{\lambda v_d/b} - 1), \quad i_R = v_R/(bR)$$

The effect on the linear element value is the same whether v or i is scaled, but the calculation of i_d is obviously altered. The solution to the nonlinear problem must then be rescaled accordingly.

3.2 SOLUTION OF NONLINEAR RESISTIVE NETWORKS

3.2.1 A Simple Resistor Diode Network

Consider the problem of solving the resistor-diode network of Figure 3.3a for the diode voltage v_d. The usual solution procedure would involve a graphical construction as shown in Figure 3.3b, with the solution being at the intersection of a load line and the diode characteristic. Graphical techniques such as this suffer from one deficiency, however: they apply principally to networks with a single nonlinearity.

The only general approach to solution of problems involving more than one nonlinearity is by iteration, i.e., the guessing of a solution and the successive refinement of that guess. As we will see, there are drawbacks to such a procedure, namely, (1) the process may in fact diverge from a solution, and (2) the user must construct a test to determine when the iteration has converged to acceptable accuracy. The best known iterative method is Newton's method, now to be described.

Using the resistor-diode network example, let us linearize the nonlinear characteristic around the "starting point" shown. This is equivalent to replacing the diode with a voltage source and a linear resistor. Analytically, the linearized diode equation is obtained by expanding i_d in a truncated Taylor's series about i_d^o, viz*

$$i_d = i_d^o + \left.\frac{\partial i_d}{\partial v_d}\right|_{v_d = v_d^o} (v_d - v_d^o) \tag{3.1}$$

---

*The superscript denotes the number of iterations that have been performed.

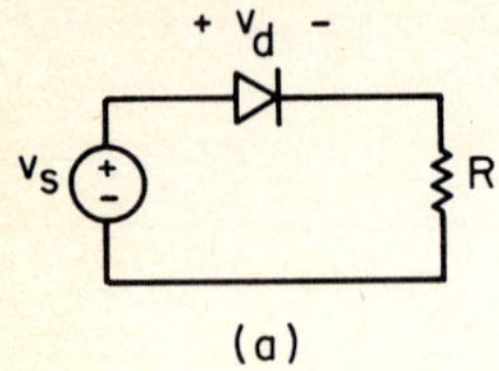

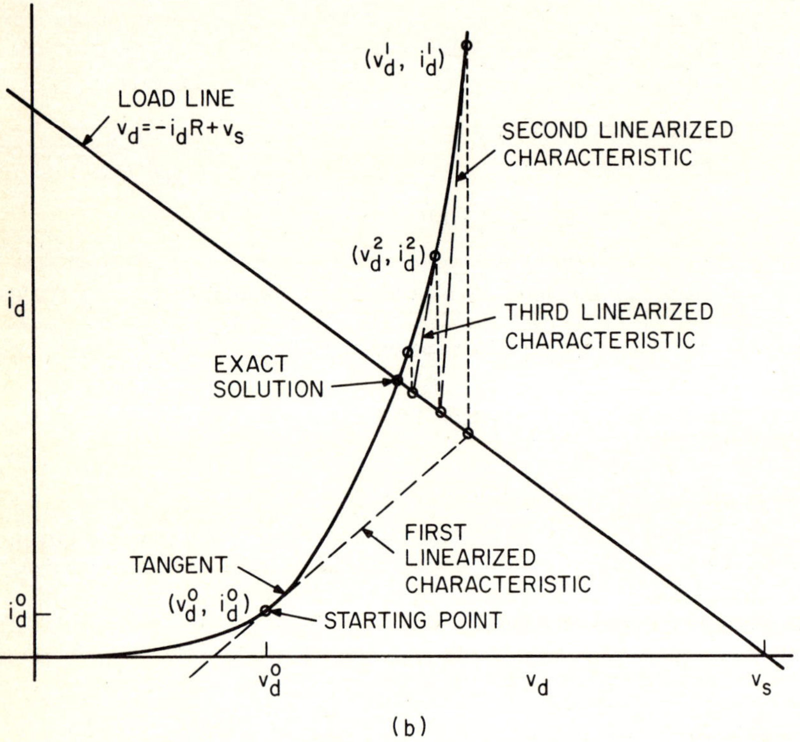

FIG. 3.3 Resistor-Diode Network and Its Iterative Solution

$$= I_s(e^{\lambda v_d^o} - 1) + \lambda I_s e^{\lambda v_d^o} \Delta v_d^o \tag{3.2}$$

where $\Delta v_d^o = v_d - v_d^o$. The reader will note that i_d of (3.1) does indeed pass through (i_d^o, v_d^o) with the correct slope $\partial i_d / \partial v_d$. The load line equation is given by

$$i_d = \frac{-v_d + v_s}{R} = \frac{-v_d^o - \Delta v_d^o + v_s}{R}$$

The intersection of this linearized diode characteristic and the load line yields our first refined guess; it is found by equating i_d of each equation, as

$$\frac{-v_d^o - \Delta v_d^o + v_s}{R} = I_s(e^{\lambda v_d^o} - 1) + \lambda I_s e^{\lambda v_d^o} \Delta v_d^o$$

from which Δv_d^o can be found. The diode voltage is then updated by forming the refined guess

$$v_d^1 = v_d = v_d^o + \Delta v_d^o$$

A second linearization at (i_d^1, v_d^1) yiends Δv_d^1, etc. In this case, it is obvious from Figure 3.3b that the process converges. The general formula for updating v_d has the form

$$\Delta v_d^m = v_d^{m+1} - v_d^m = \frac{-RI_s(e^{\lambda v_d^m} - 1) - v_d^m + v_s}{1 + \lambda RI_s e^{\lambda v_d^m}} \tag{3.3}$$

Example 3.1. Let v_s of Figure 3.3a be a function of time described by

$$v_s(t) = 1.7\,(1 + .2 \sin(t))$$

This waveform is shown in Figure 3.4. It is desired to calculate the diode current $i_d(nT)$, where $n = 0, 1, \ldots, 20$ and $T = (2\pi)/20$. We expect $i_d(t)$ to be nearly sinusoidal; however, some distortion will be introduced by the nonlinearity of the diode.

Even though this is not strictly speaking a dc problem since v_s varies with time, the absence of energy storage elements implies that the circuit will react instantaneously at each $t = nT$ as if the source were a dc source of value $v_s(nT)$. Newton's method can then be used to solve this equivalent dc problem for each n.

The program to determine i_d is given in Appendix Table 3.1 and the resultant $i_d(nT)$ in Table 3.1. The iteration is judged to have converged at each n when $|\Delta v_d^m| < 10^{-6}$, a number chosen to correspond to the six to seven significant figures carried in single precision Fortran.

The current i_d is not symmetric about its value for $n = 0$ as the voltage v_s is, i.e., the average of the peak values of i_d is .02549, while $i_d(0) = .02547$. The reader can, by choosing larger values of R, cause greater relative distortion to occur.

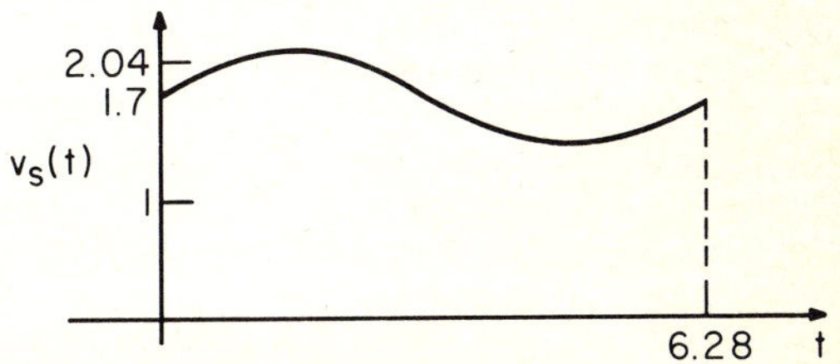

FIG. 3.4 Input Waveform

TABLE 3.1 v_s (nT) and i_d(nT) for Example 3.1

CSAT CURRENT = 0.1000E-08 RESISTANCE = 0.5000E 02

TIME	SOURCE	CURRENT
0.0	0.1700E 01	0.2547E-01
0.3142E 00	0.1805E 01	0.2754E-01
0.6283E 00	0.1900E 01	0.2940E-01
0.9425E 00	0.1975E 01	0.3088E-01
0.1257E 01	0.2023E 01	0.3183E-01
0.1571E 01	0.2040E 01	0.3216E-01
0.1885E 01	0.2023E 01	0.3183E-01
0.2199E 01	0.1975E 01	0.3088E-01
0.2513E 01	0.1900E 01	0.2940E-01
0.2827E 01	0.1805E 01	0.2754E-01
0.3142E 01	0.1700E 01	0.2547E-01
0.3456E 01	0.1595E 01	0.2341E-01
0.3770E 01	0.1500E 01	0.2156E-01
0.4084E 01	0.1425E 01	0.2009E-01
0.4398E 01	0.1377E 01	0.1915E-01
0.4712E 01	0.1360E 01	0.1882E-01
0.5027E 01	0.1377E 01	0.1915E-01
0.5341E 01	0.1425E 01	0.2009E-01
0.5655E 01	0.1500E 01	0.2156E-01
0.5969E 01	0.1595E 01	0.2341E-01
0.6283E 01	0.1700E 01	0.2547E-01
0.6597E 01	0.1805E 01	0.2754E-01

3.2.2 Iterative Solution of Large Nonlinear Networks

Companion Models. In extending the iterative approach to networks with more nonlinearities, it is clear that we must abandon attempts to find a graphical representation of the iteration, as displayed in Figure 3.3b. We could, of course, simply write and solve a set of nonlinear algebraic equations using KCL and KVL and an extension of Newton's method. A much more interesting possibility is the following, which converts the nonlinear network to a linear one that we can solve using, for example, mesh and node analysis.

The key to this approach is exemplified by the equation describing the linearization of the nonlinear diode characteristic

$$i_d^{m+1} = i_d^m + \left[\frac{\partial i_d}{\partial v_d} \bigg|_{v_d = v_d^m} \right] (v_d^{m+1} - v_d^m) \tag{3.4}$$

where m is the iteration counter (0, 1, . . .). This equation can be considered a description of the branch model of Figure 3.5, that is, a KCL current at node 1.

If this branch model is inserted into the complete circuit in place of the diode, we observe in Figure 3.6a that the new network is now linear, and therefore solvable by node analysis. We will term the linearized network a "companion" network model.

The companion network has two important properties that permit us to substitute it for the original network.

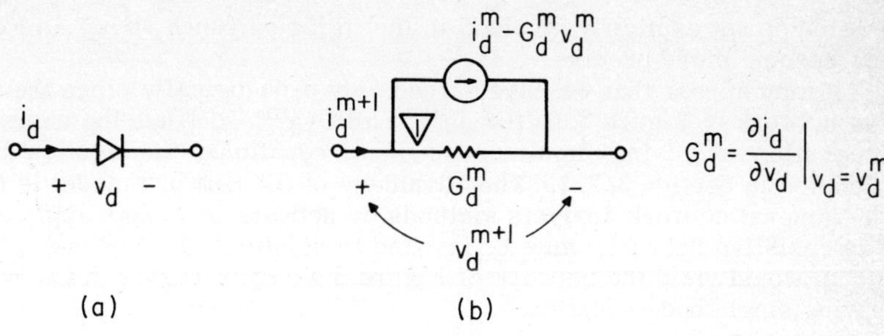

FIG. 3.5 Companion Model of Diode

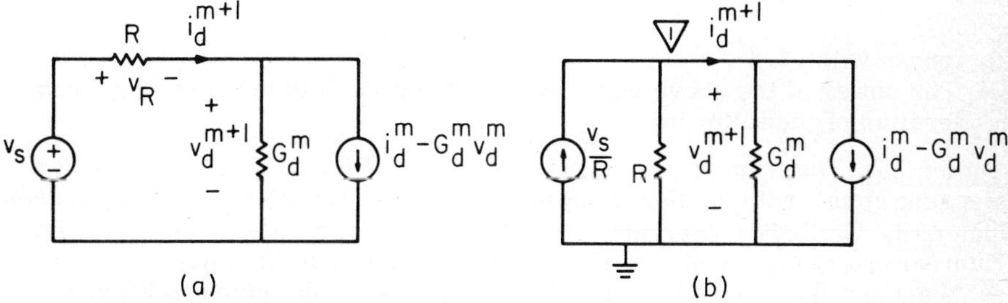

FIG. 3.6 Linearized (Companion) Network Models

Property 1: Branch currents and voltages satisfy the same KCL and KVL laws as the original network, since the branch structure of both networks is the same.

Property 2: The current and voltage relationships within a branch approximate those in the original network; for example, a diode current and voltage relationship is a linearized approximation to the exponential branch relationship.

Of morc importance is the meaning of the companion network model as a description of the iterative solution process. If we are at the m-th iteration, then v_d^m is known (or guessed if m is zero). Therefore,

(1) $$i_d^m = I_s(e^{\lambda v_d^m} - 1)$$ is known;

(2) $$G_d^m = \lambda I_s e^{\lambda v_d^m}$$ is known and is the slope of the v_d - i_d characteristic at (v_d^m, i_d^m).

Numerical values for both the current source $i_d^m - G_d^m v_d^m$ and the linearized conductance G_d^m of the companion network can thus be calculated. All the branches of the companion network are then determined, and we can solve for v_d^{m+1} of Figure 3.6a by node analysis. From v_d^{m+1}, i_d^{m+1} and G_d^{m+1} are formed and the iteration continues. If the iteration converges, branch currents and voltages of the linear companion network must approach those of the nonlinear network (e.g., $i_d^{m+1} \to i_d$, $v_d^{m+1} \to v_d$). This, the approximation involved in Property 2 becomes more precise. Even currents and voltages not explicitly involved in the iteration (such as v_R in Figure 3.6a) become more precise.

It may appear that we have gained only pedagogically since the solution of the network of Figure 3.6a for i_d^{m+1} and v_d^{m+1} defines the same iteration equation as solving simultaneous linear equations, the solution method we adopted in Section 3.2.1. The advantage of the network model is that any of the general network analysis methods or network theorems applicable to linear resistive networks may be invoked in solution. For example, Norton's theorem would yield the network of Figure 3.6b from Figure 3.6a, with the following single node equation:

$$v_d^{m+1} = \left(\frac{v_s}{R_s} + G_d^m v_d^m - i_d^m\right) \Big/ \left(\frac{1}{R_s} + G_d^m\right) \tag{3.5}$$

This agrees with (3.3).

The power of the above network model approach is more evident from consideration of the following.

Example 3.2. The transistor amplifier circuit of Figure 3.7a operates in a "quiescent state" with no signal applied ($i_s = 0$). Knowledge of the quiescent values of i_c (collector current) and v_{ce} (collector-emitter voltage) is important so that the transistor will amplify i_s without distortion.

We adopt the reduced Ebers-Moll transistor model given in Figure 3.7b, where the transistor is replaced by a diode and a current-controlled current source. For proper operation, the quiescent voltage of the diode should be near .4 volts.

The analysis of the network of Figure 3.7b can be regarded as a two-step process. The node equations are first written with all current sources on the right hand side, viz,

$$\begin{bmatrix} \frac{1}{R_1} + \frac{1}{R_2} & 0 & -\frac{1}{R_1} \\ 0 & \frac{1}{R_E} & 0 \\ -\frac{1}{R_1} & 0 & \frac{1}{R_L} + \frac{1}{R_1} \end{bmatrix} \begin{bmatrix} v_1^{m+1} \\ v_2^{m+1} \\ v_3^{m+1} \end{bmatrix} = \begin{bmatrix} -i_d^{m+1} + \alpha i_d^{m+1} \\ i_d^{m+1} \\ -\alpha i_d^{m+1} + \frac{E}{R_L} \end{bmatrix} \tag{3.6}$$

Since i_d^{m+1} is an unknown, it must be expanded into unknown node variables (v_i^{m+1}) and variables known from the previous iteration.

$$i_d^{m+1} = G_d^m v_d^{m+1} + i_d^m - G_d^m v_d^m = G_d(v_1^{m+1} - v_2^{m+1}) + i_d^m - G_d(v_1^m - v_2^m) \tag{3.7}$$

Substituting (3.7) into (3.6) and transferring expressions involving v_i^{m+1} to the left hand side, we finally have

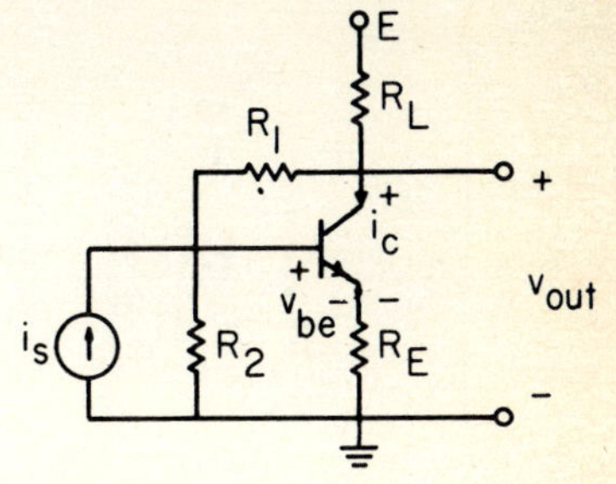

(a) Transistor Amplifier Circuit
$R_L = 5K$, $R_1 = 20K$,
$R_2 = 3K$, $R_E = .3K$
$E = 10.$

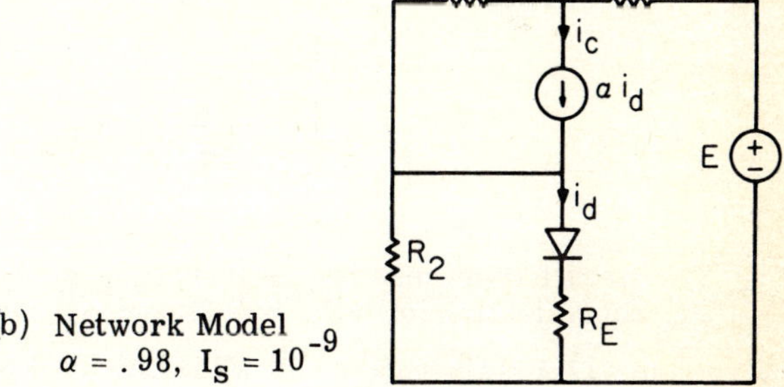

(b) Network Model
$\alpha = .98$, $I_s = 10^{-9}$

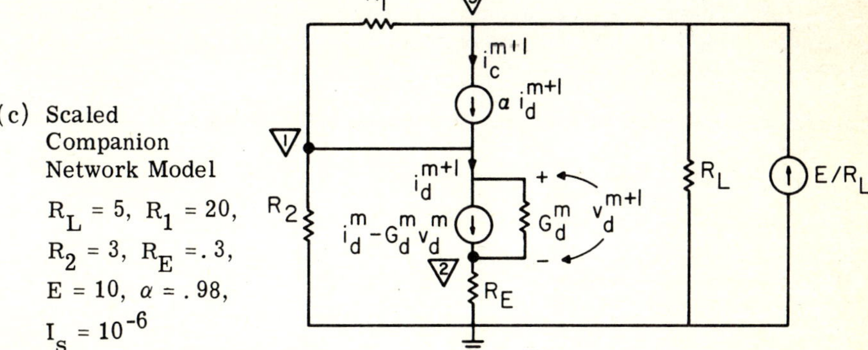

(c) Scaled Companion Network Model
$R_L = 5$, $R_1 = 20$,
$R_2 = 3$, $R_E = .3$,
$E = 10$, $\alpha = .98$,
$I_s = 10^{-6}$

FIG. 3.7 Evolution of Companion Model of Transistor Amplifier

$$\begin{bmatrix} \frac{1}{R_1} + \frac{1}{R_2} + (-\alpha + 1) G_d^m & -(1-\alpha) G_d^m & -\frac{1}{R_1} \\ -G_d^m & \frac{1}{R_E} + G_d^m & 0 \\ -\frac{1}{R_1} + \alpha G_d^m & -\alpha G_d^m & \frac{1}{R_L} + \frac{1}{R_1} \end{bmatrix} \begin{bmatrix} v_1^{m+1} \\ v_2^{m+1} \\ v_3^{m+1} \end{bmatrix}$$

$$= \begin{bmatrix} (\alpha - 1)(i_d^m - G_d^m (v_1^m - v_2^m)) \\ i_d^m - G_d^m (v_1^m - v_2^m) \\ -\alpha (i_d^m - G_d^m (v_1^m - v_2^m)) + E/R_L \end{bmatrix} \tag{3.8}$$

The iteration process begins by setting $m = 0$ and guessing a value for v_d^0, or, equivalently, for $v_1^0 - v_2^0$. From v_d^0, we calculate $i_d^0 = I_s(e^{\lambda v_d^0} - 1)$ and $G_d^0 = I_s \lambda e^{\lambda v_d}$. This calculation establishes both sides of (3.8), and we may solve for v_i^1. Then we set $m = 1$ and repeat the process.

In preparing this problem for computer solution, we current scale by selecting $b = 1000$. The resistor values of Figure 3.7c are then in kilohms and the current is milliamps, including the saturation current I_s.

A program (DNET) to solve (3.8) iteratively is given in Appendix Table 3.2. This program is arranged to have the same form as RNET of Appendix Table 2.1. The results of performing an iterative solution are shown in Table 3.2.

TABLE 3.2 Iteration Results for Bias Example

```
R1 =      20.000 R2 =       3.000 RE =       0.300
RL =       5.000 E =      10.000 ALF =       0.980
ITER NO.=  0 VBE =0.37722 VCE=4.78841 IC=3.57301
ITER NO.=  1 VBE =0.35791 VCE=4.69308 IC=1.64989
ITER NO.=  2 VBE =0.34543 VCE=4.63150 IC=1.00156
ITER NO.=  3 VBE =0.34117 VCE=4.61051 IC=0.84489
ITER NO.=  4 VBE =0.34078 VCE=4.60857 IC=0.83173
ITER NO.=  5 VBE =0.34078 VCE=4.60856 IC=0.83163
```

3.3 COMPOSITE COMPANION MODELS

As already pointed out, an outstanding feature of the companion network approach is that linear resistive network analysis techniques may be used to

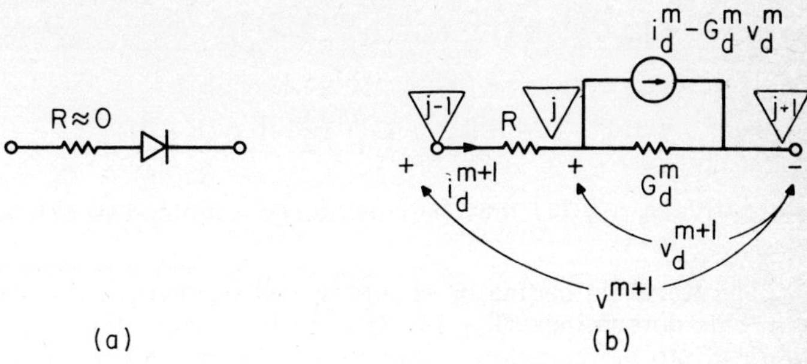

FIG. 3. 8 Augmented Diode Model

alter the form of the companion network model. We want to examine the use of this technique to avoid accuracy problems in dc analysis.

Consider the analysis of a network containing diodes, where the diode bulk resistances, although small, are felt to be important to the circuit simulation. The complete diode companion model is shown in Figure 3. 8b.

The node matrix of this circuit will appear in part as

$$
\begin{array}{llll}
 & & \text{j-th column} & \\
 & \frac{1}{R} + \dots & -\frac{1}{R} & \cdot\cdot \\
\text{j-th row} & -\frac{1}{R} & \frac{1}{R} + G_d^m & \cdot\cdot \\
 & \cdot\cdot & \cdot\cdot & G_d^m + \dots
\end{array}
$$

Reduction of this node matrix by Gauss elimination or factorization methods must result in cancellation of $1/R^2$ terms, which can cause a significant round-off error since R is small. This cancellation could be avoided by using the general network reduction methods of Section 2. 4. 1, or performing a symbolic reduction of the companion model to eliminate the interior j-th node. For example, we may invoke Thevenin's or Norton's theorem to eliminate the j-th node. In this case, we have

$$G_{eq}^m = \frac{1}{R + 1/G_d^m} = \frac{G_d^m}{1 + RG_d^m} \tag{3.9}$$

$$i_{eq}^m = \frac{\frac{1}{R}(-i_d^m + G_d^m v_d^m)}{\frac{1}{R} + G_d^m} = \frac{-i_d^m + G_d^m v_d^m}{1 + RG_d^m} \tag{3.10}$$

These formulae do not completely specify the Norton equivalent, since v_d^m - which determines G_d^m and i_d^m - has been lost in obtaining the equivalent. However, from Figure 3. 8b we can write the updating formula

$$v_d^{m+1} = v^{m+1} - R\, i_d^{m+1}$$

$$= v^{m+1} - R\,(i_d^m + G_d^m\,(v_d^{m+1} - v_d^m))$$

or (3.11)

$$v_d^{m+1} = [\,v^{m+1} - R\,(i_d^m - G_d^m\,v_d^m)]\,/\,(1 + RG_d^m)$$

Alternatively, (3.11) may be considered a method of calculating $v_d{}^m$ from v^m and $v_d{}^{m-1}$.

The iteration begins by choosing $v_d{}^0$ as before, and calculating $G_d{}^0$ and $i_d{}^0$. This determines G_{eq}^0 and i_{eq}^0 and the companion diode model may be inserted into the complete companion network. After the diode terminal voltage v^1 is calculated, the diode voltage $v_d{}^1$ is formed from (3.11) and $G_d{}^1$ and $i_d{}^1$ can be calculated.

3.4 CONVERGENCE PROBLEMS AND THEIR SOLUTION

As we mentioned previously, Newton iteration is not guaranteed to converge; also, if several solutions exist, the desired one usually cannot be specified with absolute assurance that the iteration will proceed in that direction. In practice, the method converges if the initial guess is good, and the neighborhood of a prescribed stable point often can be established from other circuit considerations sufficiently well to result in convergence to that point.

Convergence is especially difficult for biasing circuits with a significant amount of feedback, since a small error in the initial guess of one of the circuit variables can be amplified and returned as a large apparent error. For example, if a diode voltage is guessed significantly below its actual value, the first iteration tends to overshoot this value (as shown in Figure 3.3). Considerable computational effort is then necessary to compensate for this indiscretion. One way of avoiding the overshoot is simply not to take a full step, i.e., for a vector of node voltages $\underline{v}$

$$\underline{v}^{m+1} = \underline{v}^m + \alpha\,\Delta\,\underline{v}^m$$

where $\alpha < 1$. The value of α is often chosen so that no diode voltage exceeds a prescribed positive voltage $v_{d_{max}}$. This is known as damped Newton iteration; this will be considered in detail in Chapter XI. Another method involves setting to zero any elements producing feedback and then incrementing them to their normal values.

A more specialized but highly efficient scheme when exponential nonlinearities are present uses the curvature of the diode characteristic to improve the behavior of Newton iteration. When a diode voltage v_d' is calculated to be negative after solution of the node equations, then conventional updating formulae are used, viz,

$$v_d^{m+1} = v_d' \qquad\qquad v_d' < 0$$

$$i_d^{m+1} = I_s(e^{\lambda v_d^{m+1}} - 1) \tag{3.12}$$

$$G_d^{m+1} = \lambda\,I_s\,e^{\lambda v_d^{m+1}}$$

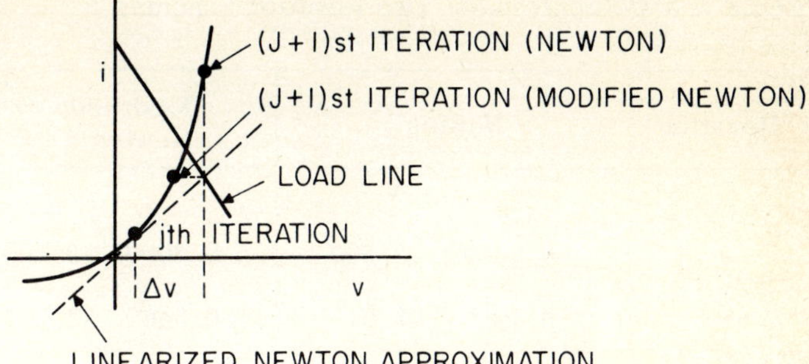

FIG. 3.9 Comparison of Iteration Strategies

This corresponds to a vertical projection from the solution of the linearized equations (Figure 3.9). For $v_d' > 0$, a horizontal projection onto the diode characteristic as shown will often result in a new guess closer to the final solution of the nonlinear problem [3.4]. This projection is achieved by determining the current flowing into the companion diode model from

$$i_d^{m+1} = i_d^m + G_d^m (v_d^{m+1} - v_d^m)$$

rather than from (3.12). Then v_d^{m+1} is re-evaluated as

$$v_d^{m+1} = \frac{1}{\lambda} \ell n \left(\frac{i_d^{m+1}}{I_s} + 1 \right)$$

and G_d^{m+1} then obtained as

$$G_d^{m+1} = \lambda(i_d^{m+1} + I_s)$$

Example 3.3. The network of Figure 3.10 is characterized by a large amount of voltage feedback. A comparison of the iterative solution using Newton iteration with and without the above modification is shown in Table 3.3. We see that the modified method offers significantly faster convergence.

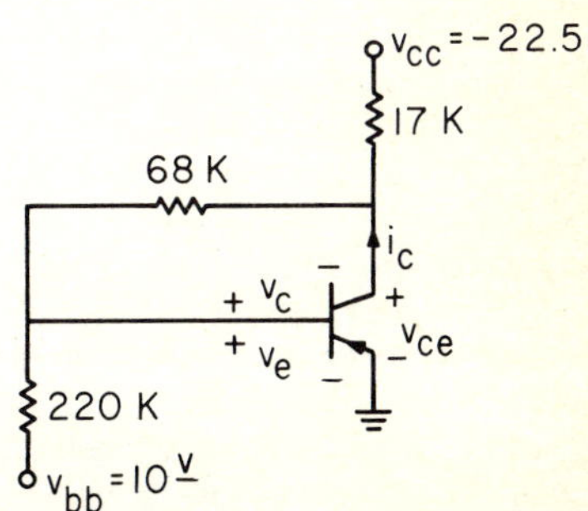

FIG. 3.10 An Ill-conditioned Biasing Problem

TABLE 3.3 Comparison of Two Iteration Schemes

Iteration	Newton		Modified Newton	
	v_e	v_c	v_e	v_c
0	0.3000	-3.500	0.3000	-3.500
1	0.4327	-3.918	0.3460	-3.918
2	0.4085	-3.912	0.3462	-3.891
3	0.3855	-3.906	0.3462	-3.896
4	0.3657	-3.898		
5	0.3522	-3.896		
6	0.4368	-3.896		
7	0.3462	-3.896		

3.5 A GENERAL DC CIRCUIT ANALYSIS PROGRAM (DCAP)

As in Chapter 2, we conclude by examining a general purpose analysis program which incorporates the solution methods of this chapter. A program (DCAP) for the dc analysis of networks with diode nonlinearities is contained in Appendix Table 3.3. The modifications in RCAP (Appendix Table 2.4) necessary to perform dc analysis are similar in part to the differences between DNET and RNET. These are summarized in the following list and can be found in the DCAP program.

(1) The main program of RCAP is adjusted similarly to DNET to permit iteration of the node equation, with updating of the node voltage array V.
(2) Convergence of the iteration is checked in the main program.
(3) The diode must be added to the model library in RCAP in subroutine NODEQ. Since the node voltages $\underline{v}_i^m$ are available in the V array, the calculation of the linearized conductance and source of the companion model of a diode is straightforward.
(4) Because most dc analysis problems involve voltage sources, a gyrator-current source of a model of a voltage source is added in subroutine PREPAR (see Problem 3.12). For every independent voltage source encountered in the list of input elements, two voltage controlled current sources and an independent current source are added to the input list. These elements are then processed in the normal manner in NODEQ.

Example 3.4. The unscaled version of the circuit of Figure 3.7c is analyzed for α = (.99,.98,.9) using the DCAP program. The input and output data for this problem are shown in Table 3.4 and 3.5. The output voltages are

TABLE 3.4 Input Data for Example 3.4

```
TRANSISTOR AMPLIFIER BIAS MODEL
R1           03 01        20000.
R2           01 00        3000.
I1   VRE     03 01   V    .003300,.0032667,.003000
D1           01 02   V    .4
RE           02 00         300.
RL           03 04         5000.
VE           04 00        10.
AMEN
```

shown at each iteration. It should be noted that the current controlled source has been converted to a voltage controlled current source (VCCS) and the value changed accordingly (justify ?).

3.6 APPROXIMATION OF DEVICE MODELS

It is increasingly common to use measured voltage-current component characteristics in dc analysis. These point-wise characteristics are commonly approximated in one of two ways:

(1) by piecewise linear approximations as shown in Figure 3.11; the analysis is then always performed on a linear network, the difficulty being to find which linear line segment of the characteristic to use [3.6], [3.7], [3.9]; it has recently been shown that Gaussian elimination can be used to solve such networks [3.15];

(2) by spline-fitting techniques (see Problem 3.10), where at least the first derivative of the approximating function is maintained continuous (unlike the piecewise linear approximations above); this permits Newton iteration to proceed in the normal manner; the derivative continuity also facilitates the numerical integration to be performed in Chapter IV.

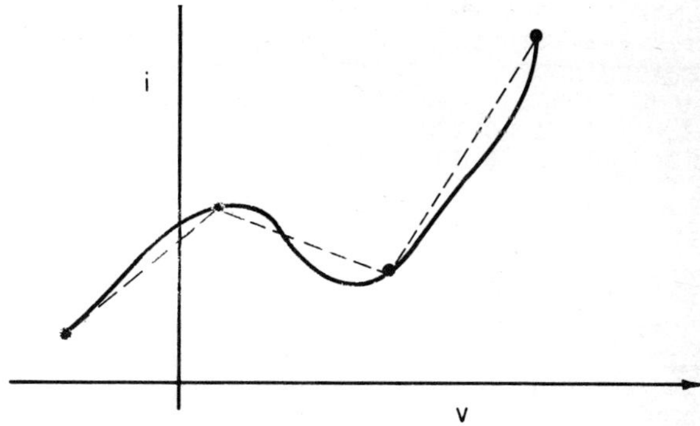

FIG. 3.11 Piecewise Linear Model

3.7 OTHER SOLUTION METHODS

Several other methods of dc analysis problems have been proposed. One approach is to view the dc analysis problem as a transient analysis problem that has reached a steady state condition. That is, we regard the constant dc sources as step inputs to a dynamic system that, with

TABLE 3.5 DC Analysis Results of Example 3.4

```
**********TRANSISTOR AMPLIFIER BIAS MODEL
BRANCH NODE      ELEMENT      BRANCH CON-
NAME    NOS.      VALUE        NAME   TROL
R1      3  1  0.2000000D 05
R2      1  0  0.3000000D 04
I1      3  1  0.3300000D-02 RE       V
D1      1  2  0.4000000D 00
RE      2  0  0.3000000D 03
RL      3  4  0.5000000D 04
VE      4  0  0.1000000D 02
OUTPUT NODES =  3  1  OUTPUT VOLTAGE =       4.2968414
OUTPUT NODES =  1  2  OUTPUT VOLTAGE =       0.3772784
OUTPUT NODES =  3  1  OUTPUT VOLTAGE =       4.2181630
OUTPUT NODES =  1  2  OUTPUT VOLTAGE =       0.3580885
OUTPUT NODES =  3  1  OUTPUT VOLTAGE =       4.1678710
OUTPUT NODES =  1  2  OUTPUT VOLTAGE =       0.3458222
OUTPUT NODES =  3  1  OUTPUT VOLTAGE =       4.1511267
OUTPUT NODES =  1  2  OUTPUT VOLTAGE =       0.3417382
OUTPUT NODES =  3  1  OUTPUT VOLTAGE =       4.1496504
OUTPUT NODES =  1  2  OUTPUT VOLTAGE =       0.3413781
OUTPUT NODES =  3  1  OUTPUT VOLTAGE =       4.1496400
OUTPUT NODES =  1  2  OUTPUT VOLTAGE =       0.3413756
**********TRANSISTOR AMPLIFIER BIAS MODEL
BRANCH NODE      ELEMENT      BRANCH CON-
NAME    NOS.      VALUE        NAME   TROL
R1      3  1  0.2000000D 05
R2      1  0  0.3000000D 04
I1      3  1  0.3266700D-02 RE       V
D1      1  2  0.4000000D 00
RE      2  0  0.3000000D 03
RL      3  4  0.5000000D 04
VE      4  0  0.1000000D 02
OUTPUT NODES =  3  1  OUTPUT VOLTAGE =       4.2676912
OUTPUT NODES =  1  2  OUTPUT VOLTAGE =       0.3407861
OUTPUT NODES =  3  1  OUTPUT VOLTAGE =       4.2676644
OUTPUT NODES =  1  2  OUTPUT VOLTAGE =       0.3407793
**********TRANSISTOR AMPLIFIER BIAS MODEL
BRANCH NODE      ELEMENT      BRANCH CON-
NAME    NOS.      VALUE        NAME   TROL
R1      3  1  0.2000000D 05
R2      1  0  0.3000000D 04
I1      3  1  0.3000000D-02 RE       V
D1      1  2  0.4000000D 00
RE      2  0  0.3000000D 03
RL      3  4  0.5000000D 04
VE      4  0  0.1000000D 02
OUTPUT NODES =  3  1  OUTPUT VOLTAGE =       5.0430492
OUTPUT NODES =  1  2  OUTPUT VOLTAGE =       0.3368108
OUTPUT NODES =  3  1  OUTPUT VOLTAGE =       5.0429806
OUTPUT NODES =  1  2  OUTPUT VOLTAGE =       0.3364718
OUTPUT NODES =  3  1  OUTPUT VOLTAGE =       5.0429742
OUTPUT NODES =  1  2  OUTPUT VOLTAGE =       0.3364695
**********TRANSISTOR AMPLIFIER BIAS MODEL
```

energy storage elements removed, is identical to the original dc problem. Since the problem is originally posed as a dc problem, we must decide upon a method of generating a related dynamic system. This problem has been studied in [3.12].

Another direction of study involves finding alternatives to Newton's method for iteration of the describing equations [3.5], [3.8]. Still other studies, which are farther removed from the development of efficient algorithms, are concerned with the question of uniqueness of a dc solution [3.9], [3.10] and related iteration methods. As a general rule, one finds that algorithms that guarantee to find all dc solutions are significantly slower than Newton iteration.

Problems

3.1 One method of biasing a transistor is illustrated in Figure P3.1a. You are asked to calculate v_{cb} and v_{be} by altering subroutine NODEQ and the input data in DNET. (It is suggested that you use the R_E symbol to represent the source E_{bb} in subroutine READ.)

For Problems 3.2 through 3.7, assume the Ebers-Moll transistor model for PNP transitors and a similar model for NPN transistors.

3.2 Find the quiescent operating points of the transistors in the amplifier circuit of Figure P3.2 (an initial guess may be made assuming zero base current flows). Repeat with all resistors 10 percent above their normal values; note the number of iterations required to calculate the new bias point from the old one.

3.3 Repeat Problem 3.2 if the Ebers-Moll diodes are modeled by a dc source of 0.3 volts in series with a 40 Ω resistor. Note that this problem does not require iteration.

3.4 Find the two stable operating points for the flip-flop of Figure P3.2b. Calculate v_{ce} and i_c for each transistor.

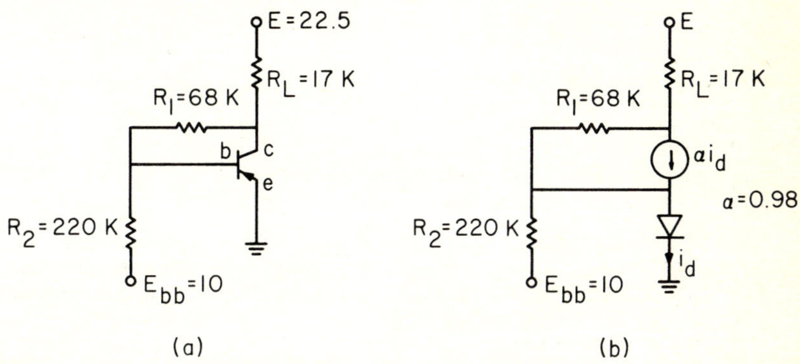

FIG. P3.1 Biasing Circuit and Model

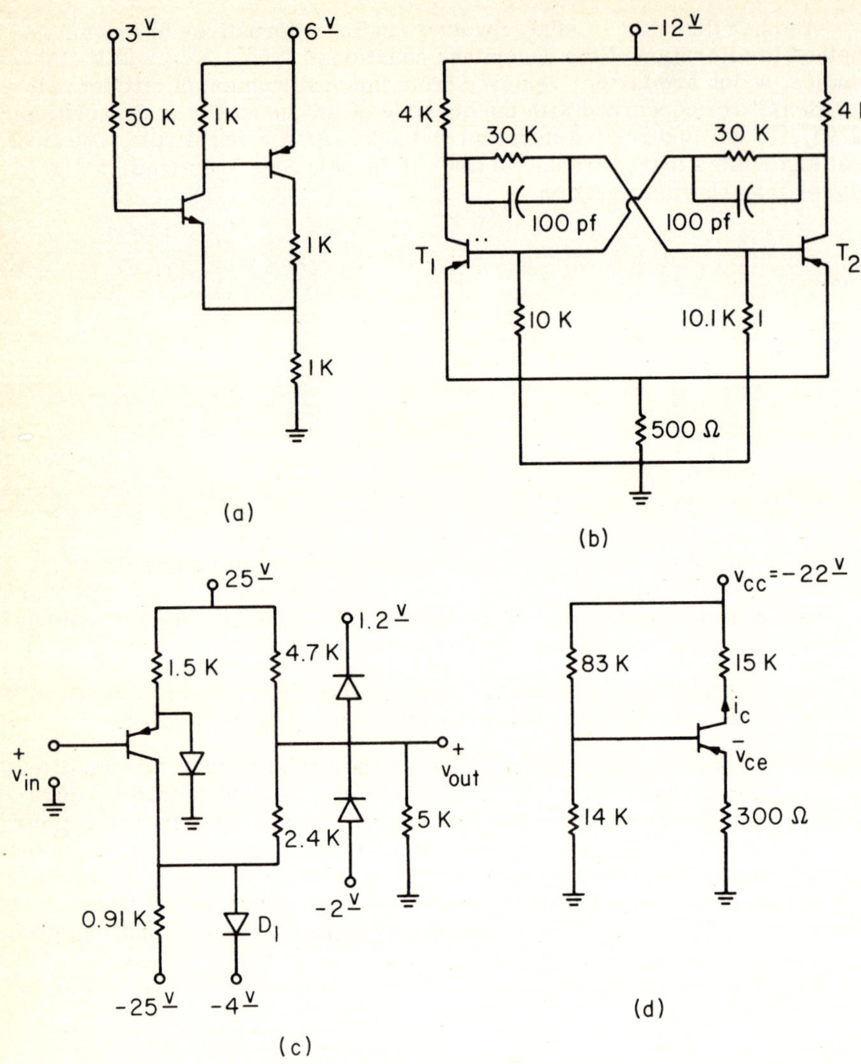

FIG. P3.2 DC Analysis Examples

3.5 Determine v_{out} for the circuit of Figure P3.2c when v_{in} = 2 volts and when v_{in} = 1.5 volts. Does diode D_1 conduct in either case?

3.6 Find the operating point for the network of Figure P3.2d. Find the partial derivatives of v_{ce} and i_c with respect to v_{cc} by perturbation.

3.7 (1) Calculate the transistor output characteristics for the

(a) CB configuration (i_c vs. v_{cb} with i_e as parameter),
(b) CE configuration (i_c vs. v_{ce} with i_b as parameter),
(c) Interchange collector and emitter terminals and repeat.

3.8 Derive a companion network diode model suitable for mesh analysis.

3.9 Use a nonzero signal input $i_s(t) = 10^{-4} \sin t$ in Figure 3.7a and calculate $v_{out}(t)$ over one cycle.

3.10 The DCAP program can be altered to accept i-v characteristics other than the exponential diode characteristic.

In Appendix Table 3.4, a "spline-fitting" interpolation algorithm [3.13] is displayed. A set of x-y data is processed in subroutine SPLINE to yield sets of polynomial coefficients. These coefficients are then consulted in subroutine FIT whenever an interpolation of the data is necessary. An estimate of the slope $\partial y/\partial x$ is also made in FIT.

You are asked to alter (1) the input language to DCAP to permit the definition of one-port measured i-v characteristics, and (2) the NODEQ subroutine to permit iterative solution of networks containing such one-ports.

3.11 A frequently occurring model of an isolation diffusion in an integrated circuit is in Figure P3.3.

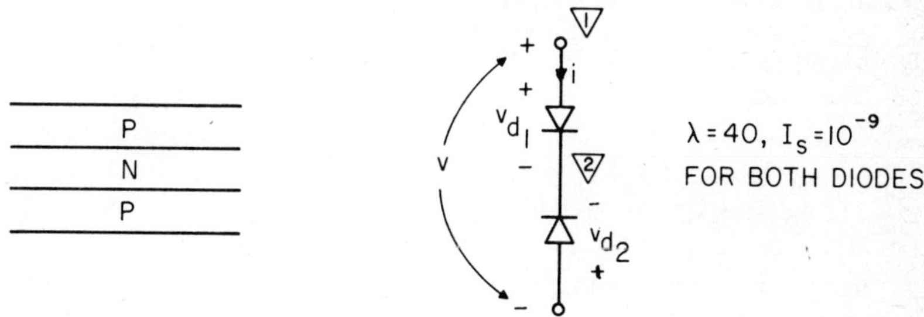

FIG. P3.3

(a) Find i as an explicit function of v.
(b) Find the companion model of the combination from (a).
(c) What is the maximum magnitude of i and of G^m, the linearized conductance of the companion model?

3.12 The gyrator is a two-port element which converts current variables at one port to voltage variables at the other port, and vice versa. Show that the connection of a gyrator and a current source as in Figure P3.4 behaves as a voltage source.

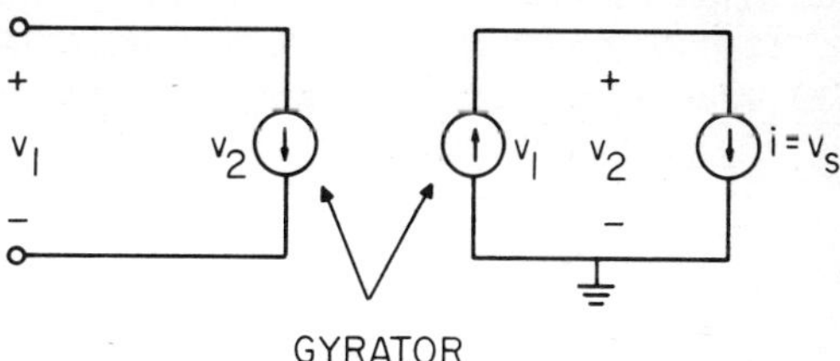

FIG. P3.4

3.13 Show that the network of Figure P3.5 behaves as a voltage source.

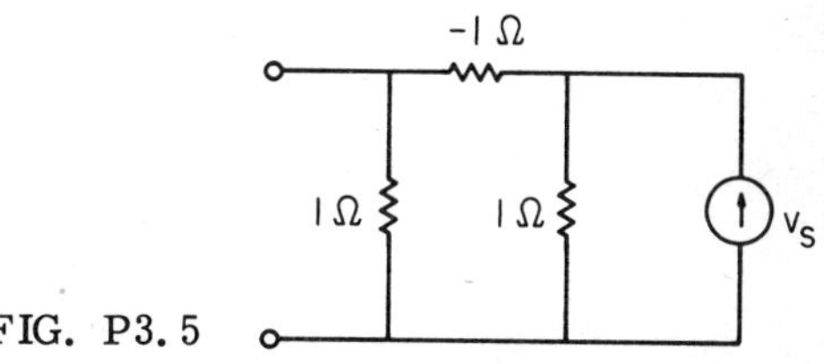

FIG. P3.5

3.14 Show that the network of Figure P3.6 satisfies the equations of an ideal transformer.

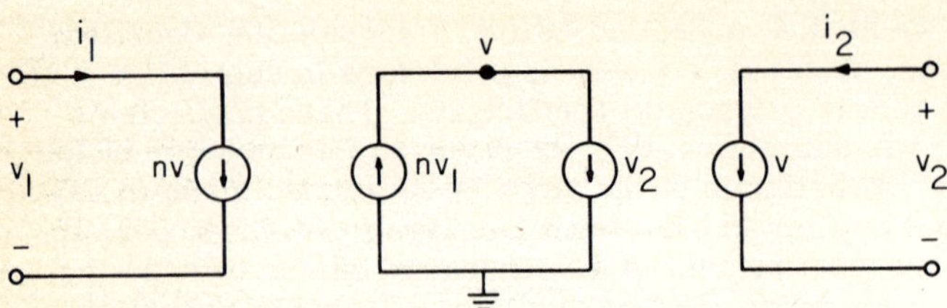

FIG. P3.6

3.15 A zero-valued resistance cannot usually be represented in a node analysis because the equivalent conductance is infinite. Show how a gyrator can be used to avoid this problem, provided pivoting is allowed on off-diagonal positions of the node matrix

3.16 The equation for a diode is more realistically given by

$$i_d = I_s\,(e^{f(v_d)\,v_d} - 1)$$

where $20 \le f(v_d) \le 40$ at room temperature. Find the companion model in terms of f and $\partial f/\partial v_d$.

3.17 A Zener diode has the exponential diode characteristic in the forward direction but i_d is multiplied by $1/(1 + v_d/v_z)^N$ in the reverse direction. Find the companion model.

3.18 It is frequently necessary to represent a nonlinear transistor β or α in the form

$$i_c = \beta(i_c)\, i_b$$

$$= \alpha(i_c)\, i_e$$

where $\alpha = \beta/(1+\beta)$. Develop a companion model that relates $\Delta i_c{}^m$ to either $\Delta i_b{}^m$ or to $\Delta i_e{}^m$, assuming $i_c{}^m$, $i_b{}^m$, and $i_e{}^m$ are known.

3.19 Add to DCAP the capability of determining the effect of temperature on diode behavior. The current through the diode is described by

$$I_s = (10^{-9}) \left(\frac{T}{T_o}\right)^3 e^{14000(T - T_o)/(TT_o)}$$

$$\lambda = (40)\, T/T_o$$

where $T_o = 290^oK$ and T is in ^oK.

REFERENCES

3.1 Broyden, C. G., "A Class of Methods for Solving Nonlinear Simultaneous Equations," Mathematics of Computation, vol. 19, pp. 577-593; October, 1965.

3.2 Katzenelson, J., "An Algorithm for Solving Nonlinear Resistive Networks," Bell System Tech. Journal, vol. 44, pp. 1605-1620; November, 1965.

3. 3 Katzenelson, J., and L. H. Seitelman, "An Iterative Method for Solution of Networks of Nonlinear Monotone Resistors," Trans. IEEE, vol. CT-13, pp. 317-322; June 1966.

3. 4 Milliman, L. D., "CIRCUS, A Digital Computer Program for Transient Analysis of Electronic Circuits," Report 346-2, Harry Diamond Laboratories, Washington, D. C.; January, 1967.

3. 5 Brown, K. M., "A Quadratically Convergent Newton-Like Method Based on Gaussian Elimination," Technical Report No. 68-23, Department of Computer Science, Cornell University, Ithaca, N. Y.; June, 1969 (also accepted for publication in the SIAM Journal of Numerical Analysis).

3. 6 Chua, L. O., "Efficient Computer Algorithms for Piecewise Linear Analysis of Resistive Nonlinear Networks," Trans. IEEE, vol. CT-18, no. 1, pp. 73-84; January, 1971.

3. 7 Ohtsuki, T., and N. Yoshida, "DC Analysis of Nonlinear Networks Based on Generalized Piecewise-Linear Characterization," Trans. IEEE, vol. CT-18, no. 1, January, 1971.

3. 8 Rohrer, R. A., "Successive Secants in the Solution of Nonlinear Network Equations" (Private communication; suggest writing to author).

3. 9 Kuh, E. S., and Ibrahim N. Hajj, "Nonlinear Circuit Theory: Resistive Networks," Proc. IEEE, vol. 59, no. 3, pp. 340-354; March, 1971.

3. 10 Sandberg, I. W., and A. N. Willson, Jr., "Some Network-Theoretic Properties of Nonlinear DC Transistor Networks," BSTJ, vol. 48, pp. 1293-1311; May-June, 1969.

3. 11 Sandberg, I. W., and A. N. Willson, Jr., "Some Theorems on Properties of DC Equations of Nonlinear Networks," BSTJ, vol. 48, no. 1, pp. 1-34; January, 1969.

3. 12 Branin, F. H., Jr., "Solution of Nonlinear DC Network Problems via Differential Equations," IBM Report TR21. 410, Kingston, N. Y.;

3. 13 Carnahan, B., H. A. Luther, and J. O. Wilkes, Applied Numerical Methods, 1969.

3. 14 McCalla, W. J., and D. O. Pederson, "Elements of Computer-Aided Circuit Analysis," Trans. IEEE, vol. CT-18, no. 1, pp. 14-26; January, 1971.

3. 15 Fugisawa, T., E. S. Kuh, and T. Ohtsuki, "Sparse Matrix Methods for Analysis of Piecewise Linear Networks," ERL Report, Dept. of Electrical Engineering and Computer Science, University of California, Berkeley, California.

4

Transient Analysis of Dynamic Networks

4.1 INTRODUCTION

Although we can formulate the integro-differential equations of complicated dynamic networks using mesh or node analysis, we can solve only small linear networks analytically (such as the network of Figure 4.1). We will find, however, that for those applications where a computer solution is acceptable, the transient response of complicated linear <u>and</u> nonlinear dynamic circuits can be calculated with a small extention of resistive network analysis techniques. The computer solution has several essential disadvantages relative to the analytic solution, however.

(1) The analytic solution is known as a continuous function of time, whereas the numerical results can be calculated only at discrete moments of time as displayed in Figure 4.1, or equivalently, in tabular form.

(2) The differential and integro-differential equations yielding the analytic solution can be only approximated numerically. Although the error made in this approximation can be estimated, we will always be left with some uncertainty in the quality of our computer solution.

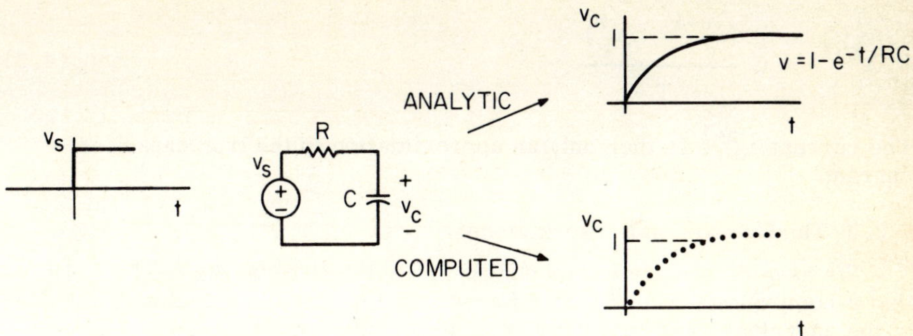

FIG. 4.1 Comparison of Analytic and Numerical Solutions

4.2 TRANSIENT ANALYSIS OF LINEAR DYNAMIC NETWORKS

4.2.1 Derivative Approximation

Although we can approximate both integrals or derivatives numerically, we will concentrate on the latter. One of many possible approximations is the backward Euler formula

$$\left.\frac{dv}{dt}\right|_{t=t^{n+1}} \approx \frac{v^{n+1} - v^{n}}{t^{n+1} - t^{n}} \tag{4.1}$$

where the "n" superscript refers to a particular time (see Figure 4.2). The term 'backward" comes from the use of past values of v to approximate the derivative. We assume that the difference between successive

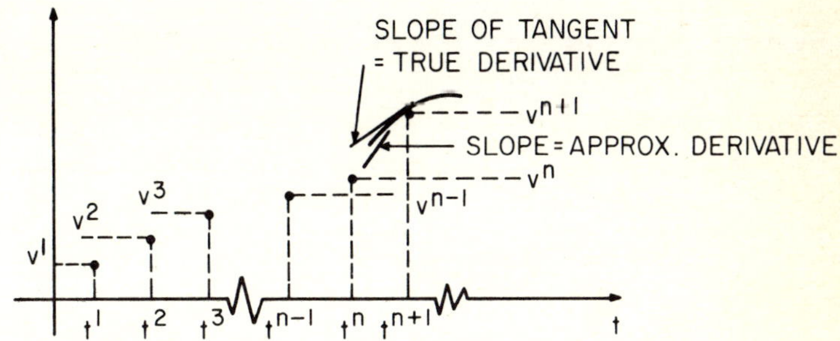

FIG. 4.2 Derivative Approximation

time steps is constant, viz,

$$t^{n+1} - t^{n} = T \tag{4.2}$$

where T is called the time step or step size.

The approximation defined, we can observe the result on the branch relationships. For example, the capacitor, defined by $i_C = C\, dv_C/dt$, is approximated by

$$i_c^{n+1} = C \frac{(v_c^{n+1} - v_c^n)}{T} \tag{4.3}$$

The current i_c^{n+1} is then only an approximation to the true capacitor current.

4.2.2 The Companion Network Model

We have demonstrated for nonlinear resistive networks the flexibility that a companion network model introduces into the numerical solution. One of the advantages was that the companion network could be analyzed using well-known methods, since it was linear and resistive. We now show that a linear resistive companion network exists which models the derivative approximation discussed above, thus avoiding both the writing and (more important) the formal numerical solution of the differential equations.

The branch representation for a capacitor in the companion network follows immediately from (4.3), by viewing this equation as a KCL sum at a branch node. Table 4.1 displays the resultant model (col. 4) which the reader should compare with the equation of col. 3. The inductor is modeled by a capacitor and gyrator (see Problem 3.12). The companion model for a resistor remains, of course, a resistor.

TABLE 4.1 Companion Network Models

	Continuous Equation	Approximate Equations	Companion Network Model
Capacitor	$i = C\frac{dv}{dt}$	$i^{n+1} = \frac{C(v^{n+1} - v^n)}{T}$	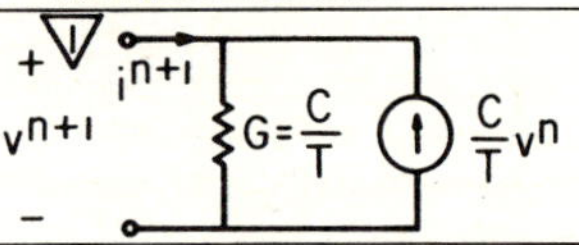 $+$ v^{n+1} $-$, i^{n+1}, $G = \frac{C}{T}$, $\frac{C}{T}v^n$
Resistor	$v = Ri$	$v^{n+1} = Ri^{n+1}$	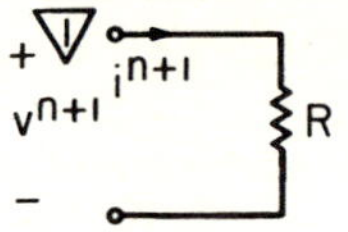 $+$ v^{n+1} $-$, i^{n+1}, R
Inductor	$v = L\frac{di}{dt}$	$v^{n+1} = \frac{L(i^{n+1} - i^n)}{T}$	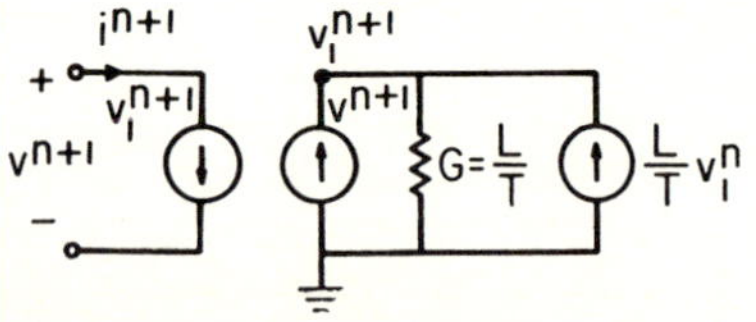 i^{n+1}, $+$ v^{n+1} $-$, v_1^{n+1}, v^{n+1}, v_1^{n+1}, $G = \frac{L}{T}$, $\frac{L}{T}v_1^n$

Returning to the simple network of Figure 4.1, we construct the companion network from the branch models of Table 4.1 (this is shown in Figure 4.3). Analysis of the companion network of Figure 4.3c in this case yields

$$v_c^{n+1} = \frac{\frac{1}{R} + \frac{C}{T}v_c^n}{\frac{1}{R} + \frac{C}{T}} \qquad n = 0, 1, \ldots \tag{4.4}$$

Note that v_c^0 need not be guessed as was necessary for iterative analysis of nonlinear networks. Instead, v_c^0 is chosen to be the initial condition $v_c(t)|_{t=0}$ which must accompany the problem description.

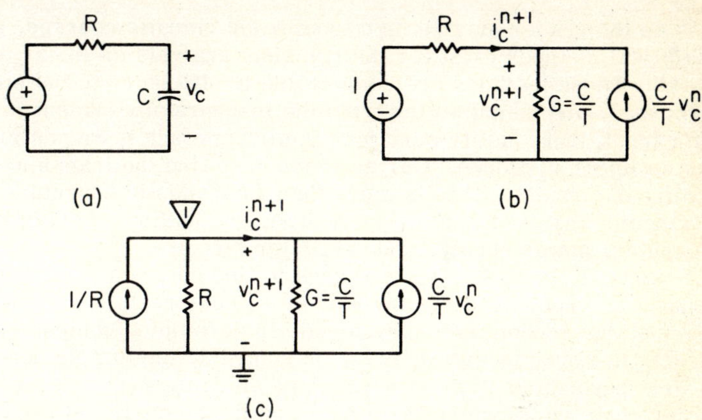

FIG. 4.3 Steps in Evaluation of Companion Model

In general, one initial condition must be specified for each capacitor and inductor in the network in order to define the sources $\frac{C}{T}v^0$ and $\frac{L}{T}i^0$ of Table 4.1, which in turn are necessary to start the step by step transient solution. Of course, this information is necessary for any transient solution - analytical or numerical.

In summary, the original and companion networks satisfy the same structural equations (KCL and KVL) and the same initial conditions. The branch relationships of the companion network are only approximate, however. The nature of this approximation can be appreciated by solving (4.4) with progressively smaller step sizes. The results for R = C = 1 are shown in Table 4.2. The error, measured with respect to the exact solution, is seen to decrease proportionally to T. This is a discouraging result; methods to be discussed later can yield an error proportional to T^2, for T sufficiently small.

TABLE 4.2 Comparison of Analytic, Numerical Solutions $(1 - e^{-1}) = .632121$

Step Size (T)	Calculated Value (t = 1)	Error
.1	.614456	.017665
.05	.623111	.009010
.02	.628472	.003649
.01	.630289	.001832
.005	.631203	.009180
.002	.631753	.000368
.001	.631937	.000184

4.3 TRANSIENT ANALYSIS OF NONLINEAR DYNAMIC NETWORKS

4.3.1 Introduction

As the reader may realize from basic circuits courses, powerful analytical methods exist for the transient analysis on linear dynamic networks. These methods are comparable in efficiency to the one we have discussed, and do not suffer from the inaccuracies evident in Table 4.2. We cannot, then, justify our algorithmic approach if we restrict our attention to linear networks. Our ambition is to find the transient behavior of nonlinear circuits. This is more than an academic adventure: the majority of circuits analyzed in the time domain are nonlinear. This includes digital computer, power supply, and oscillator circuits.

As in dc analysis, we will seek to find companion network models that reduce the problem to the analysis of a linear resistive network. A summary of our accomplishments to date in developing companion network models is shown in Figure 4.4a. In contrast, the simple network of Figure 4.4b is beyond our present capability, since the circuit contains both nonlinear resistive elements (a diode) and linear energy storage elements (a capacitor). However, the techniques for iterative analysis of resistive circuits may be combined with those for step by step transient analysis of linear dynamic circuits to yield an algorithm for solving such a circuit.

The development of our computing algorithm proceeds as follows. Ignoring the presence of nonlinear resistive elements, we replace energy storage elements by approximate resistive models (Figure 4.4c). Although the resultant network is nonlinear and so cannot be solved directly, it still offers an approximation to the original circuit. That is, as $T \to 0$, the derivative approximation for capacitances

$$C \frac{dv}{dt}\bigg|_{t=t^{n+1}} \approx C \frac{(v^{n+1} - v^{n})}{T} \tag{4.5}$$

becomes better and in the limit becomes exact. Since this is the <u>only</u> approximation involved in the resistive model, the model of Figure 4.4c may be made to represent the original circuit as closely as desired by choosing T sufficiently small.

The solution of the nonlinear resistive circuit now proceeds iteratively using Newton iteration. The solution becomes notationally cumbersome, since an iteration counter ($m = 0, 1, \ldots$) must now be used to indicate which parts of the companion model are updated at the $(m+1)$st Newton iteration and which are changed less frequently at each time step (indicated by n). We handle this notational problem by using a double superscript for variables that are dependent both on <u>m</u> and <u>n</u>. Thus, the companion model representing both the discretization and linearization of the original network is shown in Figure 4.4d. The Newton iteration formula is then given by

$$v_d^{n+1,\,m+1} = \frac{C\dfrac{v_1^{n}}{T} + \dfrac{v_s^{n+1}}{R} - i_d^{n+1,\,m} + G_d^{n+1,\,m}\, v_d^{n+1,\,m}}{\dfrac{1}{R} + G_d^{n+1,\,m} + \dfrac{C}{T}} \tag{4.6}$$

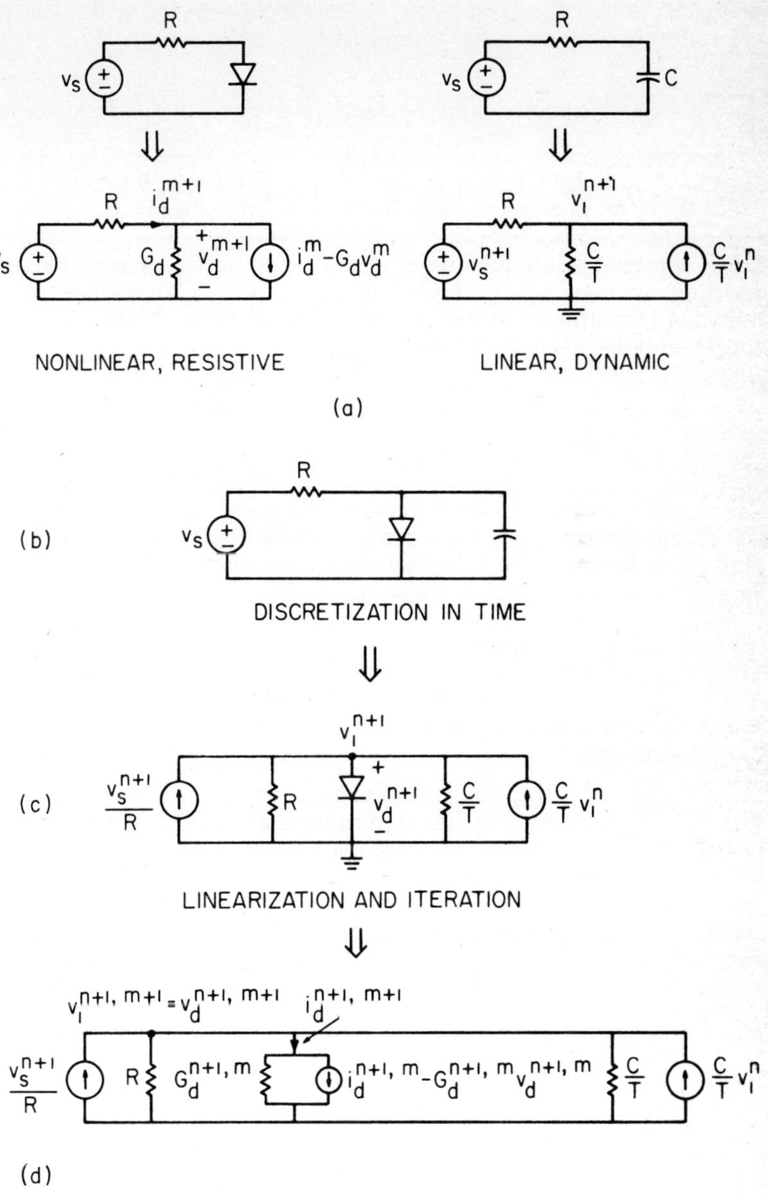

FIG. 4.4 Combination of Nonlinear and Transient Analysis Methods

where $v_d = v_1$.

The solution proceeds as follows. Let us assume that the capacitor voltage v_1^n is given at $t = 0$ ($n=0$). This voltage is equal to v_d, which we designate as $v_d^{1,0}$, i.e., the first guess in the iteration toward the calculation of v_d^1. From $v_d^{1,0}$, we can calculate

$$G_d^{1,0} = I_s \lambda e^{\lambda v_d^{1,0}} \tag{4.7}$$

$$i_d^{1,0} - G_d^{1,0} v_d^{1,0} = I_s(e^{\lambda v_d^{1,0}} - 1) - G_d^{1,0} v_d^{1,0} \tag{4.8}$$

which determines all quantities necessary to solve for node voltage $v_1^{1,1}$ and diode voltage $v_d^{1,1} = v_1^{1,1}$ in Figure 4.4d. We then set m = 1. The iteration is continued in this manner until the node voltage does not change appreciably between iterations, i.e.,

$$(v_1^{1,m_f} - v_1^{1,m_f-1})^2 \leq \epsilon$$

where m_f indicates the final iteration number. The node voltage v_1^1 then solves the nonlinear resistive network model of Figure 4.4c, for n=0. Moreover, it forms the initial guess for voltages to start a new iteration on the next time step, i.e., for any n.

$$v_1^{n+2,0} = v_1^{n+1,m_f}$$

4.3.2 Complete Solution of Nonlinear Dynamic Networks

Consider the response of the network of Figure 4.5 to a step in v_s. For $t < 0$, the circuit is assumed to have attained a dc state where no voltages or currents are changing appreciably. As $t \to \infty$, a new dc state is achieved. In either state, since $dv_c/dt = 0$, then $i_c = 0$ and the capacitor can be removed. This produces a simple resistor-diode circuit for which we have calculated the quiescent solution in several ways.

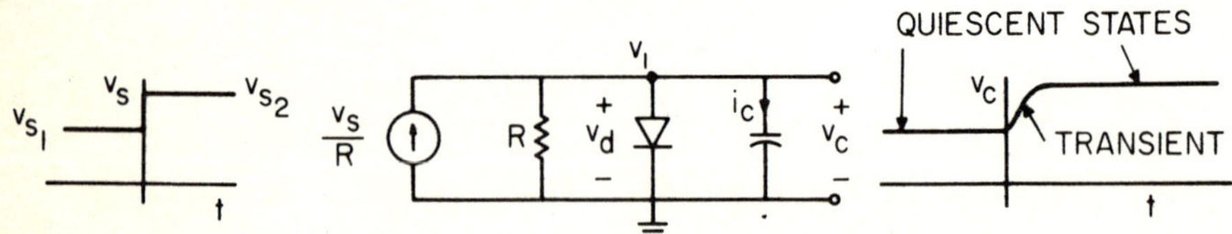

FIG. 4.5 Networks with Both DC and Transient Response

We now demonstrate that transient and dc analyses can proceed from the same equation. The iteration equation for the transient analysis is given by (4.6). Setting $T = \infty$ (or $C = 0$ in this case) and $v_s^{n+1} = v_s$, we have

$$v_c^{n+1,m+1} = \frac{\frac{v_s}{R} - i_d^{n+1,m} + G_d^{n+1,m} v_c^{n+1,m}}{\frac{1}{R} + G_d^{n+1,m}} \tag{4.9}$$

which defines the Newton iteration associated with the dc analysis (recall Equation (3.5)). Note that the superscript n+1 is superfluous and can be dropped, since (4.9) no longer contains a term to be updated with time. When this iteration in m converges, then (4.6) is used for transient response calculations, beginning with the v_c determined from (4.9).

Example 4.1. Let us return to the transistor amplifier of Chapter III, displayed again in Figure 4.6a. A capacitance C_E has been added in parallel with R_E. If we apply to this circuit a rapidly changing periodic signal, the capacitance voltage will remain essentially unchanged and the circuit will immediately attain a steady state response. We will apply a signal of value

$$i_s(t) = \begin{cases} 10^{-4}\sin(10^4 t) & t \geq 0 \\ 0 & t < 0 \end{cases} \qquad (4.10)$$

We are interested in observing v_{be} and v_{cb} over one half a period ($0 \leq t \leq \pi \times 10^{-4}$ sec), to determine if a significant transient exists.

The companion network is shown assembled in Figure 4.6b, after discretization and linearization of the model of Figure 4.6a. The node equations are written in two steps, with all sources first represented on the right hand side.

$$\begin{bmatrix} \frac{1}{R_2} + \frac{1}{R_1} + G_d^{n+1,m} & -G_d^{n+1,m} & -\frac{1}{R_1} \\ -G_d^{n+1,m} & \frac{1}{R_E} + \frac{C_E}{T} + G_d^{n+1,m} & 0 \\ -\frac{1}{R_1} & 0 & \frac{1}{R_1} + \frac{1}{R_L} \end{bmatrix} \begin{bmatrix} v_1^{n+1,m+1} \\ v_2^{n+1,m+1} \\ v_3^{n+1,m+1} \end{bmatrix}$$

$$= \begin{bmatrix} i_s^{n+1} + \alpha i_d^{n+1,m+1} - i_d^{n+1,m} + G_d^{n+1,m} v_d^{n+1,m} \\ \frac{C_E}{T} v_2^n + i_d^{n+1,m} - G_d^{n+1,m} v_d^{n+1,m} \\ -\alpha i_d^{n+1,m+1} + E/R_L \end{bmatrix}$$

Substituting for the unknown current

$$i_d^{n+1,m+1} = i_d^{n+1,m} + G_d^{n+1,m}(v_d^{n+1,m+1} - v_d^{n+1,m})$$

we finally have

$$\begin{bmatrix} \frac{1}{R_2}+\frac{1}{R_2}+(1-\alpha)G_d^{n+1,m} & (\alpha-1)G_d^{n+1,m} & -\frac{1}{R_1} \\ -G_d^{n+1,m} & \frac{1}{R_E}+\frac{C_E}{T}+G_d^{n+1,m} & 0 \\ -\frac{1}{R_1}+\alpha G_d^{n+1,m} & -\alpha G_d^{n+1,m} & \frac{1}{R_1}+\frac{1}{R_L} \end{bmatrix} \begin{bmatrix} v_1^{n+1,m+1} \\ v_2^{n+1,m+1} \\ v_3^{n+1,m+1} \end{bmatrix}$$

$$= \begin{bmatrix} i_s^{n+1} + (\alpha - 1)(i_d^{n+1,m} - G_d^{n+1,m} v_d^{n+1,m}) \\ \frac{C_E}{T} v_2^{n} + i_d^{n+1,m} - G_d^{n+1,m} v_d^{n+1,m} \\ -\alpha i_d^{n+1,m} + E/R_L \end{bmatrix} \qquad (4.11)$$

Observe that when $T = \infty$, this equation becomes identical in form to the iteration equation for dc analysis in Chapter III.

A computer program (TNET) to solve (4.11) iteratively at each time step is shown in Appendix Table 4.1a. As we have observed for the circuit of Figure 4.5, the companion network for the dc and transient solutions differ only in the presence of capacitances in the latter. Therefore, only the following minor additions need be made to the dc analysis program DNET of Chapter III to perform a transient analysis.

(1) Parameters NUMPTS, TMIN, and TMAX are used to determine the time step and the time range over which the solution is to be found.

(2) The Newton iteration - counted by the M Fortran variable in the main program - is now nested in the loop in which time is incremented - counted by the N Fortran variable.

(3) The conductance C_E/T and the current sources $C_E{}^n/T$ and $i_s{}^{n+1}$ are added to arrays G and I in NODEQ, <u>except for the dc iteration.</u>

The results of executing this program with the data of Table 4.3a are shown in Table 4.3b. Both outputs exhibit a significant transient component, since their values at the end of one half cycle are significantly different from their values at $t = 0$. It is left as a problem to determine the effect of integration accuracy on this result by using the trapezoidal integration rule.

4.3.3. Companion Model of a Nonlinear Capacitor

A nonlinear capacitor can be described by a nonlinear charge-voltage relationship

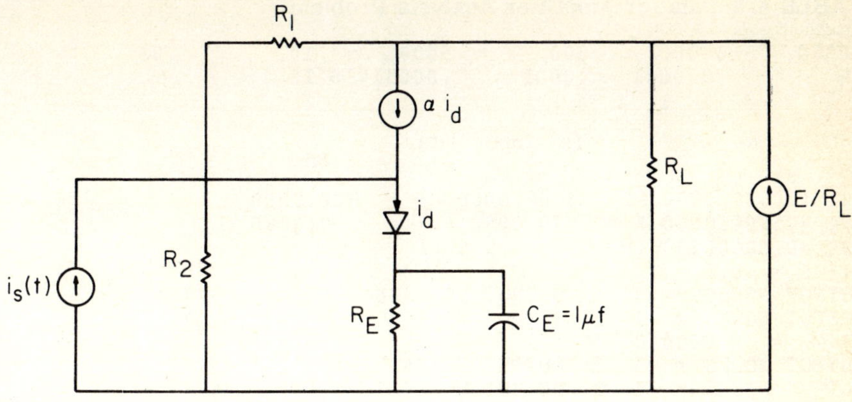

(a) LARGE SIGNAL DYNAMIC MODEL

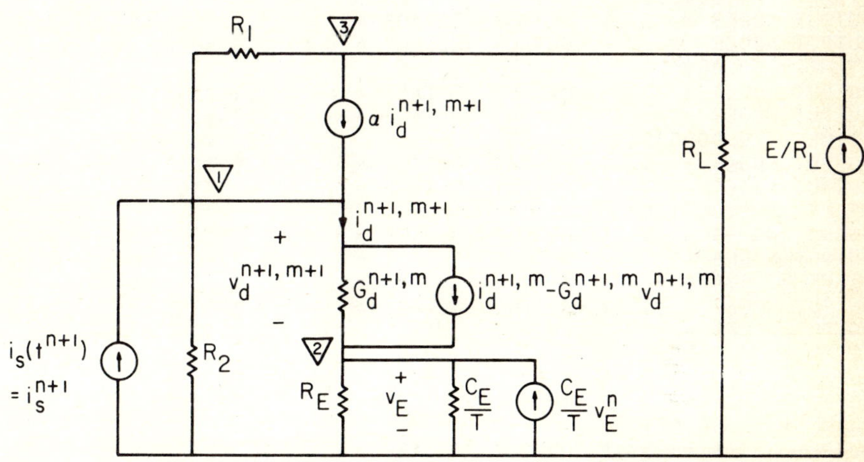

(b) COMPANION MODEL OF PART (a)

FIG. 4.6 Models for Complete Analysis of Transistor Amplifier

$$q = f(v) \tag{4.12}$$

plus the differential relationship

$$i = \frac{dq}{dt} \tag{4.13}$$

We desire to make up a library model for this element, where the components of the model are the elements we can most easily accept for node analysis - resistors, current sources, and VCCS's.

The process begins by producing a circuit model that satisfies (4.12) and (4.13), as in Figure 4.7a. The VCC's do not form a gyrator, since the current in one is proportional to the derivative of the controlling variable q. However, the node equations of the circuit do exist.

This circuit is reduced to companion network form by following the procedure of Figure 4.4c-d. First, the time derivative is discretized,

TABLE 4.3 Data for Amplifier Analysis Problem

```
20000.    3000.    300.     5000.      10.      .98
.4        .000001  .0001    .00031416 16
  3  2  3  1  1  2
```

(a) Input Data

```
 R1 =20000.0000 R2 = 3000.0000 RE =  300.0000
 RL = 5000.0000 E =    10.0000 ALF =     0.9800
 CE =0.00000100 INPUT =     0.0001
 TIME =  0.0
OUTPUT NODES =  3  1  OUTPUT VOLTAGE =      4.2677797
OUTPUT NODES =  1  2  OUTPUT VOLTAGE =      0.3407787
 TIME =  0.000020
OUTPUT NODES =  3  1  OUTPUT VOLTAGE =      3.9372215
OUTPUT NODES =  1  2  OUTPUT VOLTAGE =      0.3431716
 TIME =  0.000039
OUTPUT NODES =  3  1  OUTPUT VOLTAGE =      3.6262942
OUTPUT NODES =  1  2  OUTPUT VOLTAGE =      0.3452241
 TIME =  0.000059
OUTPUT NODES =  3  1  OUTPUT VOLTAGE =      3.3466540
OUTPUT NODES =  1  2  OUTPUT VOLTAGE =      0.3469303
 TIME =  0.000079
OUTPUT NODES =  3  1  OUTPUT VOLTAGE =      3.1087596
OUTPUT NODES =  1  2  OUTPUT VOLTAGE =      0.3482892
 TIME =  0.000098
OUTPUT NODES =  3  1  OUTPUT VOLTAGE =      2.9214535
OUTPUT NODES =  1  2  OUTPUT VOLTAGE =      0.3493028
 TIME =  0.000118
OUTPUT NODES =  3  1  OUTPUT VOLTAGE =      2.7916097
OUTPUT NODES =  1  2  OUTPUT VOLTAGE =      0.3499741
 TIME =  0.000137
OUTPUT NODES =  3  1  OUTPUT VOLTAGE =      2.7238634
OUTPUT NODES =  1  2  OUTPUT VOLTAGE =      0.3503063
 TIME =  0.000157
OUTPUT NODES =  3  1  OUTPUT VOLTAGE =      2.7204285
OUTPUT NODES =  1  2  OUTPUT VOLTAGE =      0.3503023
 TIME =  0.000177
OUTPUT NODES =  3  1  OUTPUT VOLTAGE =      2.7810113
OUTPUT NODES =  1  2  OUTPUT VOLTAGE =      0.3499645
 TIME =  0.000196
OUTPUT NODES =  3  1  OUTPUT VOLTAGE =      2.9028226
OUTPUT NODES =  1  2  OUTPUT VOLTAGE =      0.3492956
 TIME =  0.000216
OUTPUT NODES =  3  1  OUTPUT VOLTAGE =      3.0806871
OUTPUT NODES =  1  2  OUTPUT VOLTAGE =      0.3482984
 TIME =  0.000236
OUTPUT NODES =  3  1  OUTPUT VOLTAGE =      3.3072479
OUTPUT NODES =  1  2  OUTPUT VOLTAGE =      0.3469773
 TIME =  0.000255
OUTPUT NODES =  3  1  OUTPUT VOLTAGE =      3.5732581
OUTPUT NODES =  1  2  OUTPUT VOLTAGE =      0.3453401
 TIME =  0.000275
OUTPUT NODES =  3  1  OUTPUT VOLTAGE =      3.8679510
OUTPUT NODES =  1  2  OUTPUT VOLTAGE =      0.3434001
 TIME =  0.000295
OUTPUT NODES =  3  1  OUTPUT VOLTAGE =      4.1794760
OUTPUT NODES =  1  2  OUTPUT VOLTAGE =      0.3411800
 TIME =  0.000314
OUTPUT NODES =  3  1  OUTPUT VOLTAGE =      4.4953868
OUTPUT NODES =  1  2  OUTPUT VOLTAGE =      0.3387171
```

(b) Output Data

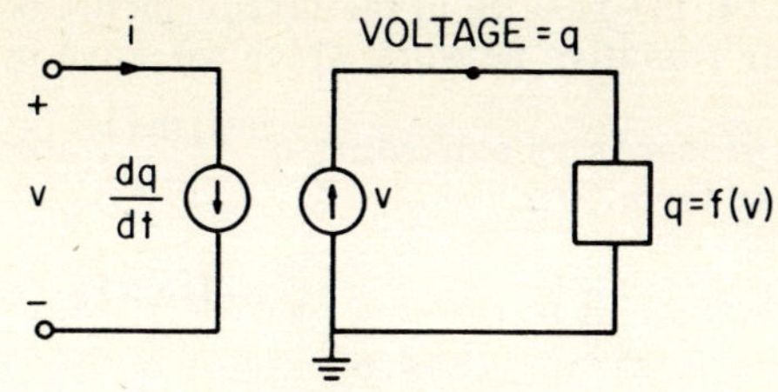

(a) CONTINOUS MODEL

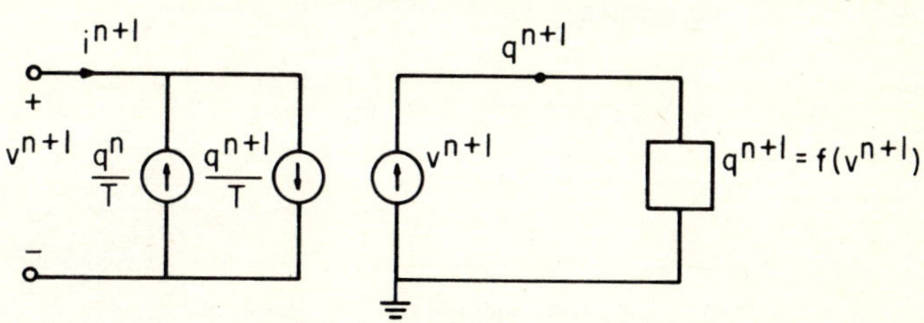

(b) DISCRETIZED MODEL

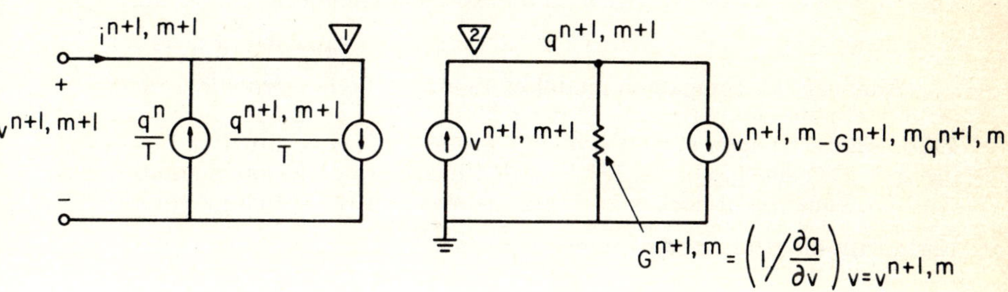

(c) COMPLETE TRANSIENT COMPANION MODEL

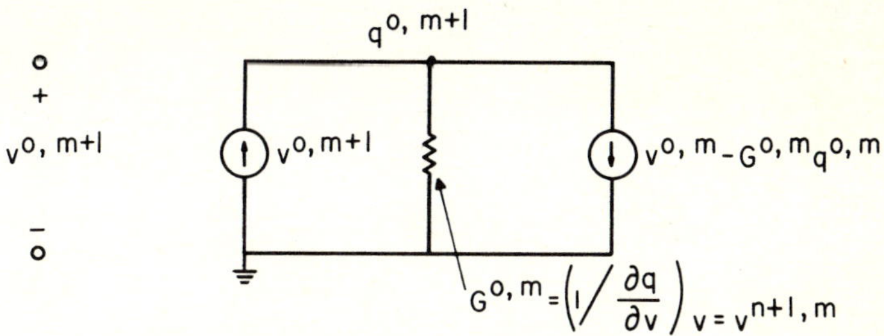

(d) DC COMPANION MODEL

FIG. 4.7 Evolution of Companion Models of Nonlinear Capacitor

yielding

$$i^{n+1} = \frac{q^{n+1} - q^n}{T} \tag{4.14}$$

Combining (4. 12) and (4. 14) results in the circuit model of Figure 4. 7b. This is still a nonlinear resistive network which we solve by linearization and iteration.

The linearization proceeds by expanding $q^{n+1,m+1}$ about $v^{n+1,m}$ and retaining only the linear term, viz,

$$q^{n+1,m+1} = q^{n+1,m} + \left.\frac{\partial q}{\partial v}\right|_{v=v^{n+1,m}} (v^{n+1,m+1} - v^{n+1,m})$$

Solving for $v^{n+1,m+1}$, we obtain

$$v^{n+1,m+1} = v^{n+1,m} + \frac{1}{\left.\frac{\partial q}{\partial v}\right|_{v=v^{n+1,m}}} (q^{n+1,m+1} - q^{n+1,m})$$

$$= v^{n+1,m} + G^{n+1,m} (q^{n+1,m+1} - q^{n+1,m})$$

This equation may be regarded as the KCL equation at node $\bigtriangledown 2$ in Figure 4. 7c, whereas (4. 14) remains the KCL equation satisfied at node $\bigtriangledown 1$.

Although the companion model of Figure 4. 7c is a transient model, a dc model may be obtained by setting $T = \infty$. The result is shown in Figure 4. 7b, where an open circuit now appears at the i-v terminals. However, the q-v relationship of (4. 12) is still iterated by successive solutions of the KCL equation at node $\bigtriangledown 2$; this insures that the transient solution will begin with the proper capacitance charge q^o.

4. 3. 4. Nonlinear Transient Model of a Diode

The exponential model of the diode given by

$$i_{dc} = I_s(e^{\lambda v_d} - 1) \tag{4. 15}$$

is of course only a resistive model. When the transient response of the diode itself is significant, the diode q-v characteristic must be considered. The minority carrier diffusion process is most simply modeled by the formula

$$q_d = q_o(e^{\lambda v_d} - 1) \tag{4. 16}$$

We immediately determine that, for the model of Figure 4. 7d,

$$G = 1/(\partial q/\partial v) = 1/(\lambda q_o e^{\lambda v_d})$$

The similarity of (4. 15) and (4. 16) permits easy inclusion of the resistive characteristic in the capacitance companion model. We write the total diode current as

$$
\begin{aligned}
i_d &= \frac{dq_d}{dt} + i_{dc} \\
&= \frac{dq_d}{dt} + I_s \left(e^{\lambda v_d} - 1\right) \\
&= \frac{dq_d}{dt} + \frac{q_o}{I_s} q_d \\
&= \frac{dq_d}{dt} + \frac{q_d}{\tau}
\end{aligned}
\qquad (4.17)
$$

where $\tau = I_s/q_o$. Linearization and discretization of this equation yields

$$
i_d^{n+1, m+1} = \frac{(q_d^{n+1, m+1} - q_d^{n})}{T} + \frac{q_d^{n+1, m+1}}{\tau}
$$

The resistive portion of the diode model therefore adds only a voltage controlled current source of value $q^{n+1, m+1}/\tau$ to the capacitance companion model. The complete model is shown in Figure 4.8b. Note that when $T = \infty$ (the dc case), the current generator $q^{n+1, m+1}/\tau$ remains to relate i_d, q_d, and v_d, thus representing the dc relationship of (4.15).

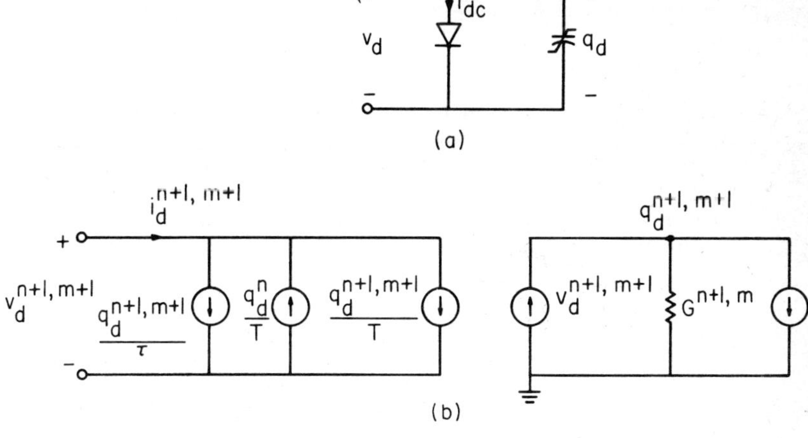

FIG. 4.8 The Diode and Its Companion Model

4.3.5. Nonlinear Transient Transistor Model

For transient analysis a transistor is commonly described by the charge control model of Figure 4.9. Basically, the model consists of two diode models that are cross-coupled by two current-controlled current sources.

The companion model is easily obtained by connecting two diode models at their common voltage terminals with the charges q_{df} and q_{dr} represented as separate node voltages as shown in Figure 4. 9. The controlled currents $\alpha_f i_{df}$ and $\alpha_r i_{dr}$ are, from (4. 16),

$$\alpha_f i_{df}^{n+1, m+1} = \alpha_f I_s (e^{qv_{df}^{n+1, m+1}} - 1) = \alpha_f \frac{q_{df}^{n+1, m+1}}{\tau_f} \tag{4.18}$$

$$\alpha_r i_{dr}^{n+1, m+1} = \alpha_r \frac{q_{dr}^{n+1, m+1}}{\tau_r} \tag{4.19}$$

Since $q_{df}^{n+1, m+1}$ and $q_{dr}^{n+1, m+1}$ are represented as node voltages, (4. 18) and (4. 19) have the circuit models of voltage-controlled current sources and so are consistent with our node analysis formulation procedure.

4. 4 A GENERAL NONLINEAR TRANSIENT CIRCUIT ANALYSIS PROGRAM (TCAP)

It should be clear by this time that both general purpose and special purpose circuit analysis programs have a similar structure, as indicated by the similarity of the main programs in RNET and RCAP and in DNET and DCAP. A general nonlinear transient analysis program can therefore be expected to have a main program similar to TNET but with subroutines similar to DCAP, the general dc analysis program. In particular, we find that the TCAP program of Appendix Table 4. 2 has the following characteristics.

(1) Variable NUMPTS, TMIN, and TMAX have the same function as in TNET.

(2) Variables M and N have the same function in the main program TNET.

(3) The model library in NODEQ of DCAP is augmented by the model for a capacitor - a conductance and current source - and by a model for the time-dependent source. Both are ignored for t = 0, the dc iteration.

(4) In PREPAR, several new elements are modeled by more familiar elements. An inductor is modeled by a gyrator and capacitor as in Table 4. 1; a current-controlled current source (CCCS) is modeled by a VCCS and a gyrator which converts a controlling current to a controlling voltage (Figure 4. 10).

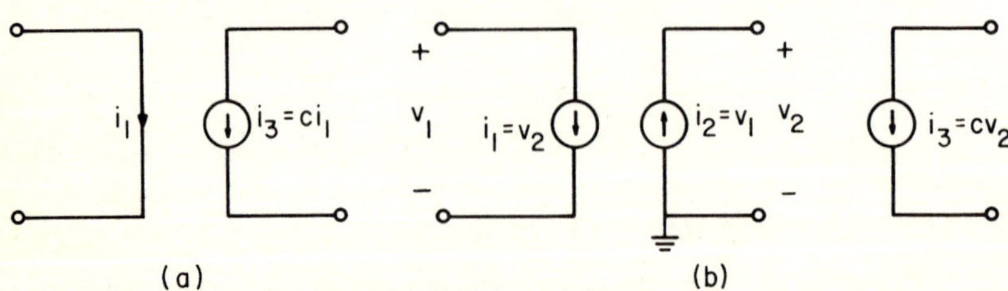

FIG. 4. 10 Conversion of a CCCS to a VCCS

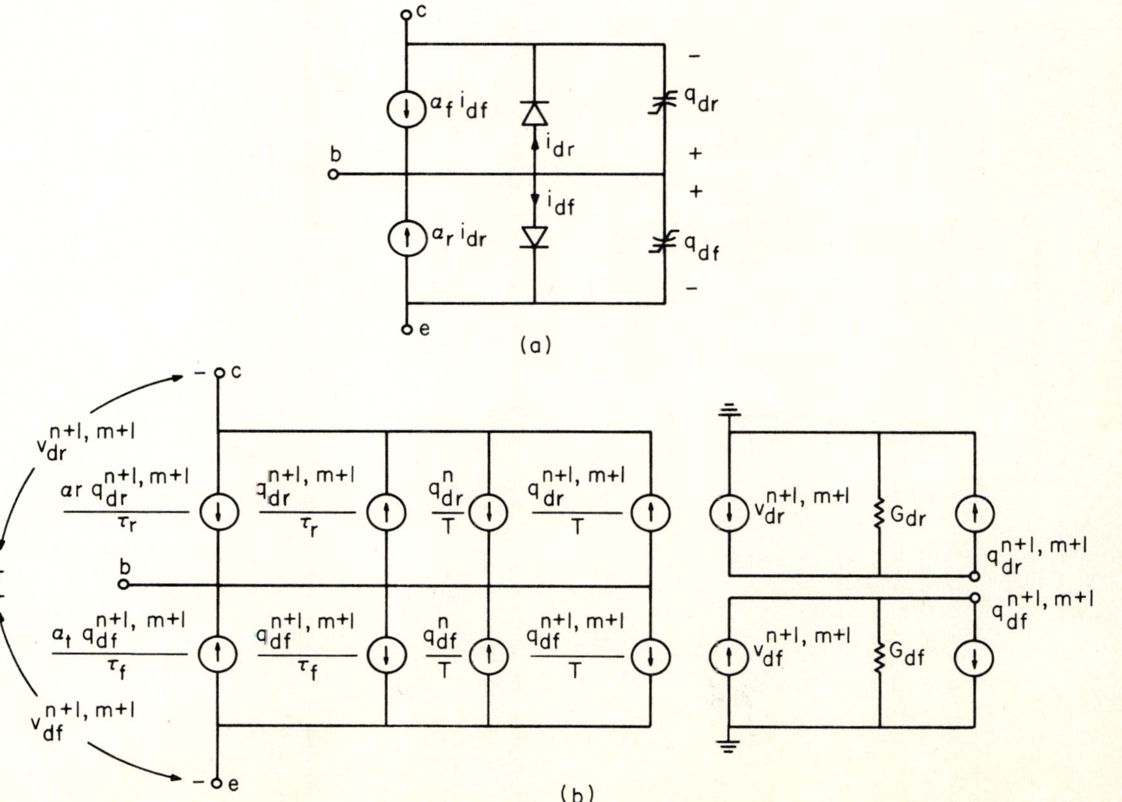

FIG. 4.9 A Dynamic Transistor Model and Its Companion Network Model

4.5 NUMERICAL CONSIDERATIONS

4.5.1 Elementary Error Analysis

The error made in the backward Euler derivative approximation

$$\left.\frac{dv}{dt}\right|_{t=t^{n+1}} \approx \frac{v^{n+1} - v^{n}}{T} \tag{4.20}$$

is easy to estimate even though the effect of this error on the response of a complicated circuit is usually difficult. This discrepancy may occur, for example, because the circuit output may be relatively insensitive to the capacitor current being approximated.

To obtain an expression for the error, we expand v^n in a Taylor's series about v^{n+1}. We have

$$v^n = v^{n+1} + (t^n - t^{n+1}) \left.\frac{dv}{dt}\right|_{t=t^{n+1}} + \frac{1}{2}(t^n - t^{n+1})^2 \left.\frac{d^2v}{dt^2}\right|_{t=t^{n+1}} + \dots$$

$$= v^{n+1} - T\left.\frac{dv}{dt}\right|_{t=t^{n+1}} + \frac{T^2}{2}\left.\frac{d^2v}{dt^2}\right|_{t=t^{n+1}} + \dots \tag{4.21}$$

Comparing (4.20) and (4.21), we find the error made in integrating from t^n to t^{n+1} with step size T is approximated by

$$\epsilon_T^{n,n+1} = \frac{T^2}{2}\left.\frac{d^2v}{dt^2}\right|_{t=t^{n+1}}$$

for T small. From this formula, we can demonstrate the linear error dependence on T shown in Table 4.2. For example, if we halve the time step T, the approximation for the error made in integrating from nT to (n + 1/2)T is

$$\epsilon_{T/2}^{n,n+1/2} = \frac{T^2}{8}\left.\frac{d^2v}{dt^2}\right|_{t=t^{n+1/2}}$$

The total error made from nT to (n+1)T is the sum of errors $\epsilon_{T/2}^{n,n+1/2}$ and $\epsilon_{T/2}^{n+1/2,n+1}$. If the second derivative term can be assumed constant, then

$$\epsilon_{T/2}^{n,n+1} = \frac{T^2}{4}\left.\frac{d^2v}{dt^2}\right|_{t=t^{n+1}}$$

This error is one half the error made with a full step size T, i.e.,

$$\epsilon_{T/2}^{n,n+1} = \frac{1}{2}\epsilon_T^{n,n+1}$$

Therefore, the error is linearly related to the step size T.

4.5.2 An Improved Approximation to the Derivative

The approximation to branch v-i relationships can be improved by permitting the branch currents to be defined at time $t^{n+1/2} = ((n+1/2)T)$. We can then write approximate formulae.

For a capacitor (Figure 4.11a):

$$i^{n+1/2} \triangleq \frac{C}{T}(v^{n+1} - v^{n})$$

For a resistor (Figure 4.11b):

$$i^{n+1/2} \triangleq \frac{i^{n+1} + i^{n}}{2} = \frac{G}{2}(v^{n+1} + v^{n})$$

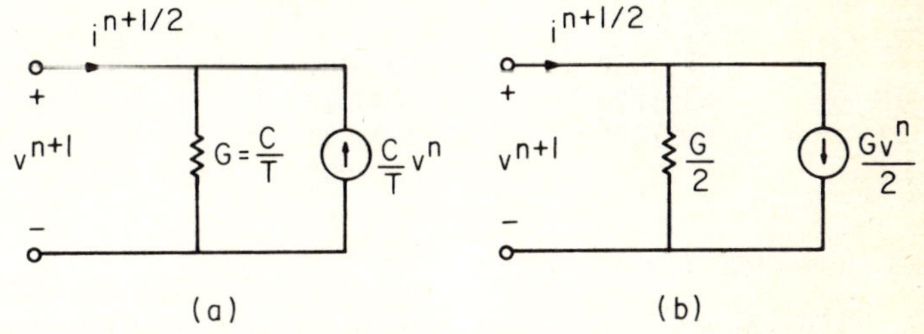

FIG. 4.11 Companion Models for Trapezoidal Integration

These turn out to be considerably better approximations than those already proposed, due to the "averaging" of currents and voltages at t^n and t^{n+1}. Mathematically they are equivalent to the trapezoidal integration rule. As a demonstration of the improved accuracy, consider again the network in Figure 4.12b. Node analysis of the companion network yields

$$v^{n+1} = \left[\frac{C}{T}v^{n} - \frac{1}{2R}v^{n} + \frac{i_s(t^{n}) + i_s(t^{n+1})}{2}\right] / \left[\frac{C}{T} + \frac{1}{2R}\right] \tag{4.22}$$

The solution with $R = C = 1$ and $i_S(t) = u(t)$ is given in Table 4.4 for different step sizes. The error decrease is now proportional to T^2, which can be proved in the manner of Section 4.5.1. Four significant figures are attained for $T = .1$; this accuracy is never attained in Table 4.1, even for $T = .001$.

A somewhat different implementation of the trapezoidal rule is explained in [4.2], where both currents and voltages in the companion model are approximations at the (n+1)st moment of time. This removes the current source in the resistance model of Figure 4.11b, but requires that i^n be calculated and saved for capacitors at each step.

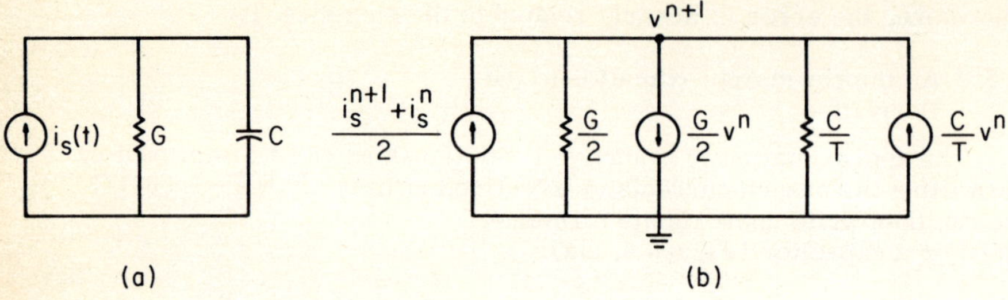

FIG. 4.12 Companion Model for Trapezoidal Rule

TABLE 4.4 Error Resulting from Improved Derivative Approximation

Step Size (T)	Calculated Value at t = 1	Error
.1	.6321972	-.0000767
.05	.6321328	-.0000123
.02	.6321328	-.0000031
.01	.6321213	-.0000008

4.5.3 Numerical Instability

Introduction. Consider the application of the forward Euler approximation

$$\left.\frac{dv}{dt}\right|_{t=t^n} \approx \frac{v^{n+1} - v^n}{T}$$

to the single equation

$$\dot{v} = -v + 1 \qquad v(0) = 0 \tag{4.23}$$

with solution $v(t) = 1 - e^{-t}$. This derivative approximation is made at $t = t^n$ rather than at $t = t^{n+1}$, so that the approximation is made forward in time. This yields the algorithm

$$v^{n+1} = v^n + T\left.\frac{dv}{dt}\right|_{t=t^n} \qquad n = 1, 2, \cdots$$

$$= v^n(1 - T) + T \qquad v^0 = 0$$

TABLE 4.5

Step Size	n = 0	n = 1	n = 2	n = 3	n = 4	n = 5
.50000	.99000	.99500	.99750	.99875	.99938	.99969
1.0000	.99000	1.0000	1.0000	1.0000	1.0000	1.0000
1.9900	.99000	1.0099	.99020	1.00970	.99039	1.00951
2.0000	.99000	1.01000	.99000	1.0100	.99000	1.0100
2.0100	.99000	1.0101	.98980	1.0103	.98959	1.0105
5.0000	.99000	1.0400	.84800	1.6400	-1.5600	11.200

TABLE 4.6

Step Size	n = 0	n = 1	n = 2	n = 3	n = 4	n = 5
.50000	.9900	.99333	.99556	.99704	.99802	.99868
1.0000	.99000	.99500	.99750	.99875	.00038	.99969
2.0000	.99000	.99667	.99889	.99963	.99988	.99996
5.0000	.99000	.99833	.99972	.99995	1.0000	1.0000

Now assume that v has been calculated to within 1% of its final value, i.e., to a value of .99. If we were demanding only 1% solution accuracy, it would seem that any step size could now be chosen if the integration were to be continued for some reason. We would require only that the solution approach unity for large n. Unfortunately, such is not the case, as Table 4.5 shows. For $T > 2$, the integration becomes unstable, i.e., on successive steps, the solution departs from the true solution.

The backward Euler derivative approximation

$$\left.\frac{dv}{dt}\right|_{t=t^{n+1}} \approx \frac{v^{n+1} - v^{n}}{T}$$

which we used previously in this chapter has far better stability properties. The result of using this formula to solve (4.23) yields the integration algorithm

$$v^{n+1} = \frac{v^{n} + T}{1 + T}$$

Table 4.6 shows that, beginning at the value $v^0 = .9900$, the solution converges to the proper value of 1 for any $T > 0$.

In most numerical analysis texts this form of instability is ignored, the assumption being that the accuracy of the forward Euler procedure usually limits the size of T more than the instability shown above. For a single equation this notion is valid; however, simultaneous equations can have a distinctly different character.

Example 4.2. In the network of Figure 4.13, we can eliminate node voltage variables to yield the two differential equations

$$\frac{d}{dt}\begin{bmatrix} v_{C_1} \\ v_{C_2} \end{bmatrix} = \begin{bmatrix} -1.457 \times 10^5 & 238. \\ -19.04 & -.0524 \end{bmatrix}\begin{bmatrix} v_{C_1} \\ v_{C_2} \end{bmatrix} + \begin{bmatrix} 5 \times 10^4 \\ 0 \end{bmatrix} u(t) \tag{4.24}$$

The natural frequencies are approximately at $s = -1.46 \times 10^5$ and $s = -.0827$ corresponding to the scaled high and low frequency cutoff frequencies of the amplifier. The step response therefore has a rise time constant $T_R = 6.85\ \mu sec$ and a fall time constant of $T_F = 12.1$ sec. The integration

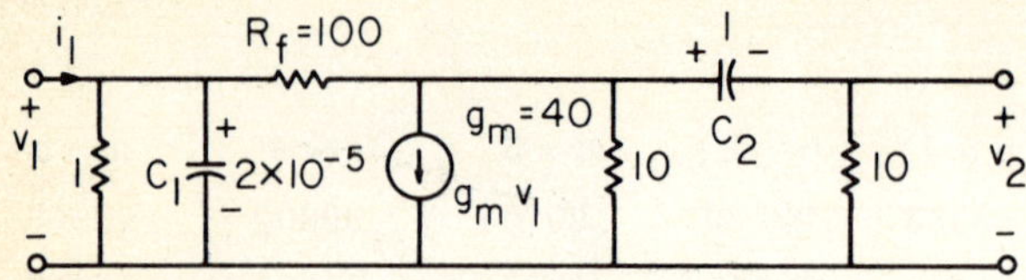

FIG. 4.13 Scaled Low and High Frequency Equivalent of RC - Coupled Amplifier

step size would seemingly be chosen small initially (say, 10^{-6}) to display the rapid transient, and then could be lengthened while the output was relatively constant. We might further conjecture that a step size of, say, 0.1 sec would be entirely adequate to view the pulse decay.

In solving (4.24) by a forward Euler derivative approximation, we would write

$$\begin{bmatrix} v_{C_1} \\ v_{C_2} \end{bmatrix}^{n+1} = \begin{bmatrix} 1 - 1.45 \times 10^5 T & 238.T \\ -19.04T & 1 - .0524T \end{bmatrix} \begin{bmatrix} v_{C_1} \\ v_{C_2} \end{bmatrix}^{n} + \begin{bmatrix} 5 \times 10^4 T \\ 0 \end{bmatrix}$$

The result of performing this integration is displayed in Figure 4.14. Two numerical experiments are performed to illustrate the problem of instability.

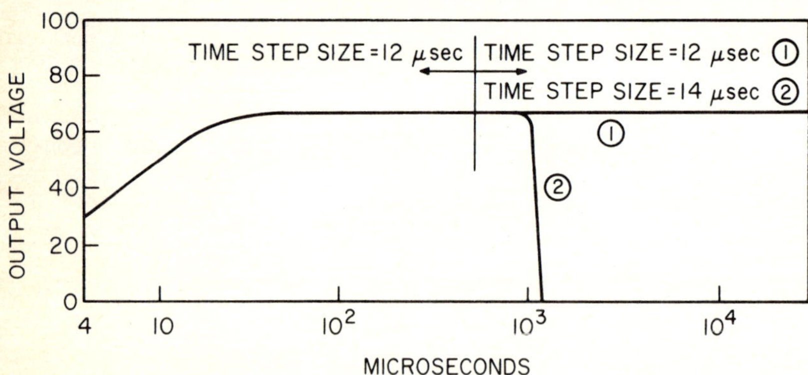

FIG. 4.14 Linear Amplifier Response by Forward Euler Integration

Experiment ①. At $t = 5 \times 10^{-4}$, the step size is maintained at 12 μsec; although this is larger than T_R, the resolution remains acceptable, due to the relatively constant response.

Experiment ②. At $t = 5 \times 10^{-4}$, the step size is increased to 14 μsec; the response tends toward infinity, and the numerical integration is unstable. (As we will show later, the numerical integration becomes unstable for $T = 13.7$ μsec.)

We could correctly conclude that with the forward Euler formula, the allowable step size is restricted by the most rapid time constant of the circuit. (We will show this in Chapter IX.) To avoid unreasonable step size limitations using forward Euler, elements producing widely separated time constants must be removed. This removal may take one of the two forms: (1) separate transient analysis of high and low frequency equivalent circuits; (2) removal of a (reactive) element during the transient analysis. For example, after 100 μsec, the current through C_1 could be set to zero; this implies $\dot{v}_{C_1} = 0$ and v_{C_1} could be eliminated from (4.24). The step size could then be lengthened to a reasonable value. This is, or course, a dangerous strategy, since the waveform could be at a local maximum or minimum. In general, automatic removal of an element is not considered a suitable strategy for general purpose programs.

In contrast to the above problems, integration of (4.24) with the backward Euler integration rule (Figure 4.14) manifests no instability problem regardless of the step size. It is left as an exercise to develop the equations for this case.

The analysis of present-day circuits is especially sensitive to instability. First, these circuits are typically large, increasing the likelihood of widely separated time constants. Second, the time constants of switching circuits such as flip-flops change radically in the process of switching between stable states. As a dramatic demonstration, the time constants of a transistor typically change by a factor of 100 or more between cutoff and saturation. Clearly, the "fast" and "slow" portion of the circuit cannot be conveniently separated in this case.

The modern philosophy in computer-aided design requires that the designer should be left free to design and should not be concerned with placating the idiosyncrasies of the numerics. For this reason, we devote the entire next section to the study of analysis of unstable integration.

Stability Analysis. We generalize the study of stability in the last section by considering the linear differential equation

$$\dot{v} = sv \qquad v(0) = 1 \tag{4.25}$$

where s is a complex constant. Then we immediately deduce the step by step integration formula to be of the form

Forward Euler: $v^{n+1} = (1 + sT)v^n$

Backward Euler: $v^{n+1} = (1 - sT)^{-1}v^n$

Since $v(t) = e^{st}$ should not grow as $t \to \infty$ if $\operatorname{Re} s \le 0$, we impose the requirement that if $v^{n+1} = k\, v^n$, then

$$|k| \leq 1 \qquad \operatorname{Re} s \leq 0$$

for stability. We may view k as a step-to-step magnification factor.

It is common to display the "stable" regions in the complex s-plane where $|k| \le 1$. We can normalize this display by recognizing that the product sT is more significant than the values of T or s individually. That is, if $s = -10^6$ and $T = 10^{-3}$, then time scaling by a constant a yields the scaled values $s = -10^6/a$ and $T = 10^{-3}a$. Clearly $Ts = -10^3$ in either case.

The stable regions corresponding to the inequalities

$$\text{Forward Euler:} \qquad |1 + sT| \leq 1 \qquad s \text{ complex} \tag{4.26}$$

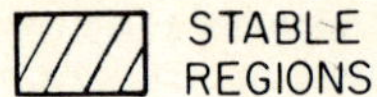

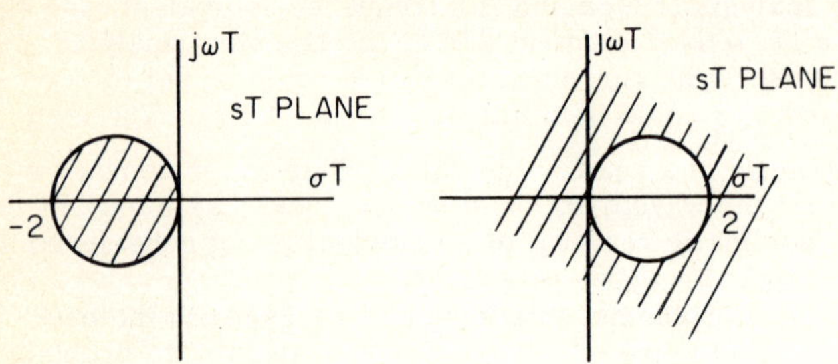

FIG. 4.15 Stability Regions of Euler Methods

$$\text{Backward Euler:} \quad \left|(1 - sT)^{-1}\right| \leq 1 \quad s \text{ complex} \tag{4.27}$$

are shown in Figure 4.15.

To illustrate the use of these figures, consider the example of (4.23). In this case, s = 1, on the negative real axis. Stability then requires

Forward Euler: $-2 \leq -T \leq 0$

or $2 \geq T \geq 0$

Backward Euler: $T\,(-1) \leq 0$

or $T \geq 0$

These check with our previous calculations showing the backward Euler procedure to be stable for all step sizes.

Conclusions. At this point, the reader should be concerned about the discrepancy between the solution techniques we have thus far presented which have been based on the companion network and the stability proofs we have presented which have been based on solutions of a single, linear, differential equation. Unfortunately, we are not in a position to discuss even the concept of stability for so difficult a problem as posed by the companion network model of a nonlinear circuit. We now state a result for linear networks, however: the integration algorithm resulting from the companion network model will be stable if and only if every natural frequency s_i of the original network satisfies the stability conditions on a single equation (we will prove this in Chapter IX). In Example 4.3 where $s_1 = -1.46 \times 10^5$, $s_2 = -.0827$, we would require for stability

Forward Euler: $-2 \leq Ts_i \leq 0$

$13.7\mu s \geq T \geq 0$

Backward Euler: $Ts_i \leq 0$

$T \geq 0$

The limitation on T for the forward Euler method is seen to lie within the limits observed in the experiments displayed in Figure 4.14. This result requires in general that the step size be limited by the largest natural frequency or smallest (fastest) time constant.

Problems

4.1 Show that the trapezoidal integration rule

$$v^{n+1/2} \triangleq \frac{\dot{v}^{n+1} + \dot{v}^{n}}{2} = \frac{v^{n+1} - v^{n}}{T}$$

has the stability region described by

$$\frac{|1 + sT/2|}{|1 - sT/2|} \leq 1$$

Sketch this region in the complex plane by setting $s = \quad + j\omega$.

4.2 Comment on the interpretation of the current $i^{n+1/2}$ used in the trapezoidal rule when a dc solution is to be found.

4.3 Obtain a companion model for the nonlinear capacitor when the trapezoidal rule is used. Comment on the dc analysis problem.

4.4 The small signal response of the amplifier of Figure 4.6 is defined by maintaining the diode conductance $G_d^{n+1,m}$ at its dc value G_d^o.

Calculate the small signal transient response in Problem 4.5 under this condition and compare with the large signal transient response.

4.5 A capacitor is sometimes placed as shown in Figure P4.1. If $i_s(t)$ is described by the sinusoid of Problem 4.4, determine the change in v_{cb} and v_{be} over the range $0 \leq t \leq 10^{-5}$.

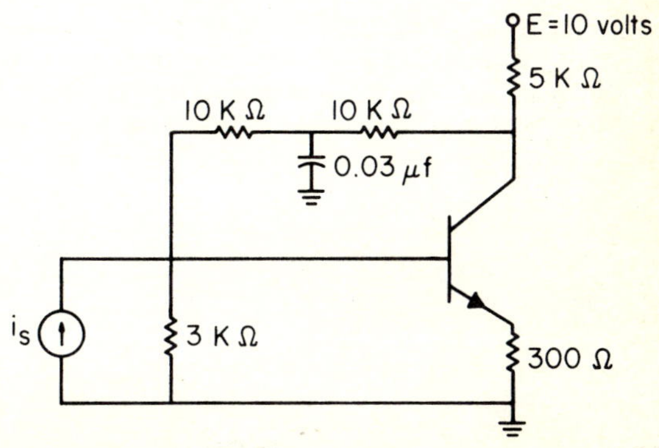

FIG. P4.1

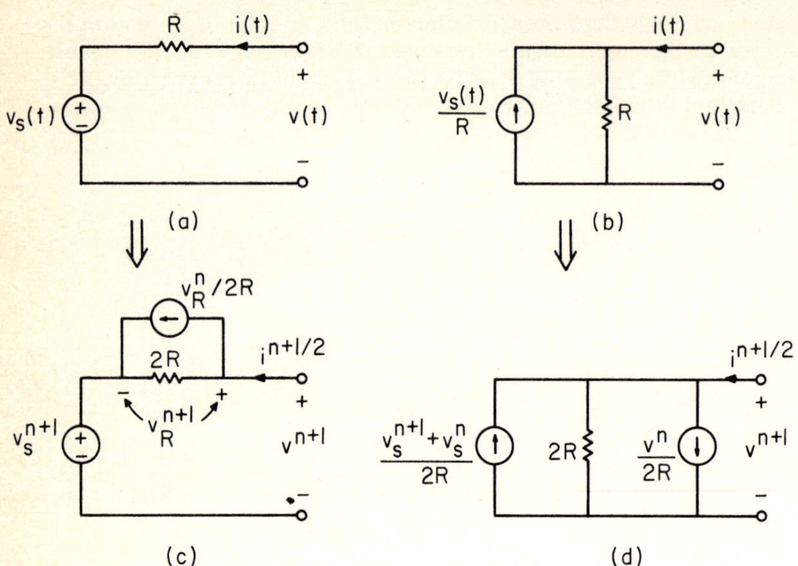

FIG. P4.2

4.6 The networks of Figure P4.2 a-b are equivalent. Show that after discretization using the trapezoidal integration rule as in Figure P4.2 c-d, the i-v characteristics remain equivalent.

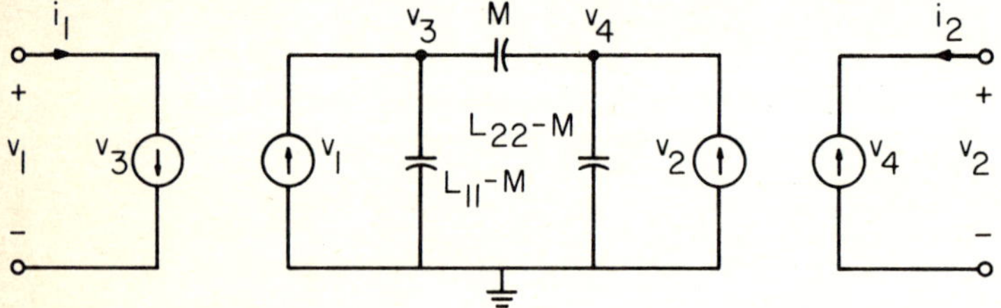

FIG. P4.3

4.7 Show that the network of Figure P4.3 satisfies the equations of a transformer. How does this transformer behave under dc conditions. Compare with the ideal transformer model of Problem 3.14.

4.8 The dynamic diode model often contains a capacitance of the form

$$q = k(v - v_p)^n$$

where $v_p \approx .9$ volts and $1/3 \leq n \leq 1/2$. Incorporate this element in the diode and transistor models of Figures 4.8 and 4.9 respectively.

REFERENCES

4.1 Jensen, R. W., and M. D. Lieberman, IBM Electronic Circuit Analysis Program, Prentice Hall; 1968.

4.2 Dommel, H. W., "Digital Computer Solution of Electromagnetic Transients in Single and Multiphase Networks," Trans. IEEE, vol. PAS - 88, no. 4, pp. 388-399; April, 1969.

4.3 Shichman, H., "Integration System of a Nonlinear Analysis Program," Trans. IEEE, vol. CT-17, no. 3, pp. 378-385; August, 1970.

5

Sensitivity Calculations

5.1 INTRODUCTION

5.1.1 Sensitivity and Computer-aided Design

There are several design reasons for wanting to calculate the sensitivity of a network output (v_o or i_o) to a set of design parameters (p_i). Among these are the need for such information in performing tolerance analysis (Chapter VII) and automatic design (Chapter VI).

This information can be obtained by stepping the parameters over a range of values and calculating the change in output. Although this may be acceptable for small networks with only one or two parameters, the computational effort can rise sharply with the number of parameters and becomes intolerable for a large network.

As an alternative to the above, consider first the possibility of calculating the symbolic partial derivative $\partial(v_o, i_o)/\partial p_i$. For example, the sensitivity of the output v_o with respect to R_1 in Figure 5.1a can be determined simply by differentiating the analytic expression for v_o as follows.

$$\frac{\partial v_o}{\partial R_1} = \frac{\partial}{\partial R_1} \frac{R_2 v_s}{R_1 + R_2} \tag{5.1}$$

$$= -\frac{R_2 v_s}{(R_1 + R_2)^2} \tag{5.2}$$

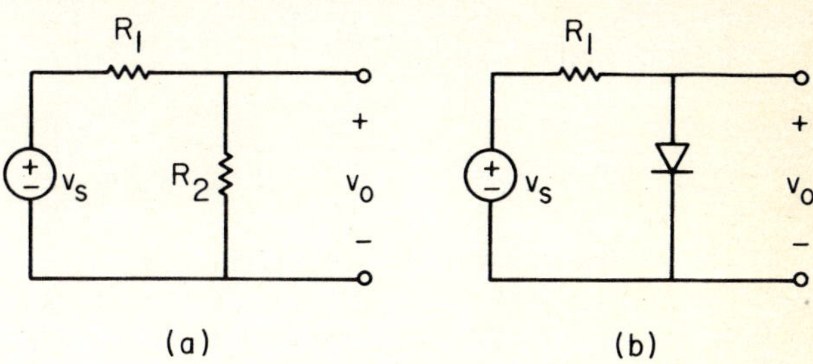

FIG. 5.1 Simple Network Examples

This procedure is not general, however; many of the networks requiring a computer solution (even the simple resistor-diode network of Figure 5.1b) do not admit an analytic solution of the form of (5.2). Instead, the solution is obtained only as the result of the iteration of a set of nonlinear equations using a method such as Newton iteration.

In this chapter we will show that sensitivity can be precisely calculated <u>without</u> an analytic expression using far less effort than would be required by pertubation. Indeed, an analysis program such as DNET or DCAP can be easily converted to yield such partial derivatives.

Before proceeding to discussion of methodology, it should be pointed out that parameters are not limited to element values as we usually know them. For example, associated with a diode defined by

$$i_d = I_s(e^{\lambda v_d} - 1) \tag{5.3}$$

we could identify I_s and λ as parameters. Physical dimensions of a transistor could also be considered parameters, provided these quantities appear somewhere in a transistor circuit model.

5.1.2 Tellegen's Theorem

The basis of our algorithm for computing sensitivities is a general network theorem by B. D. H. Tellegen [5.2]. The theorem itself has no apparent application to our topic, making discovery of its significance independently in [5.3] and [5.4] all the more remarkable. This detachment implies that the reader must be prepared to study the theorem on the promise of its later usefulness to us.

Consider two networks N and $\hat{N}$ which have the same structure but possibly different types of elements between corresponding nodes. An example is shown in Figure 5.2, where corresponding branches exist in both networks. We will even permit open and short circuits to represent corresponding branches, e.g., the capacitor in N and the short in $\hat{N}$.

We not state Tellegen's extended theorem [5.2], which relates the currents and voltages in N and $\hat{N}$. Let (i_k, v_k) and $(\hat{i}_k, \hat{v}_k)$ be (currents,

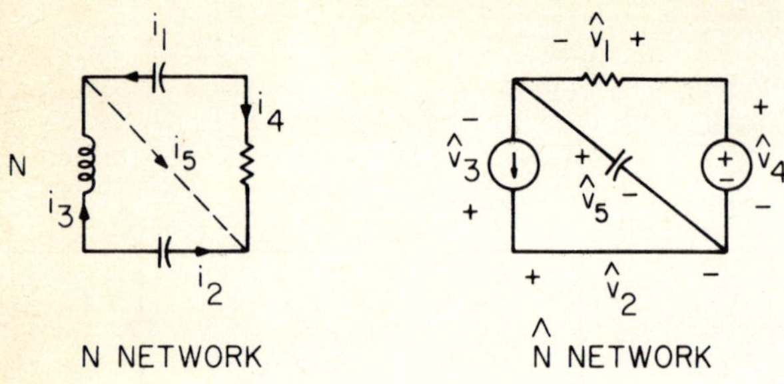

FIG. 5.2 Networks with Similar Structure

voltages) of corresponding branches in N and $\hat{N}$. Further, let every branch voltage be referenced with respect to its branch current as shown in Figure 5.3, and let corresponding currents and voltages in N and $\hat{N}$ be referenced identically as shown in Figure 5.2 and 5.3. If n_b is the number of branches, then as shown in Appendix A

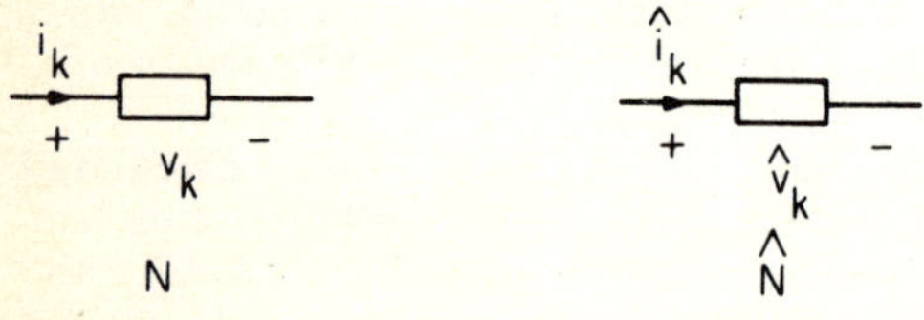

FIG. 5.3 References in N and $\hat{N}$

$$\sum_{k=1}^{n_b} i_k \hat{v}_k = 0 \qquad \sum_{k=1}^{n_b} v_k \hat{i}_k = 0 \tag{5.4}$$

A special case of this theorem occurs for $\hat{N} = N$, for then

$$\sum_{k=1}^{n_b} v_k i_k = 0 \tag{5.5}$$

which requires conservation of instantaneous power in a network.

To gain experience with the application of the theorem, consider the simple networks of Figure 5.4. The Tellegen sums of (5.4) are then

$$\sum_{k=1}^{3} i_k \hat{v}_k = (1/3)(4) + (-1/3)(4) + (-1/3)(0) = 0$$

$$\sum_{k=1}^{3} v_k \hat{i}_k = (1/3)(-4/3)+(1)(4/3)+(-2/3)(4/3) = 0$$

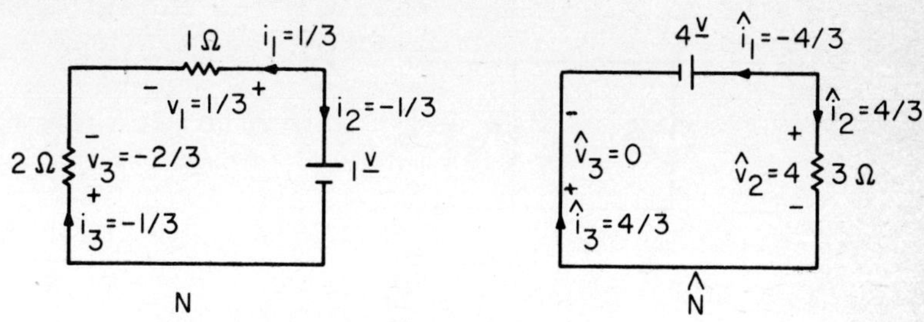

FIG. 5.4 Application of Tellegen's Theorem

5.2 SENSITIVITY CALCULATION[5.4]

5.2.1 Linear Resistive Branch Sensitivities

Let $v_k(t)$, $i_k(t)$, $\hat{v}_k(t)$, and $\hat{i}_k(t)$ represent branch variables in two networks, N and $\hat{N}$ respectively, which have identical graphs in the Tellegen sense so that (5.4) applies. Now let the voltage and current in N change by amounts $dv_k(t)$ and $di_k(t)$. Tellegen's Theorem must still be satisfied so that

$$\sum_k (v_k(t) + dv_k(t))\, \hat{i}_k(t) = 0$$

$$\sum_k \hat{v}_k(t)(i_k(t) + di_k(t)) = 0$$

which requires

$$\sum_k dv_k(t)\, \hat{i}_k(t) = 0 \qquad (5.6)$$

$$\sum_k \hat{v}_k(t)\, di_k(t) = 0 \qquad (5.7)$$

Equation (5.8) provides the key to all our sensitivity calculation methods. To see this, consider the problem of calculating the sensitivities dv_o/dR_1 and dv_o/dR_2 for the voltage divider of Figure 5.5.

The output is formally replaced by an infinite resistance branch (Figure 5.5b), and a network of the same structure is composed with as yet unknown branch relationships (Figure 5.5c). Equation (5.7) can now be written

$$(dv_s \hat{i}_s - di_s \hat{v}_s) + (dv_{R_1} \hat{i}_{R_1} - di_{R_1} \hat{v}_{R_1}) + (dv_{R_2} \hat{i}_{R_2} - di_{R_2} \hat{v}_{R_2})$$

$$+ (dv_o \hat{i}_o - di_o \hat{v}_o) = 0 \qquad (5.8)$$

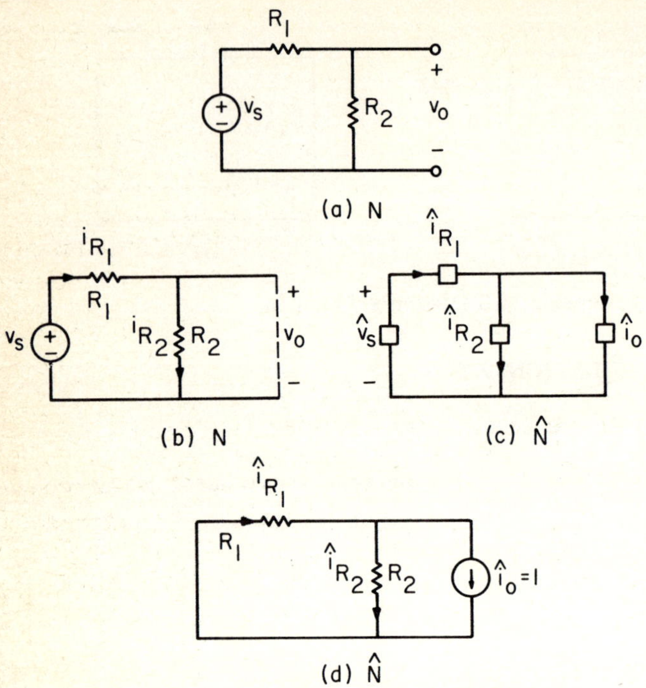

FIG. 5.5 Simple Network Example

Although (5.8) will be satisfied for any selection of elements of $\hat{N}$, we will show that a particular choice of elements will force each term - representing associated branches in N and $\hat{N}$ - individually to zero. This will lead directly to a method for sensitivity calculation [5.4].

Assume that R_1 of Figure 5.5 changes by dR_1; this will cause changes in all network branch currents and voltages, which we can evaluate as follows.

$$dv_{R_1} = \frac{\partial v_{R_1}}{\partial R_1} dR_1 + \frac{\partial v_{R_1}}{\partial i_{R_1}} di_{R_1}$$

$$= i_{R_1} dR_1 + R_1 di_{R_1}$$

Since R_2 is assumed constant, $dR_2 = 0$ and we have simply

$$dv_{R_2} = \frac{\partial v_{R_2}}{\partial i_{R_2}} di_{R_2} = R_2 di_{R_2}$$

Now we attempt to set the terms of (5.8) to zero. First, $dv_S = 0$, since the voltage v_s does not change value. Setting $\hat{v}_s = 0$ then forces the first term of (5.8) to zero. The third term is of the form

$$R_2 di_{R_2} \hat{i}_{R_2} - di_{R_2} \hat{v}_{R_2} = (R_2 \hat{i}_{R_2} - \hat{v}_{R_2}) di_{R_2} \tag{5.9}$$

which becomes zero if $\hat{N}$ contains an identical resistance R_2, so that $\hat{v}_{R_2} = R_2 \hat{i}_{R_2}$. If R_1 is also duplicated in $\hat{N}$, the remaining two terms of (5.8) become

$$(dv_{R_1} \hat{i}_{R_1} - di_{R_1} \hat{v}_{R_1}) + (dv_o \hat{i}_o - di_o \hat{v}_o) = (i_{R_1} dR_1 \hat{i}_{R_1}) + (dv_o \hat{i}_o - di_o \hat{v}_o)$$

$$= i_{R_1} dR_1 \hat{i}_{R_1} + dv_o \hat{i}_o \tag{5.10}$$

since $di_o = i_o = 0$. Finally, if we set $\hat{i}_o = 1$, we have

$$\frac{dv_o}{dR_1} = -i_{R_1} \hat{i}_{R_1} = \frac{-v_s}{(R_1 + R_2)} \; \frac{R_2}{R_1 + R_2} \tag{5.11}$$

This result can also be obtained by symbolic differentiation. Calculation of dv_o/dR_2 proceeds in an identical fashion to yield

$$\frac{dv_o}{dR_2} = -i_{R_2} \hat{i}_{R_2}$$

For both sensitivity calculations, $\hat{N}$ is identical to N, except for the shorted input in $\hat{N}$ and a unit source across the output of $\hat{N}$. Therefore, only two network analyses need be performed to compute both sensitivities.

The generalization of this procedure to <u>any</u> single-output linear resistive network is obvious, since we have invoked only Tellegen's Theorem in its development. A summary of the sensitivity calculations for general source-resistor network configurations is shown in Table 5.1. In general, two analyses suffice for computation of all sensitivities. Further, these sensitivities are as accurate as the analyses of N and $\hat{N}$, since they are formed from the <u>product</u> $i_R \hat{i}_R$; this is in contrast to calculation by perturbing R (i.e., $\Delta v_o/\Delta R$), which involves a subtraction.

For reasons to be shown in the next section, $\hat{N}$ is termed an <u>adjoint</u>* network if it is chosen so that each term of the Tellegen sum of (5.8) is zero when no parameters change (e.g., $dR_1 = 0$, $i = 1, 2, ..$). Indeed, we will reserve the $\hat{N}$ notation for the adjoint network.

---

*There is some controversy concerning the use of this term [5.8].

TABLE 5.1 Summary of Adjoint Network Relationships for Linear Resistive Networks

Branch Description	Network (N) Model	Adjoint Network Model	$dv_k \hat{i}_k - di_k \hat{v}_k$ Evaluation	Sensitivity Calculation
Voltage output	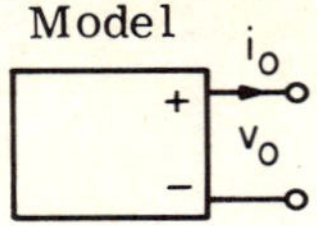	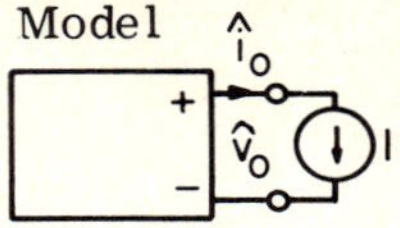	$dv_o(1) - (0)\hat{v}_o = dv_o$	
Current output	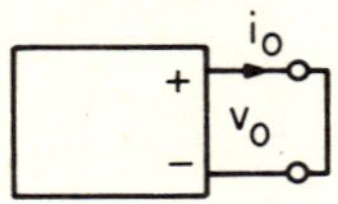	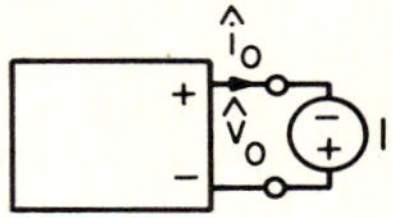	$(0)\hat{i}_o - di_o(-1) = di_o$	
Voltage source	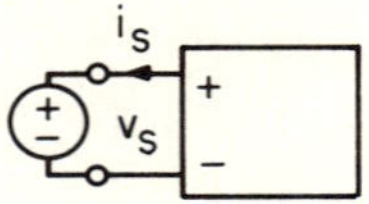	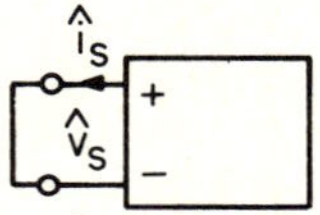	Left as a problem	
Current source	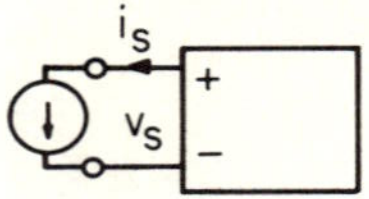	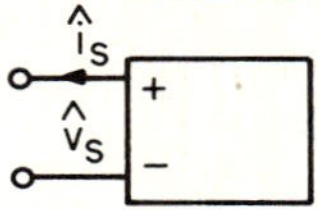	Left as a problem	
Resistance	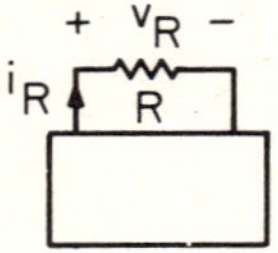	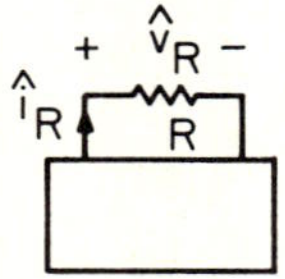	$\left(\frac{\partial v_R}{\partial i_R} di_R + \frac{\partial v_R}{\partial R} dR\right)\hat{i}_R - di_R \hat{v}_R$ $= (R di_R + i_R dR)\hat{i}_R - di_R(R\hat{i}_R)$ $= dR i_R \hat{i}_R$	$\frac{d(v_o, i_o)}{dR} = -i_R \hat{i}_R$
Conductance	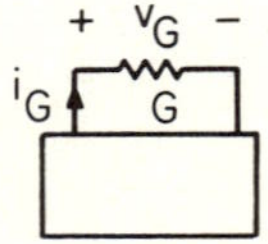	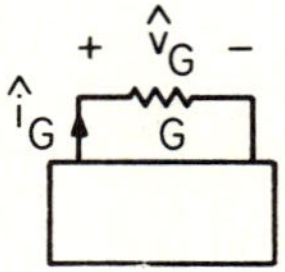	$dv_G \hat{i}_G - \left(\frac{\partial i_G}{\partial v_G} dv_G + \frac{\partial i_G}{\partial G} dG\right)\hat{v}_G$ $= dv_G(G\hat{v}_G) - (G dv_G + v_G dG)\hat{v}_G$ $= -dG v_G \hat{v}_G$	$\frac{d(v_o, i_o)}{dG} = v_G \hat{v}_G$

5.2.2 Multi-port Network Models

It is not necessary to develop adjoint network models at the branch level. We can as well model at the port level, as the following demonstrates.

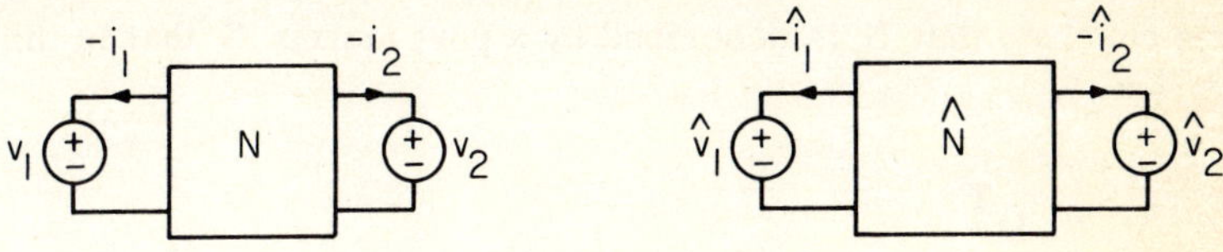

FIG. 5.6 A Two-port and Its Adjoint Model

Consider the linear resistive network of Figure 5.6, described by the equation

$$\underset{\sim}{G}\begin{bmatrix} v_1 \\ v_2 \end{bmatrix} = \begin{bmatrix} i_1 \\ i_2 \end{bmatrix}$$

or

$$\underset{\sim}{G}\underline{v} = \underline{i}$$

where $\underset{\sim}{G}$ is the short circuit port conductance matrix. The Tellegen summation for the entire network including source branches v_1 and v_2 must be zero. In addition, each term of the sum must vanish for every branch within N and $\hat{N}$, provided that N has no internal parameters. This leaves only the terms for branches v_1 and v_2, which must then sum pair-wise to zero. The terms in the Tellegen summation for the source branches are

$$\begin{aligned}
&(-\hat{i}_1)\, dv_1 - \hat{v}_1\, d(-i_1) + (-\hat{i}_2)\, dv_2 - \hat{v}_2\, d(-i_2) \\
&= -[\hat{i}_1 \;\; \hat{i}_2]\, d\begin{bmatrix} v_1 \\ v_2 \end{bmatrix} + [\hat{v}_1 \;\; \hat{v}_2]\, d\begin{bmatrix} i_1 \\ i_2 \end{bmatrix} \\
&= -\hat{\underline{i}}^T d\underline{v} + \hat{\underline{v}}^T\, d\underline{i} \\
&= -\hat{\underline{i}}^T\, d\underline{v} + \hat{\underline{v}}^T \underset{\sim}{G}\, d\underline{v} \\
&= (-\hat{\underline{i}}^T + \hat{\underline{v}}^T \underset{\sim}{G})\, d\underline{v}
\end{aligned} \tag{5.12}$$

where it is assumed that $d\underset{\sim}{G} = \underset{\sim}{0}$ (in contrast, see Example 5.2a). This expression can be made zero by setting

$$\hat{\underline{i}}^T = \hat{\underline{v}}^T \underset{\sim}{G}$$

or, taking the transpose of both sides,

$$\hat{\underline{i}} = \underset{\sim}{G}^T \hat{\underline{v}}$$

We conclude that N is described by a port matrix G that is the transpose of $\underset{\sim}{G}$, i.e.,

$$\hat{\underset{\sim}{G}} = \underset{\sim}{G}^T \tag{5.13}$$

This result can be applied to deduce the adjoint network model for a voltage-controlled current source (Table 5.2) described by

$$\underset{\sim}{G}\,\underline{v} = \begin{bmatrix} 0 & 0 \\ g_m & 0 \end{bmatrix} \begin{bmatrix} v_1 \\ v_2 \end{bmatrix} = \begin{bmatrix} i_1 \\ i_2 \end{bmatrix} \tag{5.14}$$

The adjoint model is then described by

$$\underset{\sim}{G}^T \hat{\underline{v}} = \begin{bmatrix} 0 & g_m \\ 0 & 0 \end{bmatrix} \begin{bmatrix} \hat{v}_1 \\ \hat{v}_2 \end{bmatrix} = \begin{bmatrix} \hat{i}_1 \\ \hat{i}_2 \end{bmatrix}$$

which is represented by interchanging the port connections of the original source (Table 5.2).

The derivation of (5.13) is obviously extendable to any number of ports. Further, a similar result can be ovtained when the ports are excited by voltage sources or by a mixed set of current and voltage sources. For this purpose, we now introduce the notation of a hybrid matrix $\underset{\sim}{H}$.

For a linear network N, the superposition principle allows the response at each port to be written as a weighted summation of excitations, as for example

$$i_2 = (g_m)\,v_1 + (0)\,v_2$$

in (5.14). If the port voltage excitations are collected into a vector $\underline{v}_1$ and the current excitations into a vector $\underline{i}_2$, then the respective response vectors $\underline{i}_1$ and $\underline{v}_2$ can be written as

$$\underset{\sim}{g}_{11}\underline{v}_1 + \underset{\sim}{\alpha}_{12}\underline{i}_2 = \underline{i}_1 \tag{5.15}$$

$$\underset{\sim}{\mu}_{21}\underline{v}_1 + \underset{\sim}{r}_{22}\underline{i}_2 = \underline{v}_2 \tag{5.16}$$

where $\underset{\sim}{g}_{11}$ is a matrix of transfer conductances
$\underset{\sim}{\alpha}_{12}$ is a matrix of current transfer functions
$\underset{\sim}{\mu}_{21}$ is a matrix of voltage transfer functions
$\underset{\sim}{r}_{22}$ is a matrix of transfer resistances

TABLE 5.2 Adjoint Models for Common Two-port Elements

Branch Description	Network (N) Model	Adjoint Network ($\hat{N}$) Model
Voltage-controlled current source	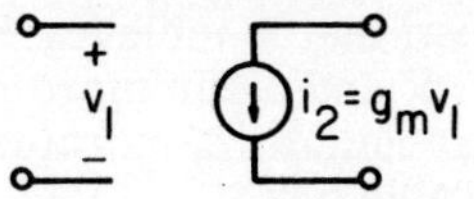 v_1; $i_2 = g_m v_1$	 $\hat{i}_1 = g_m \hat{v}_2$; $\hat{v}_2$
Current-controlled current source	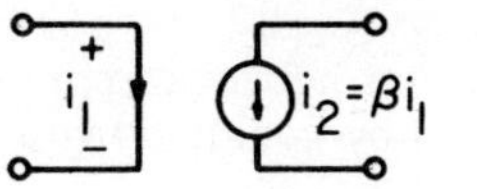 i_1; $i_2 = \beta i_1$	 $\hat{v}_1 = -\beta \hat{v}_2$; $\hat{v}_2$
Current-controlled voltage source	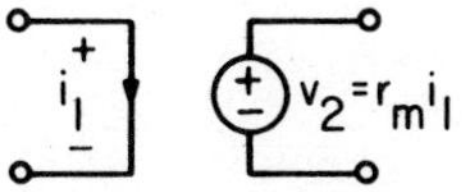 i_1; $v_2 = r_m i_1$	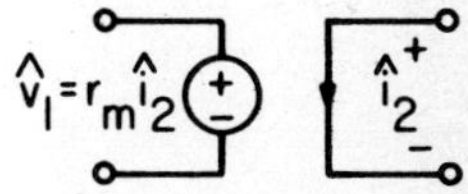 $\hat{v}_1 = r_m \hat{i}_2$; $\hat{i}_2$
Voltage-controlled voltage source	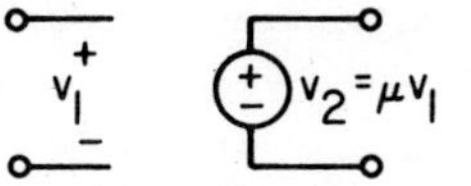 v_1; $v_2 = \mu v_1$	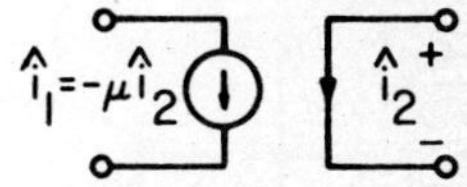 $\hat{i}_1 = -\mu \hat{i}_2$; $\hat{i}_2$
Ideal transformer	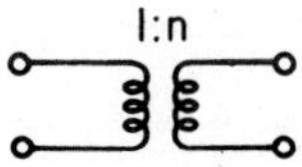 1:n	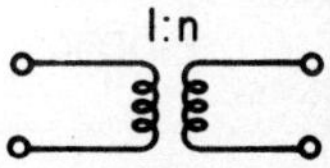 1:n

Collecting (5.15) and (5.16) into a single matrix equation, we have

$$\underset{\sim}{H}\,\underline{x} = \underline{y}$$

where $\underline{y}$ is an output vector $\begin{bmatrix} \underline{i}_1 \\ \underline{v}_2 \end{bmatrix}$

$\underline{x}$ is an input vector $\begin{bmatrix} \underline{v}_1 \\ \underline{i}_2 \end{bmatrix}$

$$\underset{\sim}{H} = \begin{bmatrix} \underset{\sim}{g}_{11} & \underset{\sim}{\alpha}_{12} \\ \underset{\sim}{\mu}_{21} & \underset{\sim}{r}_{22} \end{bmatrix}$$

Then it may be shown (left as a problem) that $\hat{N}$ is described by

$$\begin{bmatrix} \underset{\sim}{g}^T_{11} & -\underset{\sim}{\mu}^T_{21} \\ -\underset{\sim}{\alpha}^T_{12} & \underset{\sim}{r}^T_{22} \end{bmatrix} \hat{\underline{x}} = \hat{\underline{y}}$$

where $\hat{\underline{y}} = \begin{bmatrix} \hat{\underline{i}}_1 \\ \hat{\underline{v}}_2 \end{bmatrix}$, etc.

This result is so general that it includes all four types of controlled sources, ideal transformers, and other more pedogogical types of network multi-port elements such as nullators, norators, etc. [5. 11]. Some of these are summarized in Table 5. 2.

5. 2. 3 An Example

The elegance of the adjoint network approach to linear network sensitivity calculation is enhanced by considering a practical application.

Example 5. 1. A sensitivity analysis of the small-signal transfer function v_o/v_i for the differential amplifier of Figure 5. 7a is to be performed.

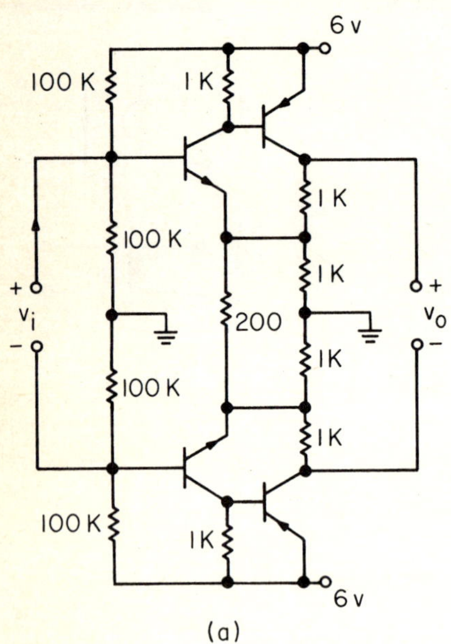

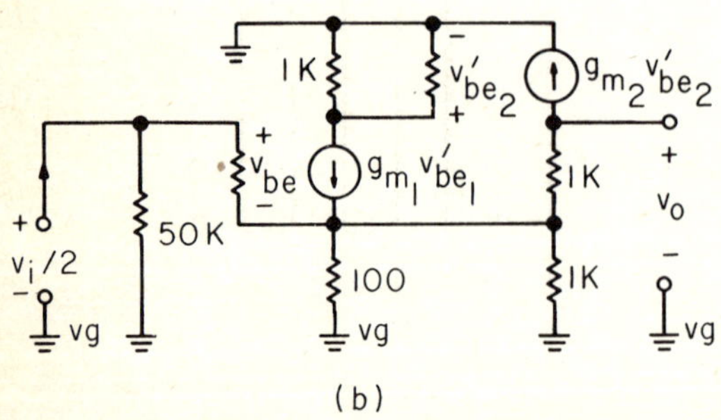

FIG. 5.7 Sensitivity Calculation for Differential Amplifier

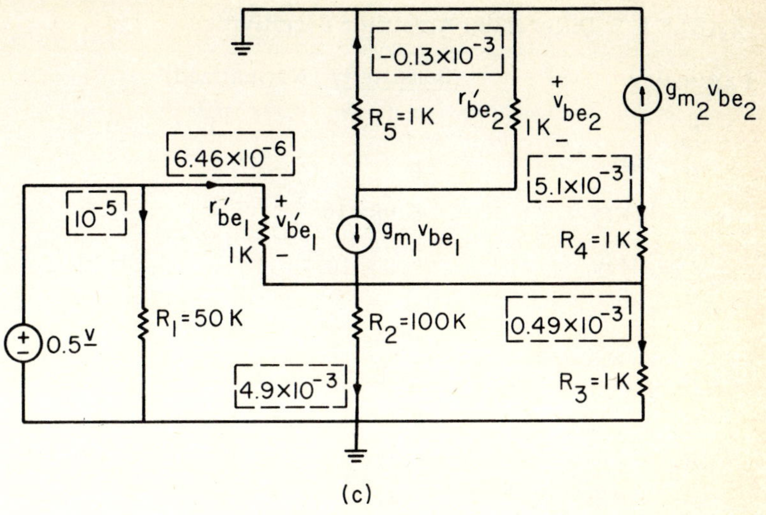

(c)

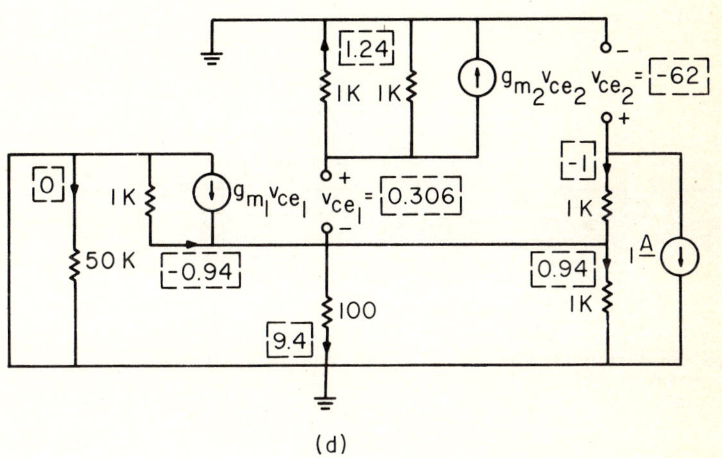

(d)

FIG. 5.7 Sensitivity Calculation for Differential Amplifier (Continued)

First, the symmetry of the amplifier necessitates analysis to be performed on only half of the circuit, so that the small signal model of Figure 5.7b becomes the basis for analysis.

Second, the transfer function sensitivity to p_j is the sensitivity dv_o/dp_j when a unit input is applied. The process is displayed in Figure 5.7c, where a .5 volt input is applied across only half of the circuit. All branch currents are calculated in preparation for computing sensitivity with respect to resistance values.

Finally, branch currents in the adjoint network are computed, after controlled sources are reversed (Figure 5.7d). The sensitivities of Table 5.3 are then computed directly from Figure 5.7c-d. The sensitivities for the half circuit must be properly interpreted to determine the sensitivities of the original circuit. This is left as a problem.

TABLE 5.3 Differential Amplifier Sensitivities

Parameter (p_j)	Sensitivity (Normalized) $p_j\,(\partial(v_o/v_i))/\partial p_j)$
R_1	0
$r_{b'e_1}$ (=1KΩ)	6.08×10^{-3}
g_{m_1} (=.04℧)	7.9×10^{-5}
R_2	-4.6
R_3	-4.6
R_4	5.1
R_5	.161
$r_{b'e_2}$ (=1KΩ)	.161
g_{m_2} (=.04℧)	.322

FIG. 5.8 Branch Variables in the Phasor Network N and $\hat{N}$

5.3 FREQUENCY DOMAIN SENSITIVITY CALCULATIONS

As we have observed in Chapter II, in sinusoidal steady state analysis it is common to represent resistive and dynamic elements as admittances in a phasor network model. Tellegen's Theorem, which depends only upon KVL and KCL, must be satisfied for the phasor networks N and $\hat{N}$. Accordingly, an adjoint phasor network may be defined with the property that the terms of the Tellegen sum

$$\sum_k (dV_k\hat{I}_k - dI_k\hat{V}_k) = 0$$

must individually sum to zero,

The calculation of sensitivities $d(V_o, I_o)/dY_k$ and $d(V_o, I_o)/dZ_k$ now proceeds similarly to the calculations $d(v_o, i_o)/dG_k$ and $d(v_o, i_o)/dR_k$ for resistive networks. It immediately follows that branch admittances appear identically in N and $\hat{N}$. Some new considerations are introduced in the actual sensitivity calculation by the phasor branch model, however.

(1) The sensitivity of V_o or I_o to a capacitance or inductance must be found from

$$\frac{d(V_o, I_o)}{dC} = \frac{d(V_o, I_o)}{dY_C}\,\frac{dY_C}{dC} = j\omega\,\frac{d(V_o, I_o)}{dY_C}$$

$$\frac{d(V_o, I_o)}{dL} = \frac{d(V_o, I_o)}{dZ_L}\,\frac{dZ_L}{dL} = j\omega\,\frac{d(V_o, I_o)}{dZ_L}$$

(2) Sensitivity of magnitude and phase can be found from

$$\begin{aligned}\frac{d}{dC}\,20\log_{10}|(V_o, I_o)| &= 8.6458\,\frac{d}{dC}\log_e|(V_o, I_o)|\\ &= 8.6458\,\frac{d}{dC}\,\mathrm{Re}[\log_e(V_o, I_o)]\\ &= 8.6458\,\mathrm{Re}\left[\frac{d}{dC}\log_e(V_o, I_o)\right]\\ &= 8.6458\,\mathrm{Re}\left[\frac{1}{(V_o, I_o)}\,\frac{d(V_o, I_o)}{dC}\right]\end{aligned}$$

$$\begin{aligned}\frac{d}{dC}\arg(V_o, I_o) &= \frac{d}{dC}\,\mathrm{Im}[\log_e(V_o, I_o)]\\ &= \mathrm{Im}\left[\frac{1}{(V_o, I_o)}\,\frac{d(V_o, I_o)}{dC}\right]\end{aligned}$$

Example 5.2a. Consider again the high frequency amplifier of Figure 2.8, repeated in Figure 5.9a. It is desired to compute the sensitivity of the output voltage with respect to the lengths of the transmission lines.

The adjoint network $\hat{N}$ has the form of Figure 5.9b. It is distinguished from N only by the transposed transistor port description $\underset{\sim}{Y}^{(t)}$, in accordance with Section 5.2.2. The transmission lines, being represented by symmetric two-port matrices, are described identically in N and $\hat{N}$.

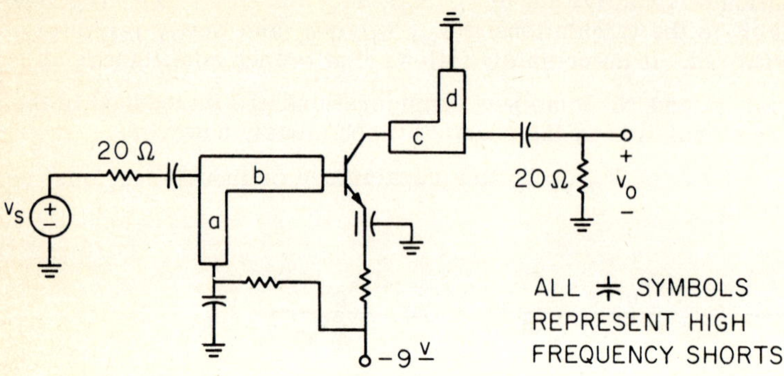

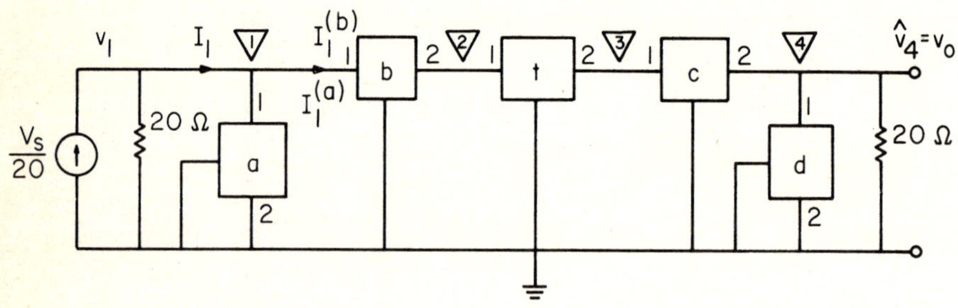

(a) Amplifier Schematic and Model

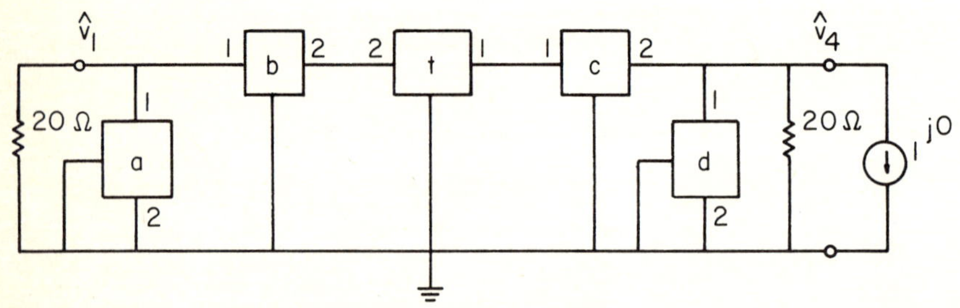

(b) Adjoint Network Model

FREQUENCY HTZ	GAIN DB	LINE LENGTH SENSITIVITIES (DB/MM) LINE A	LINE B	LINE C	LINE D
0.1500D 10	-4.62981	0.89111	-0.64371	0.50160	1.32302
0.2000D 10	-8.43481	1.10958	-0.27115	0.25147	0.42373
0.2500D 10	-5.62281	1.14548	-0.00624	-0.31525	-0.01393
0.3000D 10	-6.56583	1.12337	0.23408	-0.48842	-0.12255

(c) Partial Output Data

FIG. 5.9 Microwave Amplifier Sensitivity Calculation

The sensitivity with respect to any line length is obtained by expanding the expression of (5.12) when $d\tilde{G} \neq \tilde{0}$. In phasor terminology, we have

$$-\hat{\underline{I}}^T d\underline{V} + \hat{\underline{V}}^T d\underline{I} = -\hat{\underline{I}}^T d\underline{V} + \hat{\underline{V}}^T d\tilde{Y}\,\underline{V} + \hat{\underline{V}}^T \tilde{Y}\, d\underline{V} \tag{5.17}$$

$$= \hat{\underline{V}}^T d\tilde{Y}\,\underline{V}$$

since $\hat{\tilde{Y}} = \tilde{Y}^T$. It is readily shown that (5.17) can be expanded into

$$\hat{\underline{V}}^T d\tilde{Y}\,\underline{V} = -j\,(\omega_i \sqrt{\epsilon}/3 \times 10^8)\, Y_o \ \mathrm{csch}\,\beta$$

$$\times \begin{bmatrix} \hat{V}_i & \hat{V}_j \end{bmatrix} \begin{bmatrix} \mathrm{csch}\,\beta & -\coth\beta \\ -\coth\beta & \mathrm{csch}\,\beta \end{bmatrix} \begin{bmatrix} V_i \\ V_j \end{bmatrix} d\,(d_k)$$

$$= (a + jb)\, d(d_k)$$

where d_k is the line length, and V_i, V_j, $\hat{V}_i$, and $\hat{V}_j$ are port voltages. We can conclude that the sensitivity to a line length is given by

$$\frac{dV_o}{d(d_k)} = a + jb$$

5.4 NUMERICAL SOLUTION OF $\hat{N}$

5.4.1 Introduction

By phrasing the calculation of elemental sensitivities in terms of the analysis of an additional network $\hat{N}$, the adjoint network method exposes the range of options open to us as we consider computer implementation of a general sensitivity calculation algorithm. For example, N could be solved by node analysis and $\hat{N}$ by mesh analysis or vice versa.

In this section, we will show that if N and $\hat{N}$ are analyzed using the same equation formulation procedure, a considerable efficiency can be realized. In particular, we will show that by using node analysis to solve both N and $\hat{N}$, the solution of $\hat{N}$ can be obtained from the solution of N with small additional effort [5.9].

5.4.2 Relation Between $\tilde{Y}_N$ and $\hat{\tilde{Y}}_N$

We have established in Section 5.2.2 that any multi-port represented by a port description $\tilde{Y}$ in N is described in $\hat{N}$ by $\hat{\tilde{Y}} = \tilde{Y}^T$. But the node matrix $\tilde{Y}_N$ of the entire network N may also be viewed as a form of port matrix $\tilde{Y}$. We can establish this rather artificial result by applying port current sources between each node and the ground node and defining port voltages similarly (Figure 5.10). For this special port connection, we can write either

$$\underset{\sim}{Y}_N \underline{V} = \underline{I}$$

or

$$\underset{\sim}{Y} \, \underline{V} = \underline{I}$$

so that $\underset{\sim}{Y}_N = \underset{\sim}{Y}$. This, the node matrix is a port matrix as well.

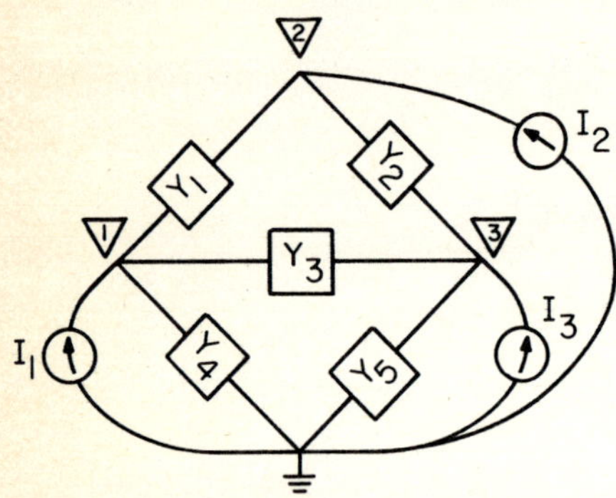

FIG. 5.10 Three-port Connection of a Four-node Network

The adjoint network $\hat{N}$, with similar port definitions, yields the relation $\hat{\underset{\sim}{Y}}_N = \hat{\underset{\sim}{Y}}$. We can therefore write

$$\hat{\underset{\sim}{Y}}_N = \hat{\underset{\sim}{Y}} = \underset{\sim}{Y}^T = \underset{\sim}{Y}_N^T \tag{5.18}$$

This is a highly useful result: <u>the node matrix of N is the transpose of the node matrix of N.</u> This transpose relationship applies as well to mesh and other forms of analysis (see Section 5.2.2).

The solutions of N and N may now be closely related as follows. As suggested in Chapter II, we could solve for $\underline{V}$ of N by factoring $\underset{\sim}{Y}_N$ into LU form, viz,

$$\underset{\sim}{Y}_N \underline{V} = \underset{\sim}{L} \, \underset{\sim}{U} \, \underline{V} = \underline{I}$$

Now consider $\hat{N}$, excited with a unit current source at the output terminals. We write

$$\hat{\underline{I}} = \hat{\underset{\sim}{Y}}_N \hat{\underline{V}} = \underset{\sim}{Y}_N^T \hat{\underline{V}}$$

$$= (\underset{\sim}{L} \, \underset{\sim}{U})^T \hat{\underline{V}}$$

$$= \underset{\sim}{U}^T (\underset{\sim}{L}^T \hat{\underline{V}}) \tag{5.19}$$

$$= \underset{\sim}{U}^T \underline{X} \tag{5.20}$$

Now assume that N has been solved, so that $\underset{\sim}{L}$ and $\underset{\sim}{U}$ have been determined. Equation (5.20) can be written as

$$\begin{bmatrix} 1 & 0 & 0 & \cdot & \cdot & 0 \\ U_{12} & 1 & 0 & \cdot & \cdot & 0 \\ U_{13} & U_{23} & 1 & \cdot & \cdot & 0 \\ \cdot & \cdot & \cdot & \cdot & \cdot & \cdot \\ \cdot & \cdot & \cdot & \cdot & \cdot & \cdot \\ U_{1n} & U_{2n} & \cdot & \cdot & \cdot & 1 \end{bmatrix} \begin{bmatrix} X_1 \\ X_2 \\ X_3 \\ \cdot \\ \cdot \\ X_n \end{bmatrix} = \begin{bmatrix} 0 \\ 0 \\ \vdots \\ 1 \\ \cdot \\ 0 \end{bmatrix} \longleftarrow \tag{5.21}$$

where complex quantities are shown in upper case. A forward substitution step now yields $\underline{X}$. We then write, from (5.19), $\underset{\sim}{L}^T \underline{V} = \underline{X}$, or

$$\begin{bmatrix} L_{11} & L_{21} & L_{31} & \cdot\cdot & L_{n1} \\ 0 & L_{22} & L_{32} & \cdot\cdot & L_{n2} \\ 0 & 0 & L_{33} & \cdot\cdot & \cdot \\ \cdot & \cdot & \cdot & \cdot\cdot & \cdot \\ \cdot & \cdot & \cdot & \cdot\cdot & \cdot \\ 0 & 0 & 0 & & L_{nn} \end{bmatrix} \begin{bmatrix} \hat{V}_1 \\ \hat{V}_2 \\ \cdot \\ \cdot \\ \hat{V}_n \end{bmatrix} = \begin{bmatrix} X_1 \\ X_2 \\ \cdot \\ \cdot \\ \cdot \\ X_n \end{bmatrix} \tag{5.22}$$

A back substitution completes the solution of $\hat{N}$.

Recalling that the cost of factorization varies as $n^3/3$ for a full matrix while that for two substitutions varies as n^2, we can conjecture that the adjoint network solution adds trivially to the solution of N. This economy is only partially realized due to the matrix sparsity. In practice, the two substitutions together typically require 1/3 the effort of the factorization. Of more importance to the coding of a general algorithm, however, is that only one node matrix need be constructed. In Example 5.2a, for instance, we need not explicitly insert $\underset{\sim}{Y}^{(t)}$ into $\underset{\sim}{Y}_N$ and $\lfloor Y^{(t)} \rfloor^T$ into $\underset{\sim}{\hat{Y}}_N$. This process is subsumed in the two substitution steps of (5.21) and (5.22).

<u>Example 5.2b.</u> A program to evaluate the parameter sensitivities of Example 5.2a is given in Appendix Table 5.1. Using the input data of Table 2.7, the results of Figure 5.9c are obtained. We observe that, except for line A, changing a line length will increase gain at one frequency only at the expense of gain at another frequency.

5.5 DC SENSITIVITY ANALYSIS OF NONLINEAR NETWORKS

5.5.1 Adjoint Models

The presence of resistive nonlinearities does not seriously complicate the sensitivity computation procedure. Let us represent a two-terminal resistive nonlinearity in the form

$$i = g(v, p) \tag{5.23}$$

or

$$v = r(i,p)$$

where p is a parameter. For example, a diode dependent on the temperature T is represented as

$$i_d = I_s(e^{qv_d/kT} - 1)$$

where p = T. The contribution of a branch described by (5.23) to the Tellegen sum then has the form

$$dv\,\hat{i} - di\,\hat{v} = dv\,\hat{i} - \left(\frac{\partial g}{\partial v}dv + \frac{\partial g}{\partial p}dp\right)\hat{v} \tag{5.24}$$

This term will be independent of dv if

$$\hat{i} = \left(\frac{\partial g}{\partial v}\right)\hat{v}$$

The adjoint network therefore contains a linearized conductance of value $G = \partial g/\partial v$ in place of the nonlinear diode branch. Moreover, the sensitivity of the network to p is easily calculated from (5.8) and (5.24) as

$$\frac{d(v_o, i_o)}{dp} = \left(\frac{\partial g}{\partial p}\right)\hat{v} \tag{5.25}$$

Example 5.3. The sensitivity of the current in the resistor-diode network of Figure 5.11a to changes in R and in T is to be found.

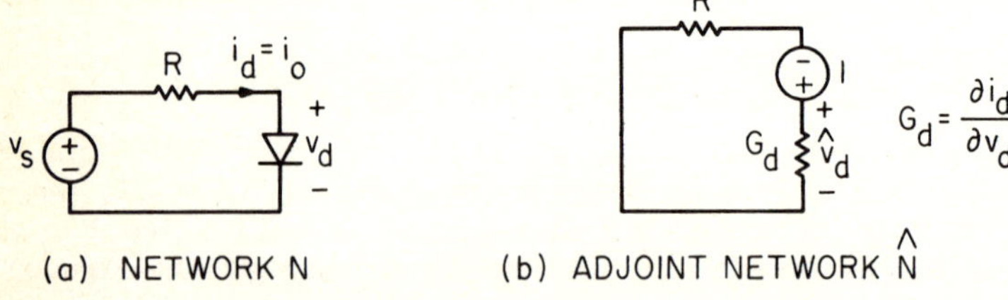

FIG. 5.11 Network Models for Calculation of di_d/dT

The diode is represented in $\hat{N}$ by a linearized conductance of value

$$\frac{\partial i_d}{\partial v_d} = \frac{qI_s}{kT}e^{qv_d/kT}$$

where v_d is the diode voltage in N after convergence of the dc iteration. Also, since the output is the current i_d, we conclude from Table 5.1 that a unit voltage source is placed in series with G_d. The sensitivity is then calculated from (5.25) as

$$\frac{di_o}{dT} = \left(\frac{\partial i_d}{\partial T}\right) \hat{v}_d$$

$$= \frac{-qv_d I_s e^{qv_d/kT}}{kT^2} \hat{v}_d \tag{5.26}$$

In this case, the diode voltage $\hat{v}_d$ can be calculated by hand by regarding $\hat{N}$ as a voltage divider. In particular, we have

$$\hat{v}_d = \frac{1}{1 + RG_d}$$

Substituting this expression into (5.26) yields

$$\frac{di_d}{dT} = \frac{-q\, v_d\, I_s\, e^{qv_d/kT}}{(kT^2)\,(1 + RG_d)}$$

The above procedure may be extended to multi-port nonlinearities, e.g., a transistor described by

$$i_e(v_{eb}, v_{cb}) = I_{es}(e^{\lambda v_{eb}} - 1) - \alpha_r I_{cs}(e^{\lambda v_{cb}} - 1)$$

$$i_c(v_{eb}, v_{cb}) = -\alpha_f I_{es}(e^{\lambda v_{eb}} - 1) + I_{cs}(e^{\lambda v_{cb}} - 1)$$

In this case, the summation would be of the form

$$dv_{eb}\hat{i}_e - di_e(v_{eb}, v_{cb})\hat{v}_{eb} + dv_{cb}\hat{i}_c - di_c(v_{eb}, v_{cb})\hat{v}_{cb}$$

$$= dv_{eb}\hat{i}_e - \left(\frac{\partial i_e}{\partial v_{eb}} dv_{eb} + \frac{\partial i_e}{\partial v_{cb}} dv_{cb}\right)\hat{v}_{eb} + dv_{cb}\hat{i}_c$$

$$- \left(\frac{\partial i_c}{\partial v_{eb}} dv_{eb} + \frac{\partial i_c}{\partial v_{cb}} dv_{cb}\right)\hat{v}_{cb}$$

which is equal to zero for all dv_{eb} and dv_{cb} if

$$\hat{i}_e - \frac{\partial i_e}{\partial v_{eb}} \hat{v}_{eb} - \frac{\partial i_c}{\partial v_{eb}} \hat{v}_{cb} = 0$$

$$\hat{i}_c - \frac{\partial i_e}{\partial v_{cb}} \hat{v}_{eb} - \frac{\partial i_c}{\partial v_{cb}} \hat{v}_{cb} = 0$$

In matrix form this becomes

$$\begin{bmatrix} \hat{i}_e \\ \hat{i}_c \end{bmatrix} = \begin{bmatrix} \dfrac{\partial i_e}{\partial v_{eb}} & \dfrac{\partial i_c}{\partial v_{eb}} \\ \dfrac{\partial i_e}{\partial v_{cb}} & \dfrac{\partial i_c}{\partial v_{cb}} \end{bmatrix} \begin{bmatrix} \hat{v}_{eb} \\ \hat{v}_{cb} \end{bmatrix} \tag{5.27}$$

If for a transistor we define

$$\underline{i}^{(t)} = \begin{bmatrix} i_e \\ i_c \end{bmatrix}, \quad \hat{\underline{i}}^{(t)} = \begin{bmatrix} \hat{i}_e \\ \hat{i}_c \end{bmatrix}, \quad \underline{v}^{(t)} = \begin{bmatrix} v_{eb} \\ v_{cb} \end{bmatrix}, \quad \hat{\underline{v}}^{(t)} = \begin{bmatrix} \hat{v}_{eb} \\ \hat{v}_{cb} \end{bmatrix}$$

then (5.27) may be written succinctly as

$$\hat{\underline{i}}^{(t)} = \left(\frac{\partial \underline{i}^{(t)}}{\partial \underline{v}^{(t)}} \right)^T \hat{\underline{v}}^{(t)}$$

We observe that the matrix representation of a transistor in $\hat{N}$ is a linearized, transposed version of the nonlinear equations which described it in N.

In general, if a nonlinear multi-port in N is described by the vector equation

$$\underline{i} = \underline{h}(\underline{v})$$

then the representation in $\hat{N}$ is given by

$$\hat{\underline{i}} = \left(\frac{\partial \underline{h}}{\partial \underline{v}} \right)^T \hat{\underline{v}}$$

$$= \underset{\sim}{J}_{\underline{v}}^{T} \hat{\underline{v}} \tag{5.29}$$

where $\underset{\sim}{J}_{\underline{v}}$ is a Jacobian matrix.

5.5.2 Numerical Solution of $\hat{N}$

As in the case of linear networks, we need expend neither programming nor computational effort in explicitly forming the matrix description of $\hat{N}$ from the linearized, transposed matrix representative of all its elements. Instead, we now show a familiar relationship between the network matrix of $\hat{N}$ and the matrix of the companion network used in dc analysis of N.

Consider a multi-port nonlinear element represented by (5.28). Using the technique of Chapter II, we develop a companion model by linearizing $\underline{h}$ around $\underline{v} = \underline{v}^m$, viz,

$$\begin{aligned} \underline{i}^{m+1} &= \underline{i}^m + \left.\frac{\partial \underline{h}}{\partial \underline{v}}\right|_{\underline{v}=\underline{v}^m} (\underline{v}^{m+1} - \underline{v}^m) \\ &= \underset{\sim}{J}_{\underline{v}}^{m} \underline{v}^{m+1} + (\underline{i}^m - \underset{\sim}{J}_{\underline{v}}^{m} \underline{v}^m) \end{aligned} \tag{5.30}$$

The matrix $\underset{\sim}{J}_{\underline{v}}$ relates currents and voltages at the yet to be performed (m + 1)st iteration, whereas the second term is a source term of known currents from the m-th iteration.

After convergence of the iteration to obtain the dc solution, currents and voltages in the companion network are solutions of the original nonlinear dc network. If the adjoint network model of this multi-port element is now desired, we need only remove the source term of (5.30) and transpose $\underset{\sim}{J}_{\underline{v}}$ to obtain the adjoint matrix representation of (5.29)

Since each element is represented in $\hat{N}$ by a transposition of its companion network description, the arguments of Section 5.3.2 can be invoked to conclude that the entire network node matrix of $\hat{N}$ is the transposed version of the companion network node matrix. The sources in each network are of course different. The adjoint network is excited by a unit source across the output, the companion network by sources obtained from (5.30).

In summary, to compute the voltages and currents of $\hat{N}$ after convergence of the dc iteration, we need save only the most recent LU factorization of the companion network matrix. A forward and back substitution of the form of (5.21) and (5.22) then yield the solution of $\hat{N}$.

Example 5.4a. A sensitivity analysis of the biasing circuit of Figure 5.12a is to be performed. (This was previously analyzed in Chapter III.) In particular, the sensitivities

$$\frac{\partial v_{ce}}{\partial R_E}, \quad \frac{\partial v_{ce}}{\partial \alpha}, \quad \frac{\partial v_{ce}}{\partial T}, \quad \frac{\partial v_{ce}}{\partial E} \tag{5.31}$$

are to be calculated, where T is the temperature in degrees Kelvin in the diode relationship $\lambda = q v_d/kT$.

The companion and adjoint networks are shown in Figure 5.12b-c. The adjoint network is described by the transposed node matrix of Equation (3.8), viz,

$$\begin{bmatrix} \frac{1}{R_2} + \frac{1}{R_1} + (1-\alpha)G_d & -G_d & -\frac{1}{R_1} + \alpha G_d \\ (\alpha - 1)G_d & \frac{1}{R_E} + G_d & -\alpha G_d \\ -\frac{1}{R_1} & 0 & \frac{1}{R_1} + \frac{1}{R_2} \end{bmatrix} \begin{bmatrix} \hat{v}_1 \\ \hat{v}_2 \\ \hat{v}_3 \end{bmatrix} = \begin{bmatrix} 1 \\ 0 \\ -1 \end{bmatrix}$$

where the superscript m is dropped because the dc solution has already

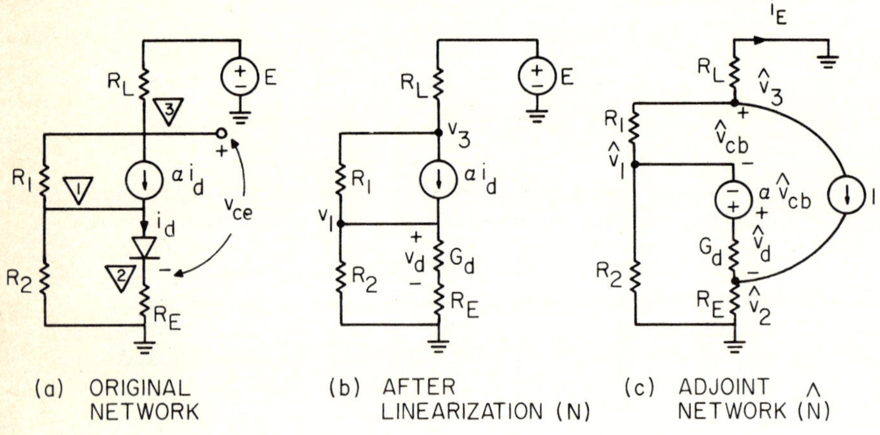

FIG. 5.12 Two-Step Generation of $\hat{N}$ for Nonlinear Network

been found. The sensitivities of (5.31) are determined from the node voltages of N and $\hat{N}$ as follows.

(1) From Table 5.1,

$$\frac{dv_{ce}}{dR_E} = - i_{R_E} \hat{i}_{R_E}$$

$$= - v_2 \hat{v}_2 / R_E^2$$

$$= .003137$$

(2) From Table 5.2,

$$\frac{dv_{ce}}{d\alpha} = i_d(\hat{v}_3 - \hat{v}_1)$$

$$= I_s(e^{\lambda(v_1 - v_2)} - 1)(\hat{v}_3 - \hat{v}_1)$$

$$= -11.47$$

(3) From Table 5.1,

$$\frac{dv_{ce}}{dE} = -\hat{i}_E$$

$$= -\hat{v}_3/R_L$$

$$= .3102$$

(4) From (5.26),

$$\frac{dv_{ce}}{dT} = \frac{d(I_s(e^{qv_d/kT} - 1)\,\hat{v}_d}{dT}$$

$$= \frac{-qv_d I_s\,(e^{qv_d/kT})}{kT^2}\,\hat{v}_d \tag{5.32}$$

where it is assumed that I_S is independent of temperature. Examining the adjoint network model of Figure 5.12c, we observe that

$$v_d = v_1 - v_2$$

whereas

$$\hat{v}_d = \hat{v}_1 - \hat{v}_2 + \alpha(\hat{v}_3 - \hat{v}_1)$$

The sensitivity calculated from (5.32) is then

$$dv_{ce}/dT = .005607$$

Note that the diode voltage in the adjoint network must include the effect of the voltage controlled voltage source shown in Figure 5.12c.

The program DSNET from which these calculations were made is shown in Appendix Table 5.2; the output is displayed in Table 5.4.

TABLE 5.4 Results of DC Sensitivity Analygis of Amplifier

```
 R1 = 20000.000 R2 =  3000.000 RE =   300.000
 RL =  5000.000 E =    10.000 ALF =     0.980
OUTPUT NODES =  3  1  OUTPUT VOLTAGE =      4.4111864
OUTPUT NODES =  1  2  OUTPUT VOLTAGE =      0.3772230
OUTPUT NODES =  3  1  OUTPUT VOLTAGE =      4.3351727
OUTPUT NODES =  1  2  OUTPUT VOLTAGE =      0.3579055
OUTPUT NODES =  3  1  OUTPUT VOLTAGE =      4.2860692
OUTPUT NODES =  1  2  OUTPUT VOLTAGE =      0.3454267
OUTPUT NODES =  3  1  OUTPUT VOLTAGE =      4.2693350
OUTPUT NODES =  1  2  OUTPUT VOLTAGE =      0.3411740
OUTPUT NODES =  3  1  OUTPUT VOLTAGE =      4.2677915
OUTPUT NODES =  1  2  OUTPUT VOLTAGE =      0.3407817
OUTPUT NODES =  3  1  OUTPUT VOLTAGE =      4.2677797
OUTPUT NODES =  1  2  OUTPUT VOLTAGE =      0.3407787
 RE SEN =  0.3137D-02
 ALF SEN = -0.1147D 02
 TEMP SEN =  0.5607D-02
 E SEN =  0.3102D 00
```

5.6 A GENERAL DC SENSITIVITY ANALYSIS PROGRAM

The form of the DSNET program suggests the ease with which a general network analysis program such as DCAP can be converted to compute sensitivities. Indeed, the only difficulty is determination of which of the network variables in N and $\hat{N}$ are necessary for calculation of $d(v_o, i_o)/dp_j$. For example, we could expect node analysis to yield the node voltage vectors $\hat{\underline{v}}$ and $\hat{\underline{v}}$. Unfortunately, $\hat{N}$ may contain nodes in addition to those represented in $\hat{\underline{v}}$. The network of Figure 5.12c shows an example. Also, branch currents in $\hat{N}$ are sometimes necessary to determine parameter sensitivities; the calculation dv_{ce}/dE of Example 5.4 is an example.

Fortunately, the set of basic elements used in DCAP - conductances VCCS's, current sources, and diodes - all have adjoint network representations that maintain the same number of nodes; also all require only the node voltages of N and $\hat{N}$ for sensitivity evaluations.

The DSAP program of Appendix Table 5.3 will calculate the sensitivity of a dc solution to any resistance value, to any diode temperature and saturation current, and to the strength of any controlled or independent source. The user indicates by a tag in column 20 of the branch description those branches to which the sensitivity is to be determined. After each element of the input list has been represented in terms of the four basic elements in subroutine PREPAR, the dc analysis proceeds in the normal manner, i.e., at each iteration, the library of four linearized element values is consulted. After convergence and determination of $\hat{\underline{v}}$, the tagged branches are processed as follows. Depending on which of the four types of elements is involved, an appropriate sensitivity calculation using $\underline{v}$ and $\hat{\underline{v}}$ is carried out in subroutine SENN. For example, a branch originally described as a voltage source is converted in PREPAR to a gyrator and a current source. The sensitivity of the expanded network to the current source is then identical to the sensitivity of the original network to the voltage source.

Example 5.4b. The sensitivity of analysis of Example 5.4a is repeated with the DSAP program. The input and output data are displayed in Table 5.5 showing results nearly identical with Table 5.5. The discrepancy in temperature sensitivity is considered in Problem 5.2. As a check on the temperature sensitivity calculation, the temperature is stepped by 10° K and v_{ce} recalculated.

TABLE 5.5 DSAP Results

```
TRANSISTOR AMPLIFIER SENSITIVITY CALCULATION
R1           03 01        20000.
R2           01 00        3000.
I1   ID1     03 01      F .98
D1           01 02      D  .4
RE           02 00      N  300.
RL           03 04         5000.
VE           04 00      Q 10.
IOUT         03 02  V
TEMP                      290.,300.
AMEN
```

(a) Input

```
**********TRANSISTOR AMPLIFIER SENSITIVITY CALCULATION
 BRANCH NODE     ELEMENT      BRANCH CON-
 NAME    NOS.    VALUE        NAME    TROL
 R1     3  1  0.2000000E 05
 R2     1  0  0.3000000E 04
 I1     3  1  0.9800000E 00 D1      I
 D1     1  2  0.4000000E 00
 RE     2  0  0.3000000E 03
 RL     3  4  0.5000000E 04
 VE     4  0  0.1000000E 02
 IOUT   3  2  0.0
 TEMP   0  0  0.2900000E 03
 NODES  =  3  2 VOLTAGE =     4.788410
 NODES  =  3  2 VOLTAGE =     4.608717
 NODES  =  3  2 VOLTAGE =     4.608559
SENSITIVITY OF VOLTAGE  3- 2 W.R.T.
 DEPENDENT SOURCE I1    =  -0.1147D 02
 TEMP. OF DIODE D1    =  -0.1549D-01
 RESISTANCE RE    =    0.3137D-02
 SOURCE VE    =    0.3102D 00
*****TEMP =    0.3000E 03
 NODES  =  3  2 VOLTAGE =     4.776693
 NODES  =  3  2 VOLTAGE =     4.454074
 NODES  =  3  2 VOLTAGE =     4.453563
SENSITIVITY OF VOLTAGE  3- 2 W.R.T.
 DEPENDENT SOURCE I1    =  -0.1197D 02
 TEMP. OF DIODE D1    =  -0.1552D-01
 RESISTANCE RE    =    0.3273D-02
 SOURCE VE    =    0.3100D 00
DIODE VOLTAGES,CURRENTS
 D1
   0.341 ,    0.000832
   0.309 ,    0.000867
```

(b) Output

5.7 SECOND ORDER SENSITIVITIES IN LINEAR NETWORKS*

Observing that first order sensitivities $\partial(v_o, i_o)/\partial p_i$ are so easily calculated, we are led to ask how complicated the calculation of second order sensitivities $\partial^2(v_o, i_o)/\partial p_i\, \partial p_j$ might be. Although their utility may be open to question at present, it is still an interesting exercise.

To estimate the magnitude of the computational problem, we can observe that with r parameters p_j the matrix of second partials is of the form

$$H = \begin{bmatrix} \dfrac{\partial^2(v_o, i_o)}{\partial^2 p_1} & \dfrac{\partial^2(v_o, i_o)}{\partial p_2 \partial p_1} & \cdots & \dfrac{\partial (v_o, i_o)}{\partial p_r \partial p_1} \\ \dfrac{\partial^2(v_o, i_o)}{\partial p_1, \partial p_2} & \dfrac{\partial^2(v_o, i_o)}{\partial p_2^2} & \cdots & \vdots \\ \vdots & \vdots & \ddots & \vdots \\ \dfrac{\partial^2(v_o, i_o)}{\partial p_1 \partial p_r} & & & \dfrac{\partial^2(v_o, i_o)}{\partial^2 p_r} \end{bmatrix} \tag{5.33}$$

The symmetry of H requires the calculation of only the upper triangular matrix shown in (5.33). If the partial derivatives are calculated by successive differencing of p_j, i.e.,

$$\frac{\partial^2 (v_o, i_o)}{\partial p_i\, \partial p_j} \approx \frac{\left.\dfrac{\partial (v_o, i_o)}{\partial p_j}\right|_{p_i = p_i^o + \Delta p_i} - \left.\dfrac{\partial (v_o, i_o)}{\partial p_j}\right|_{p_i = p_i^o}}{\Delta p_i}$$

a total of $r(r+1)/2$ separate analyses are required <u>after</u> the first derivatives are estimated. We will show in this section that only $\overline{r}$ additional analyses are required.

The model we propose for development of our calculation procedure is shown in Figure 5.13b, namely, a linear resistive single-input, single-output network, where sensitivities $\partial^2(v_o, i_o)/\partial R_i\, \partial R_j$ or $\partial^2(v_o, i_o)/\partial G_j \partial G_j$ are to be found. The resistive network example of Figure 5.13a will be carried through the procedure.

To begin, we write the first order sensitivity in the form

$$\frac{\partial (v_o, i_o)}{\partial G_j} = v_{G_j} \hat{v}_{G_j}$$

*S. W. Director, private communication.

where j=1 in the Figure 5.13c. Since both v_{G_j} and $\hat{v}_{G_j}$ are dependent on both G_i and G_j, we would expand the second order sensitivity as

$$\frac{\partial (v_o, i_o)}{\partial G_i G_j} = \frac{\partial v_{G_j}}{\partial G_i} \hat{v}_{G_j} + v_{G_j} \frac{\partial \hat{v}_{G_j}}{\partial G_i} \tag{5.34}$$

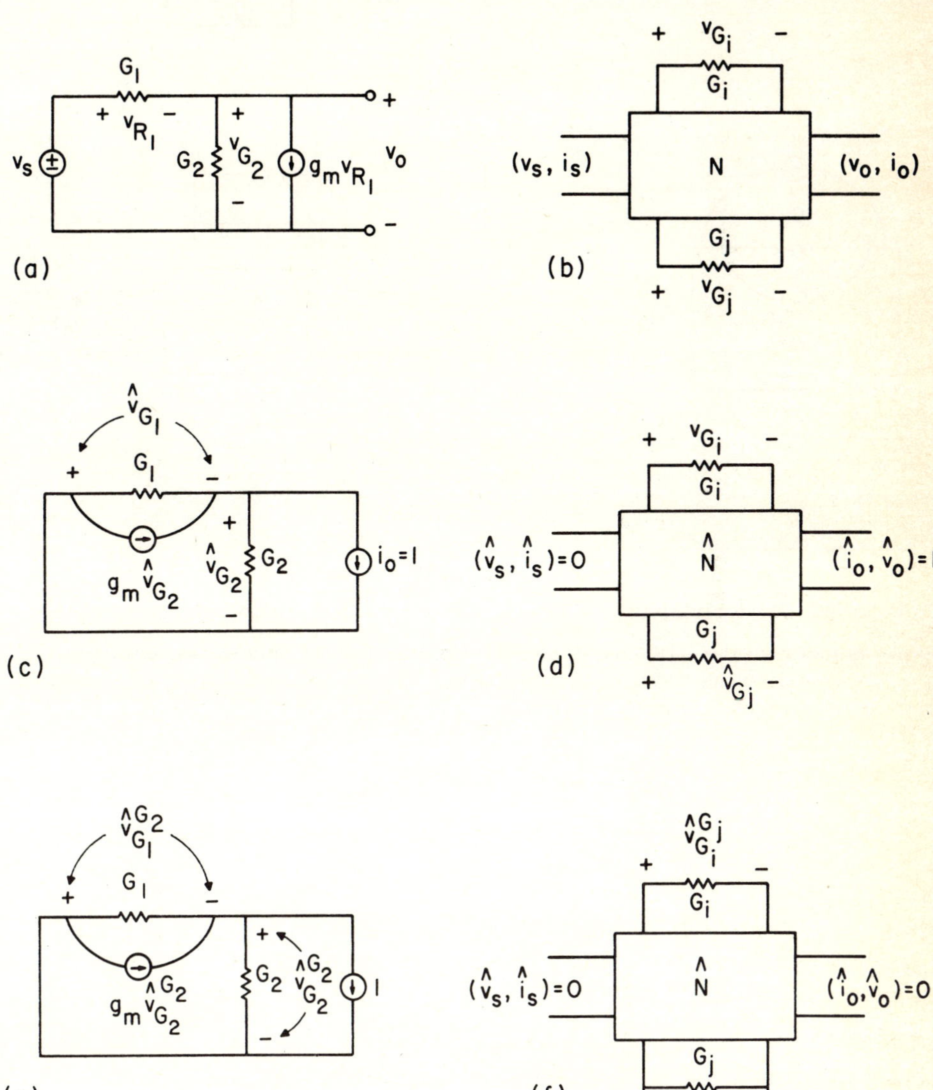

FIG 5.13 Calculation of Second Order Sensitivity (i = 1, j = 2 for a-c-e-g-i)

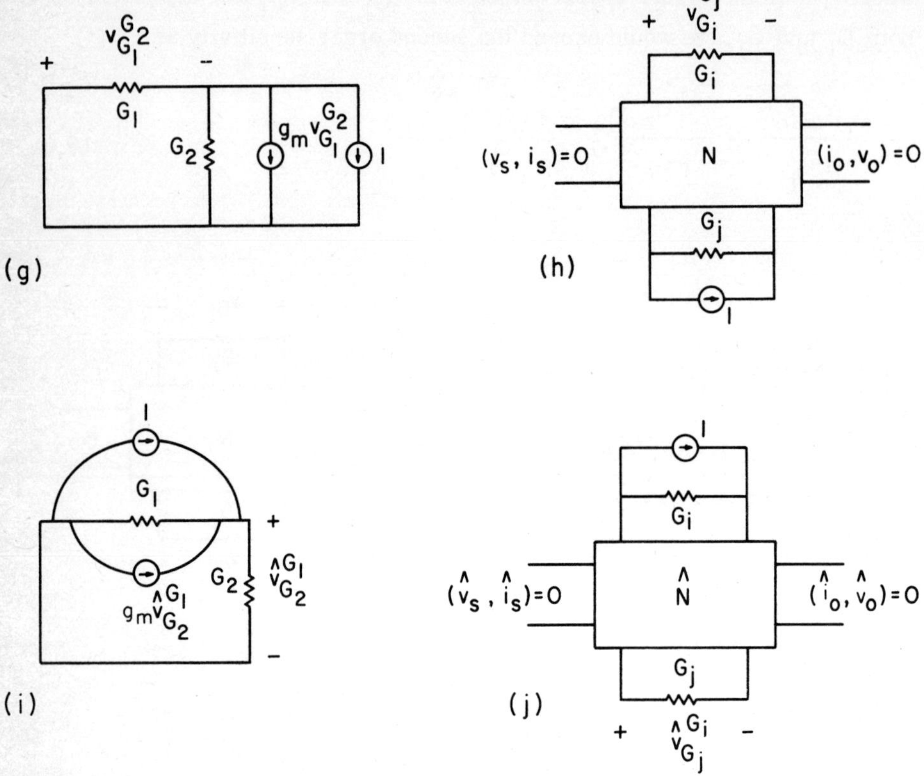

FIG. 5.13 (Continued)

Now the partial derivatives $\partial v_{G_j}/\partial G_i$ and $\partial \hat{v}_{G_j}/\partial G_i$ can be calculated in the same manner as the first order sensitivity — as the product of two voltages.

To visualize the process of calculating $\partial v_{G_j}/\partial G_i$, we temporarily regard v_{G_j} as an "output" voltage. Then

$$\frac{\partial v_{G_j}}{\partial G_i} = v_{G_i} \hat{v}_{G_i}^{G_j}$$

where $\hat{v}_{G_i}^{G_j}$ is measured across branch G_i of $\hat{N}$ but with the excitation (a unit current source) in parallel with G_j (the "output branch"). This is shown in Figure 5.13e-f.

The term $\partial \hat{v}_{G_j}/\partial G_i$ of (5.34) can be similarly developed. The voltage $\hat{v}_{G_j}$ is now viewed as an output variable; we then write

$$\frac{\partial \hat{v}_{G_j}}{\partial G_i} = \hat{v}_{G_i} v_{G_i}^{G_j}$$

where $\hat{v}_{G_i}$ is shown in Figure 5.13c-d and $v_{G_i}^{G_j}$ is measured in the adjoint of $\hat{N}$, i.e., the "adjoint of the adjoint." But, from the rules of forming the elements of $\hat{N}$ for linear networks, we realize that this is simply the original linear network N.

Collecting all terms involved in the derivative calculation, we can now write

$$\frac{\partial^2(v_o, i_o)}{\partial G_i \partial G_j} = (v_{G_i} \hat{v}_{G_i}^{G_j}) \hat{v}_{G_j} + v_{G_j} (\hat{v}_{G_i} v_{G_i}^{G_j}) \tag{5.35}$$

The simultaneous calculation of all r second order sensitivities must now be carefully considered. For fixed j, the calculation of all $\hat{v}_{G_i}^{G_j}$ and v_{G_i} $(i=1,2,\ldots,r)$ can be made simultaneously from two analyses, with a unit current source across the $\hat{G}_j$ branch in N and $\hat{N}$ as shown. This would appear to require a total of 2r+2 analyses to compute second order sensitivities for all j $(j=1,2,\ldots,r)$.

It may be shown (left as a problem) that the calculations made with N in Figure 5.13g-h can in fact be made with the adjoint network $\hat{N}$ of Figure 5.13i-j. In particular, it can be shown [5.8] that

$$\hat{v}_{G_j}^{G_i} = v_{G_i}^{G_j} \tag{5.36}$$

For example, the reader can verify that $v_{G_1}^{G_2} = v_{G_2}^{G_1}$ in Figure 5.13i-j. The property of (5.36) is known as interreciprocity.

Substituting (5.36) into (5.35) yields

$$\frac{\partial^2(v_o, i_o)}{\partial G_i \partial G_j} = v_{G_i} \hat{v}_{G_j} \hat{v}_{G_i}^{G_j} + v_{G_j} \hat{v}_{G_i} \hat{v}_{G_j}^{G_i} \tag{5.37}$$

This formula is symmetrical in i and j as indeed it should be.

The procedure for determining $\underset{\sim}{H}$ of (5.34) can now be summarized.

(1) All v_{G_i} and v_{G_j} are determined in the original network analysis (Figure 5.13b);

(2) all $\hat{v}_{G_i}$ and $\hat{v}_{G_j}$ are determined in the single analysis of $\hat{N}$ in Figure 5.13d; all first order sensitivities can then be found;

(3) for any j, all $\hat{v}_{G_i}^{G_j}$ $(i=1,2,\ldots,r)$ can be determined in one analysis (Figure 5.13c) with a unit source across G_j; for $j=1,2,\ldots,r$, a total of r analyses are therefore necessary; all second order sensitivities can then be found using (5.37).

Thus, a total of r+2 analyses suffice to calculate the original network response as well as all first and second order sensitivities.

Problems

5.1 Show that for an independent source

$$\frac{d(v_o, i_o)}{dE} = -\hat{i}_E \qquad \frac{d(v_o, i_o)}{dJ} = \hat{v}_J$$

5.2 Show that for a controlled source of strength k,

$$\frac{d(v_o, i_o)}{dk} = \pm \begin{array}{l}\text{(controlling branch variable in } N) \times \\ \text{(controlling branch variable in } \hat{N})\end{array}$$

5.3 Show that the sensitivity of power in an output branch can be found from two analyses, using the adjoint network concept. Is this algorithm valid

(a) for the linear resistive networks
(b) for networks in the sinusoidal steady state,
(c) for nonlinear dc resistive networks ?

5.4 How many matrix factorizations and substitutions are required to find the sensitivities of m outputs to r parameters for a linear network?

5.5 Suppose we want to find the zeros of a network function T(s) by Newton iteration. We might define

$$\frac{dT}{ds}\Big|_{s=s^m} \Delta s^m = -T(s^m)$$

Indicate how we could calculate both dT/ds and T for a given value s^m in only two node analyses!

5.6 A linear two-port N is connected in parallel with its adjoint network $\hat{N}$. Show that the resultant two-port is now reciprocal.

5.7 The DSAP program of Appendix Table 5.3 assumes the following temperature dependence.

$$i_d = I_s(e^{qv_d/kT} - 1)$$

where

$$I_s = 10^{-9}(T/T_o)^3\, e^{B(T-T_o)/TT_o}$$

$$T_o = 290^oK$$

Develop the expression for $d(v_o, i_o)/dT$ used in subroutine

5.8 Show that the scattering matrix $\hat{\underset{\sim}{S}}$ of $\hat{N}$ is given by [5.12]

$$\hat{\underset{\sim}{S}} = \underset{\sim}{S}^T$$

5.9 Show how the determinantal identity

$$\Delta_{ij}\,\Delta_{k\ell} - \Delta\,\Delta_{ijk\ell} = \Delta_{i\ell}\,\Delta_{kj}$$

can be used to obtain

$$\frac{d(v_o, i_o)}{dG} = v_G \hat{v}_G$$

(The subscripts refer to rows and columns deleted from a matrix, e.g., the node matrix $\tilde{G}_N$.)

5.10 Consider the problem of calculating $d\Delta/ds$ and $d^2\Delta/d^2s$ where

$$\Delta = \det(G + s\tilde{C})$$

Suppose we write

$$(\tilde{G} + s\tilde{C})\underline{X} = \underline{b}$$

where $\underline{X} = [x_1 \ldots x_n]^T$ and $\underline{b}$ is chosen such that $x_j = 1/\Delta$ (if indeed this is possible). Show that

(a) dx_j/ds can be found with a forward and back substitution
(b) d^2x_j/d^2s can be found with two additional forward and back substitutions (a more difficult problem).

5.11 If a dc i-v characteristic is given by

$$i = f(v, p)$$

find the adjoint network representation of this element for small signal sinusoidal steady-state analysis. Give the expression for finding dV_o/dp where V_o is a phasor quantity.

REFERENCES

5.1 Bordewijk, J. L., "Reversal Theorems for Linear Electrical Triode Networks," HET PTT-Bedriff deel III, no. 2; August, 1950.
5.2 Tellegen, B. D. H., "A General Network Theorem, with Applications," Phillips Res. Rpt., no. 7, pp. 259-269; 1952.
5.3 Penfield, P., R. Spence, and S. Duinker, "Tellengen's Theorem," Memo. 153, Department of Electrical Engineering, M. I. T.; 1968.
5.4 Director, S. W., and R. A. Rohrer, "A Generalized Adjoint Network and Network Sensitivities," IEEE Trans., vol. CT-16, pp. 330-336; August, 1969.
5.5 Penfield, P., R. Spence, and S. Duinker, Tellegen's Theorem and Electrical Networks, M. I. T. Press; 1970.
5.6 Bordewijk, J. L., "Inter-reciprocity Applied to Electrical Networks," Appl. Sci. Res., vol. 6, sec. B, pp. 1-74; 1956.
5.7 Spence, R., Linear Active Networks, Wiley-Interscience; 1970.
5.8 Bordewijk, J. L., "Comments on 'Automated Network Design and the Intereciprocity Concept'," Trans. IEEE, vol. CT-18, no. 1, p. 179; January, 1971.

5.9 Director, S. W., "LU Factorization in Network Sensitivity Calculations," Trans. IEEE, vol CT-18, no. 1, pp. 184-185; January, 1971.

5.10 Parker, S. R., "Sensitivity: Old Questions, Some New Answers," Trans. IEEE, vol. CT-18, no. 1 pp. 27-34; January, 1971.

5.11 Kuh, E. S., and R. A. Rohrer, Theory of Linear Active Networks, Holden-Day; 1967.

5.12 Monaco, V. A., and P. Tiberio, "On Linear Network Scattering Matrix Sensitivity," Alta Frequenza, vol. 39, no. 2; 1970.

6

Automatic Design

6.1 INTRODUCTION

6.1.1 Iterative Design

One of the more controversial concerns in design regards the extent to which the digital computer can replace human insight. To develop a reasonable attitude to this question, let us study a typical design situation. A circuit often fails to deliver adequate performance because circuit parasitics (e.g., stray capacitance, lead inductance, leakage resistance) are not accounted for by the idealized component models necessary to render the design humanly tractable. When these parasitics are included, then the designer, with computer assistance, can alter

(1) the values of design parameters, such as bias resistances, coupling coefficients, etc., or if necessary
(2) the circuit structure itself.

These two corrective measures are evidently in increasing order of severity and require an increasingly higher level of judgment.

If we are to evolve a methodology of "automated" or "hands-off" design, where "algorithm" replaces "insight," then it is reasonable to consider first the parameter adjustment problem of (1) above. As we shall see in this chapter, certain common design problems can be easily re-formulated as mathematical problems. The solution of these mathematical problems by iteration of the design parameters then yields the required element values. The iteration technique will turn out to be so powerful that we will sometimes be able to regard the initial design by the engineer as more of a "guess" than the result of an organized design procedure. At a minimum, we can expect that at least part of the design procedure can be automated and that, by having such an option available, the designer can educate himself on its value to his particular problem.

Before proceeding, we should caution the reader that only elementary iteration techniques will be used in our present study; advanced methods are discussed in Chapter XI.

The present discussion will be to show the results achieveable by automatic design with small and somewhat contrived examples. We will, however, feel sufficiently comfortable with the reliability of our algorithms to develop a problem-oriented language for automatic dc design, based on DSAP of Chapter V.

6.1.2 Mathematical Description

We will examine three common classes of network design problems for possible use of automatic design:

(1) s-plane or pole-zero design,
(2) frequency domain design,
(3) nonlinear dc design.

With discussion of each, we will introduce a new iteration technique that is suggested by characteristics of the design problem.

In general, the goal of this iteration will be to match a calculated and a desired response. For example, a set of n bias voltages and currents may be required in a dc circuit model, or it may be desired to match a frequency response $|T(j\omega_i)|$ to a specified response $|T_s(j\omega_i)|$ at frequencies ω_i, $i = 1, 2, \ldots n$.

Mathematically, the problem may be stated as follows. Let

$$\underline{p} = \begin{bmatrix} p_1 \\ p_2 \\ \cdot \\ \cdot \\ p_r \end{bmatrix}, \quad \underline{z} = \begin{bmatrix} z_1 \\ z_2 \\ \cdot \\ \cdot \\ z_n \end{bmatrix}, \quad \hat{\underline{p}} = \begin{bmatrix} \hat{p}_1 \\ \hat{p}_2 \\ \cdot \\ \cdot \\ \hat{p}_r \end{bmatrix}, \quad \hat{\underline{z}} = \begin{bmatrix} \hat{z}_1 \\ \hat{z}_2 \\ \cdot \\ \cdot \\ \hat{z}_n \end{bmatrix}, \quad \underline{f} = \begin{bmatrix} f_1 \\ f_2 \\ \cdot \\ \cdot \\ f_n \end{bmatrix} \tag{6.1}$$

where $\underline{p}$ contains the design parameters,
$\underline{z}$ contains the calculated response description and is a function of $\underline{p}$,
$\hat{\underline{z}}$ contains the desired response description and so is a vector of constants,
$\hat{\underline{p}}$ contains the unknown optimal parameter values to be achieved after iteration.

If we now define

$$\underline{f}(\underline{p}) = \underline{z}(\underline{p}) - \hat{\underline{z}} \tag{6.2}$$

then the solution of the design problem occurs at $\underline{f} = \underline{0}$.

6.2 S-PLANE OPTIMIZATION

6.2.1 The Rationale

In the design of filters and equalizers, the specification to be realized is usually in the frequency domain, i.e., magnitude, phase, and/or delay. The iteration is then commonly carried out in the same domain (Section 6.3).

In the adjustmentof active filters, wideband amplifiers, and oscillators, new considerations are introduced which make frequency domain optimization less appealing than an equivalent s-plane iterative scheme:

1. the active network may be unstable initially; to iterate toward a stable network requires the natural frequencies to cross the $j\omega$-axis in the s-plane, at which point the magnitude response becomes infinite: thus, the iteration must be based on the phase or delay until the crossing is completed; then the iteration could seemingly return to the magnitude response; the reverse may be true in the design of oscillators from stable works; in contrast, s-plane optimization methods are insensitive to $j\omega$-axis crossing;
2. the coefficients of a network function are linear functions of any element value, i.e.,

$$T(s) = \frac{N_1(s) + y_j N_2(s)}{D_1(s) + y_i D_2(s)}$$

This linear dependence is, of course, in contrast to the corresponding frequency domain design criteria functions; it can be shown that this can result in local minima in the latter case which are avoided by s-plane optimization [6.7].

6.2.2 The Coefficient Matching Method*

For simplicity, assume that a network function $T(s)$ describing a network is of the all-pole form

$$T(s) = \frac{1}{\sum_{i=0}^{n-1} a_i s^i} \tag{6.3}$$

It is desired to adjust $r = n$ network elements to realize a set of coefficients $\hat{a}_i$. If we define error functions

$$f_{i+1} = a_i - b_o \hat{a}_i \qquad i = 0, 1, 2, \ldots, n-1 \tag{6.4}$$

where b_o is unknown, then the solution $\underline{f} = \underline{0}$ defines a class of networks having the desired poles, i.e., the set $\hat{a}_i$ is realized within a multiplicative

---

* The general features of this method were first suggested to this author by H. J. Orchard.

constant b_o. Clearly, this matching process can be altered to realize (1) both zeros and poles or (2) T(s) precisely, including any gain constant.

Equation (6.4) may be viewed as representing n nonlinear equations in n unknown coefficients a_i and b_o. But these coefficients may be regarded as functions of the network element values p_j. We can therefore linearize (6.4) and write the Newton iteration formula

$$\sum_{j=1}^{r} \frac{\partial f_i}{\partial p_j} \Delta p_j + \frac{\partial f_i}{\partial b_o} \Delta b_o = -f_i \qquad i = 1, 2, \dots n \tag{6.5}$$

or

$$\sum_{j=1}^{r} \frac{\partial a_i}{\partial p_j} \Delta p_j - \hat{a}_i \Delta b_o = -a_i + b_o \hat{a}_i \qquad i = 0, 1, \dots n-1 \tag{6.6}$$

where the iteration counter m has been deleted for convenience.

Example 6.1. The unterminated filter of Figure 6.1 with poles at $s = 1 \pm j1$ is to have a 0.5 Ω source resistance inserted while maintaining the same poles.

Equation (6.6) becomes in matrix form

$$\begin{bmatrix} \frac{\partial a_o}{\partial \Gamma} & \frac{\partial a_o}{\partial C} & -\hat{a}_o \\ \frac{\partial a_1}{\partial \Gamma} & \frac{\partial a_1}{\partial C} & -\hat{a}_1 \\ \frac{\partial a_2}{\partial \Gamma} & \frac{\partial a_2}{\partial C} & -\hat{a}_2 \end{bmatrix} \begin{bmatrix} \Delta\Gamma \\ \Delta C \\ \Delta b_o \end{bmatrix} = \begin{bmatrix} -a_o + b_o \hat{a}_o \\ -a_1 + b_o \hat{a}_1 \\ -a_2 + b_o \hat{a}_2 \end{bmatrix} \tag{6.7}$$

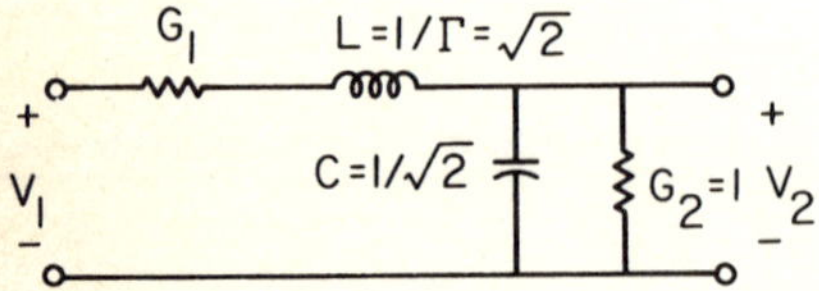

FIG. 6.1 LC Filter of Example 6.1

The network function, in terms of admittances, is

$$\frac{V_2}{V_1}(s) = \frac{\Gamma G_1}{C G_1 s^2 + (\Gamma C + G_1 G_2)s + (G_1 + G_2)\Gamma} \tag{6.8}$$

$$\begin{bmatrix} 0 & G_1 & -1 \\ C & \Gamma & -\sqrt{2} \\ G_1 + G_2 & 0 & -1 \end{bmatrix} \begin{bmatrix} \Delta\Gamma \\ \Delta C \\ \Delta b_o \end{bmatrix} = \begin{bmatrix} -a_o + b_o \\ -a_1 + \sqrt{2}b_o \\ -a_2 + b_o \end{bmatrix} \qquad (6.9)$$

so that for the first iteration ($b_o = 1$)

$$\begin{bmatrix} 0 & 2 & -1 \\ 0.707 & 0.707 & -\sqrt{2} \\ 3 & 0 & -1. \end{bmatrix} \begin{bmatrix} \Delta\Gamma \\ \Delta C \\ \Delta b_o \end{bmatrix} = \begin{bmatrix} 0 \\ -0.353 \\ -0.5 \end{bmatrix} \qquad (6.10)$$

The results of subsequent iterations are shown in Table 6. 1.

TABLE 6. 1 Iteration Table for Example 6. 1

Iteration	L	C
0	1.4140	0.70700
1	1.6490	0.90917
2	1.6727	0.89674
3	1.6725	0.89668

The initial value chosen for b_o is immaterial. This follows from an inspection of (6. 9), since both b_o and Δb_o are multiplied by the same constants; therefore $b_o + \Delta b_o$ will be independent of b_o.

It is possible to add simple side constraints to the iteration, as the following shows.

Example 6. 2. Let it be desired to insert loss ($Q = Q_L = Q_C = 20$) in the double terminated design of Example 6. 1.

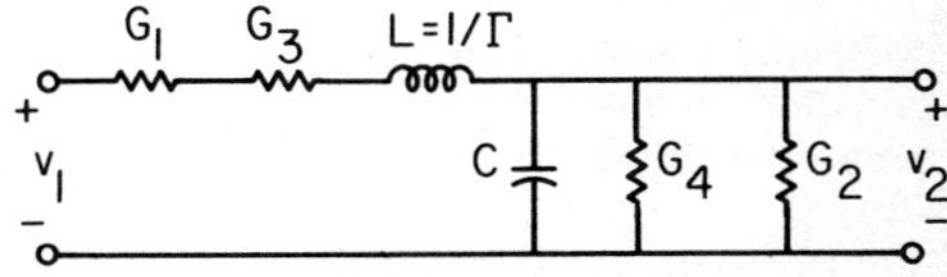

FIG. 6. 2 LC Filter of Example 6. 2

We insert two conductances representing loss as shown in Figure 6. 2. Beginning with the final values of L and C in Table 6. 1, we again apply

(6.6) with

$$G_3 = Q\Gamma \tag{6.11}$$

$$G_4 = C/Q \tag{6.12}$$

maintained throughout the iteration. Equations (6.11) and (6.12) imply that the terms associated with G_3 and G_4 in the Newton iteration are constrained by

$$\frac{\partial a_i}{\partial G_3} \Delta G_3 = Q \frac{\partial a_i}{\partial G_3} \Delta\Gamma \tag{6.13}$$

$$\frac{\partial a_i}{\partial G_4} \Delta G_4 = \frac{1}{Q} \frac{\partial a_i}{\partial G_4} \Delta C \tag{6.14}$$

Thus the columns of the partial derivative matrix corresponding to ΔG_3 and ΔG_4 may be weighted (by Q and 1/Q) and added to the columns associated with $\Delta\Gamma_1$ and ΔC_1. The resultant matrix is then the same order as that of Example 6.1. The results of iteration are given in Table 6.2.

TABLE 6.2 Iteration Table for Example 6.2

Iteration	L	C
0	1.6725	0.8967
1	1.6041	0.9494
2	1.6173	0.9949
3	1.6177	0.9951

6.2.3 The Dominant Pole Case

In many practical applications, it is required to specify only certain dominant poles, while the remaining poles are of little concern. For example, the bandpass tuned amplifier of Figure 6.3 is obviously stable if $C_c = 0$, since the tuned circuit N is then isolated from the input by a controlled source. Moreover, there is one pair of complex poles near the $j\omega$-axis due to the tuned circuit N.

When C_c is increased from zero, a negative real pole appears which does not materially affect the frequency response; also, the poles near the $j\omega$-axis move toward the right half plane and, unless the other elements are adjusted, may result in instability. Thus, we might pose the problem of adjusting the other elements to maintain the original poles near the $j\omega$-axis in position while C_c is increased.

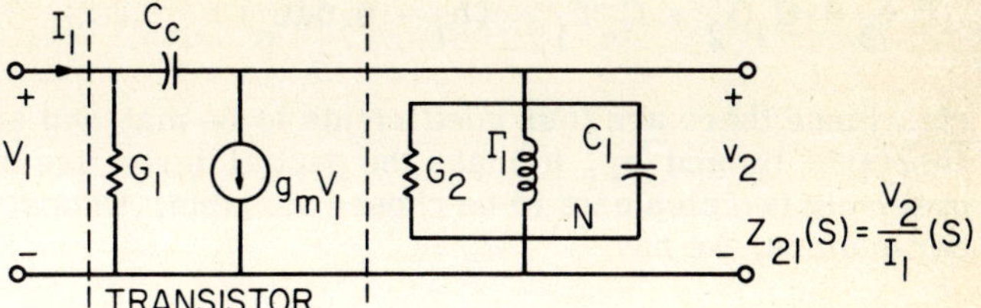

FIG. 6.3 Tuned Amplifier Example

The previous iterative procedure is applicable here if we note that the denominator polynomial can in general be written

$$D(s) = (b_o + b_1 s + \dots b_k s^k)\ (a_o + a_1 s + \dots a_n s^n) \tag{6.15}$$

$$= B(s)\ A(s) \tag{6.16}$$

where A(s) contains the dominant roots and B(s) the nondominant. Then D(s) can be expanded

$$D(s) = \sum_{i=0}^{k+n} d_i s^i \tag{6.17}$$

where d_i is a linear function of the unknown b_i's. Thus, it plays the same role as the b_o of (6.4).

Example 6.3. It is desired to keep the dominant poles of the transfer impedance of the amplifier of Figure 6.3 at $s = -0.02 \pm j1$ as C_c is increased. Initial guesses of element values are

$$C_1 = 1. \quad L_1 = 1. \quad R_2 = 25. \quad R_1 = 1. \quad g_m = 1.$$

which yield the exact design when $C_c = 0$.

There is one nondominant pole, so that we have the denominator of the transfer impedance as

$$D(s) = (C_1 C_c)\, s^3 + (G_1(C_1 + C_c) + C_c G_2 + g_m C_c)\, s^2$$
$$+ (G_1 G_2 + C_c \Gamma_1)\, s + \Gamma_1 G_1 \tag{6.18}$$

$$= (b_o + b_1 s)\ (1 + 0.04\, s + s^2) \tag{6.19}$$

$$= b_o + (b_1 + 0.04 b_o)\, s + (0.04 b_1 + b_o)\, s^2 + b_1 s^3 \tag{6.20}$$

The f_i similar to (6.4) are formed by equating (6.18) to (6.20)

$$f_1 = \Gamma_1 G_1 - b_o \tag{6.21}$$

$$f_2 = G_1G_2 + C_cT_1 - (b_1 + 0.04b_o) \qquad (6.22)$$

etc. Since there are four coefficients to be matched and two undetermined constants b_o and b_1, to make the partial derivative matrix square requires precisely two elements to be chosen variable. Choosing C_1 and G_2 as parameters, we have

$$\begin{bmatrix} 0 & 0 & -1. & 0 \\ 0 & G_1 & -0.04 & -1. \\ G_1 & C_c & -1. & -0.04 \\ C_c & 0 & 0 & -1. \end{bmatrix} \begin{bmatrix} \Delta C_1 \\ \Delta G_2 \\ \Delta b_o \\ \Delta b_1 \end{bmatrix} = \begin{bmatrix} b_o - F_1G_1 \\ b_1 + 0.04b_o - G_1G_2 - C_c\Gamma_1 \\ 0.04b_1 + b_o - G_1(C_1+C_c) - C_c(G_2+g_m) \\ b_1 - C_1C_c \end{bmatrix} \qquad (6.23)$$

The results for $C_c = .0001$ are given in Table 6.3.

TABLE 6.3 Iteration Table for Example 6.3

Iteration	C_1	R_2
0	1.000	25.00
1	0.8012	49.50
2	0.8012	49.70

6.2.4 Related Topics

The use of Newton iteration to match coefficients is a rather obvious ploy so long as the partial derivative matrix is square. When there are more variable elements than required, it is questionable how this freedom may be employed to advantage. One answer to these questions follows in Section 6.3 and 6.4.

Large examples require special attention be given to the efficient generation of the coefficient derivatives. This topic is considered in Problem 6.4.

As a final example, we present the following design problem.

Example 6.4. The active RC amplifier of Figure 6.4 is initially unstable, with a dominant pole at $s = 0.0169 \pm j0.566$. The design is to be altered so as to realize a pole at $s = 0.05 \pm j1$.

After six iterations, requiring four seconds of computing time on an IBM 360/67, the final element values shown in parentheses in Figure 6.4 were obtained. Note the large percentage change in certain of the elements, indicating the wide range of convergence in parameter space.

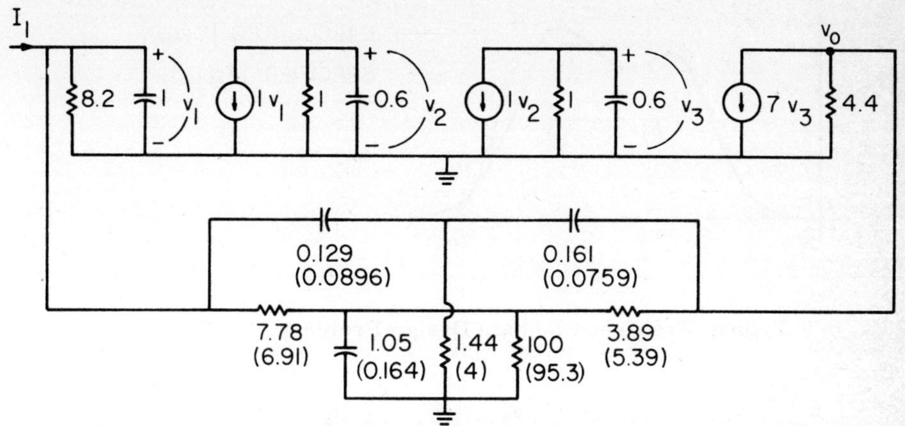

FIG. 6.4 Active Filter, with Element Values Shown Before and After Adjustment of Dominant Poles

6.3 FREQUENCY DOMAIN DESIGN

6.3.1 Statement of the Problem

By far the most common specification in frequency domain design is the magnitude function given over a range of frequencies as shown in Figure 6.5. We can convert the problem to a tractable matching problem by discretizing the frequency variable. That is, we define

$$e_i = |T_s(j\omega_i)| - |T(j\omega_i)| \qquad i = 1, 2, \dots n \tag{6.24}$$

and $\underline{f} = [e_1 e_2 \dots]^T$. Then if $f = 0$, the specification is considered realized. Since $|T(j\omega_i)|$ is a complicated function of the network element values, solution of $\underline{f} = \underline{0}$ requires iteration. Accordingly, we linearize the equation $\underline{f} = \underline{0}$ to obtain the Newton-type formula

$$\left(\frac{\partial \underline{f}}{\partial \underline{p}}\right)^m \Delta\underline{p}^m = -\underline{f}^m \tag{6.25}$$

or in component form

$$\sum_{j=1}^{r} \frac{\partial e_i}{\partial p_j} \Delta p_j = -e_i \qquad i = 1, 2, \dots n \tag{6.26}$$

where the iteration counter m has been dropped for notational convenience.

Example 6.5. The conductance G of Figure 6.6a is to be changed to match the frequency domain specification given at five frequencies in Figure 6.6b.

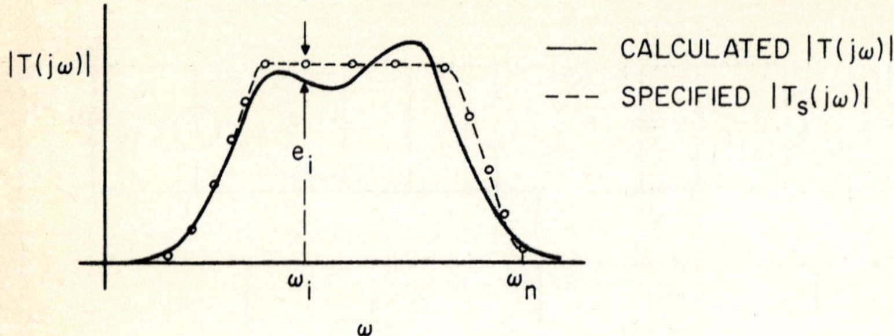

FIG. 6.5 Typical Frequency Domain Design Problem

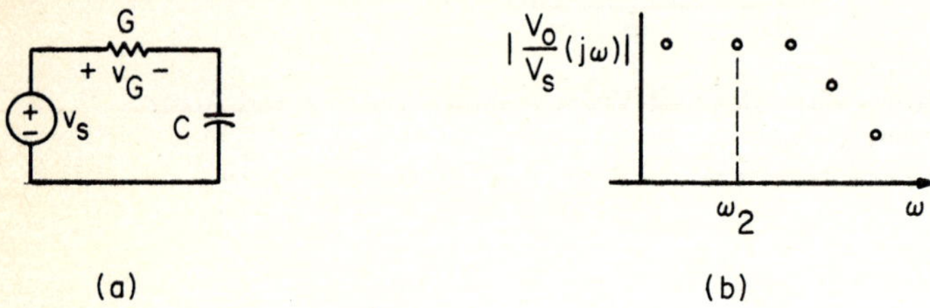

FIG. 6.6 Simple Frequency Domain Design Problem

We write (6.25) in component form as

$$\frac{\partial e_i}{\partial G}\Delta G = \frac{\partial |T(j\omega_i)|}{\partial G}\Delta G = -e_i \qquad i = 1, 2, \ldots n \tag{6.27}$$

$$= |T_s(j\omega_i)| - |T(j\omega_i)|$$

We are left with five equations but only one unknown! This suggests that we explore other mathematical formulations of the design problem which can be solved for $r < n$.

6.3.2 Least Squares Iteration [6.2]

In this procedure, we replace the problem of solving $\underline{f} = \underline{0}$ by the problem of minimizing

$$P(\underline{p}) = \sum_{i=1}^{n} f_i^{\,2}(\underline{p}) \tag{6.28}$$

The minimum of P would ideally be zero if $|T(j\omega_i)| = |T_s(j\omega_i)|$ for all i, implying an exact realization of the specifications. However, if this is not possible, the above formulation would generate at least an approximate match.

A necessary condition for this minimum to occur is

$$\frac{\partial P}{\partial p_j} = 0 \qquad j = 1, 2, \ldots r \tag{6.29}$$

or more succinctly

$$\text{grad } P = \underline{\nabla}P = \left(\frac{\partial P}{\partial \underline{p}}\right)^T = \underline{0} \tag{6.30}$$

Substituting (6.28) into (6.30) gives in component form

$$\frac{\partial P}{\partial p_j} = \sum_{i=1}^{n} 2f_i \frac{\partial f_i}{\partial p_j} \qquad j = 1, 2, \ldots r$$

or, similarly to (6.30),

$$\underline{\nabla}P = 2\underset{\sim}{\phi}^T \underline{f} = \underline{0} \tag{6.31}$$

where the (i, j) entry of $\underset{\sim}{\phi}$ is $\partial f_i/\partial p_j$.

To find p that satisfies $\underline{\nabla}P(\underline{p}) = \underline{0}$, we can apply the Newton procedure to (6.31). This requires calculation of the partial derivative of each component of $\underline{\nabla}P$ with respect to each p_j, viz,

$$\frac{\partial}{\partial p_k}\left(\frac{\partial P}{\partial p_j}\right) = 2\sum_{i=1}^{n} \frac{\partial f_i}{\partial p_k} \frac{\partial f_i}{\partial p_j} + 2\sum_{i=1}^{n} f_i \frac{\partial^2 f_i}{\partial p_k \partial p_j} \qquad \begin{matrix} j = 1, 2, \ldots r \\ k = 1, 2, \ldots r \end{matrix} \tag{6.32}$$

The matrix of second partial derivatives is commonly ignored, since its calculation is time consuming (see Chapter V). Consequently, the summation is usually approximated by

$$\frac{\partial}{\partial p_k}\left(\frac{\partial P}{\partial p_j}\right) = 2\begin{bmatrix} \frac{\partial f_i}{\partial p_k} & \cdots & \frac{\partial f_n}{\partial p_k} \end{bmatrix} \begin{bmatrix} \frac{\partial f_i}{\partial p_j} \\ \cdot \\ \cdot \\ \frac{\partial f_n}{\partial p_j} \end{bmatrix} \tag{6.33}$$

The calculation of all partial derivatives can now be written as

$$\frac{\partial \underline{\nabla}P}{\partial \underline{p}} = 2\underset{\sim}{\phi}^T \underset{\sim}{\phi} \tag{6.34}$$

so that the Newton iteration becomes, for the m-th iteration,

$$\underset{\sim}{\phi}^T \underset{\sim}{\phi}\, \Delta\underline{p} = -\underset{\sim}{\phi}^T \underline{f} \tag{6.35}$$

We may note that $\underset{\sim}{\phi}^T\underset{\sim}{\phi}$ is square, but $\underset{\sim}{\phi}$ is not unless r = n. In that case, if $\underset{\sim}{\phi}$ is invertible, then (6.35) becomes

$$\underset{\sim}{\phi}\,\Delta\underline{p} = -\underline{f} \qquad (6.36)$$

which is identical to our original iteration formula (6.25).

6.3.3 Weighting Functions

The inadequacy of (6.28) as a design criterion is evident from examination of typical filter design problem, as shown in Figure 6.7a. If we define $P = f_1^2 + f_2^2 = e_1^2 + e_2^2$, then an error of $|e_1| = .1$ at ω_1 and an error of $|e_2| = .1$ at ω_2 will affect P identically. This equal weighting of error would probably be impractical since a stopband error (at ω_2) would be far greater percentage of the value of $|T(j\omega_i)|$ and would usually be regarded as a more serious error.

A useful generalization of least square is the <u>weighted</u> least squares method, where

$$P = \sum_{i=1}^{n} \lambda_i (|T(j\omega_i)| - |T_s(j\omega_i)|)^2$$

By choosing λ_i sufficiently large - as, for example, at ω_2 - the effect of the response at ω_i on P can be enhanced.

<u>Example 6.6.</u> The voltage transfer function of the active filter of Figure 6.7b has the form

$$\frac{V_o}{V_i}(s) = \frac{(KC_1G_1)s}{C_1C_2s^2 + [G_1(C_1+C_2)+G_3(C_1+C_2-KC_1)+G_2C_1]s + [G_2(G_1+G_3)]}$$

where K represents the gain of a voltage-controlled voltage source. With $R_1 = R_3 = R_3 = 1000$, $C_1 = C_2 = 10^{-6}$, K=10, the "initial" frequency response of Figure 6.7c is obtained. We require that the response be adjusted to attain (in the least square sense) the level of 6 db at frequencies f=100, 150, 200 htz as shown. C_1 and C_2 are available as parameters.

After iteration using unweighted least squares ($\lambda_i = 1$), the values $C_1 = 1.583 \times 10^{-6}$f, $C_2 = 1.627 \times 10^{-6}$f yield the "final" response shown in Figure 6.7c. As might be expected the error is approximately equally distributed among the three frequencies.

6.4 DC DESIGN

6.4.1 Introduction

Although there are occasions when linearized models of nonlinear dc networks are adequate for design, it would be misleading to purposely so restrict our interest when the computing power of modern machines and algorithms permits solution and iterative design of the nonlinear networks themselves. The skeleton of such a design process is given by the following example.

Consider the resistor-diode circuit of Figure 6.8a. Besides solving the network using the methods of Chapter III, we require the voltage v_o to be a

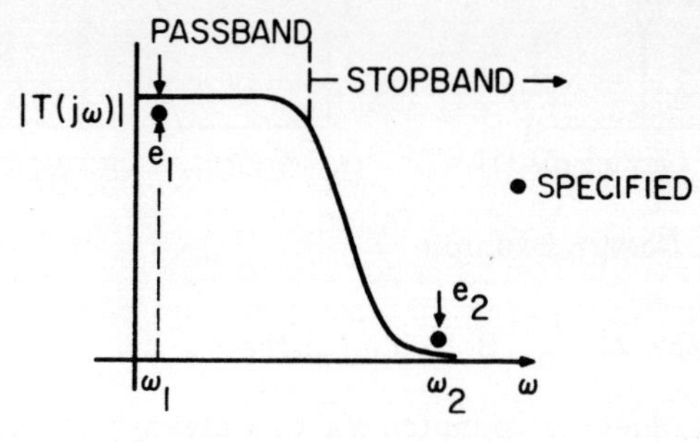

(a) Filter Design Problem

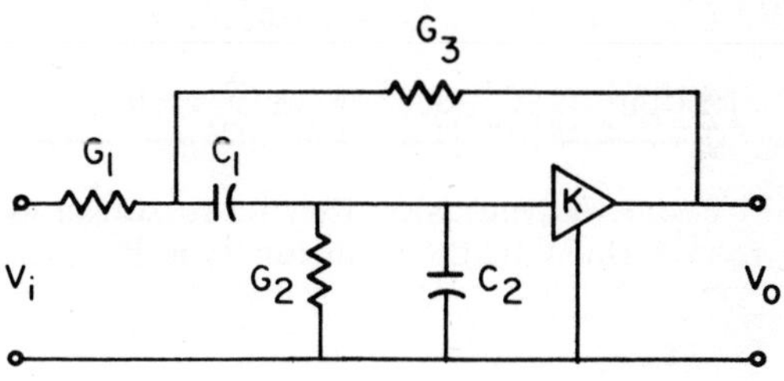

(b) Active Filter

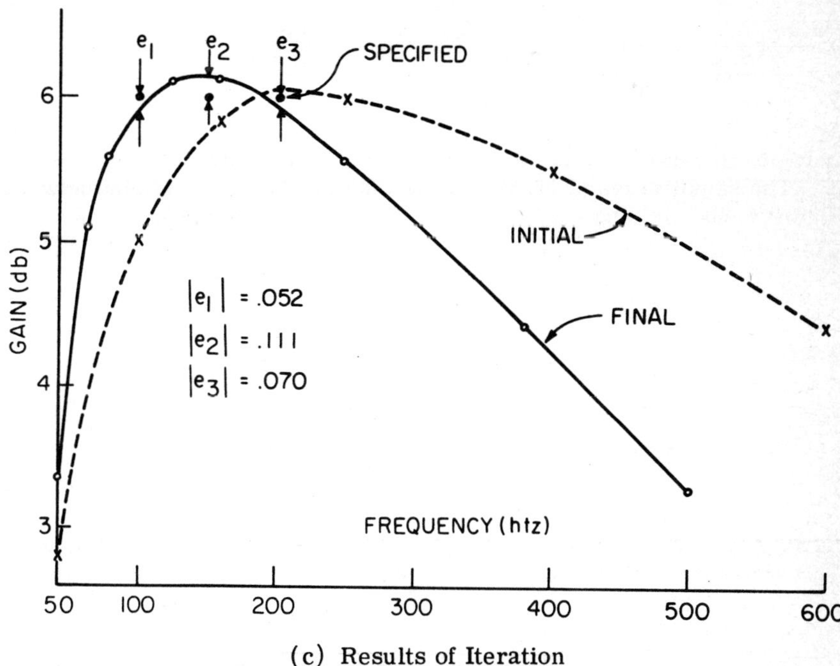

(c) Results of Iteration

FIG. 6.7 Illustrations of Automated Design in the Frequency Domain

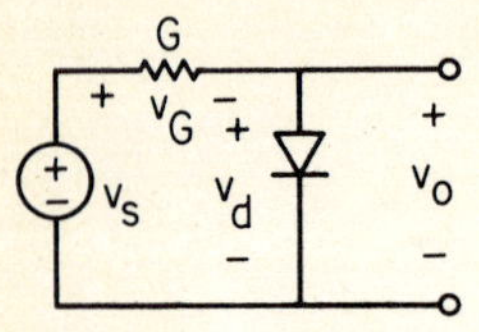

(a) ORIGINAL NETWORK (N)

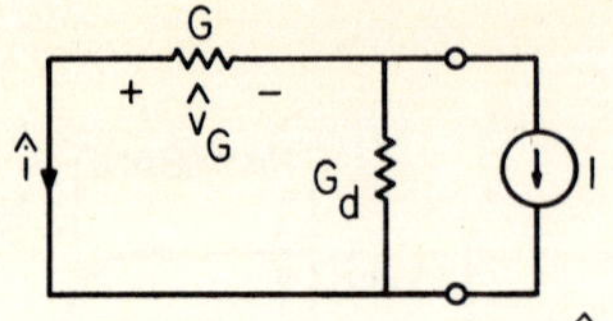

(b) ADJOINT NETWORK ($\hat{N}$)

FIG. 6. 8 DC Design Example

prescribed value of v_o. Both the source v_s and G are available as design parameters.

Since the object is to match v_o to a given voltage, we write two types of equations.

$$\text{circuit equation:}\quad f_1(v_d) = G(v_s - v_d) - I_s(e^{\lambda v_d} - 1) = 0 \tag{6.37}$$

$$\text{matching equation:}\quad f_2(v_s, G) = v_d - \hat{v}_o = 0 \tag{6.38}$$

Equation (6. 37) can be solved using Newton iteration to find v_d. Linearization of (6. 38) then defines another Newton-type iteration as

$$\frac{\partial f_2}{\partial v_s}\Delta v_s + \frac{\partial f_2}{\partial G}\Delta G = -f_2$$

or

$$\frac{\partial v_d}{\partial v_s}\Delta v_s + \frac{\partial v_d}{\partial G}\Delta G = \hat{v}_o - v_d \tag{6.39}$$

Only one iteration step is taken, and (6. 37) is re-solved.

The sensitivities of (6. 39) can be obtained from the adjoint network of Figure 6. 8b, yielding

$$\hat{i}\,\Delta v_s + v_G\,\hat{v}_G \Delta G = \hat{v}_o - v_d \tag{6.40}$$

where

$$\hat{i} = -\frac{G}{G + G_d} \qquad \hat{v}_G = \frac{1}{G + G_d}$$

Here, G_d in $\hat{N}$ is evaluated from the solution of the dc network equation. This process is summarized in Figure 6. 9.

6. 4. 2 A Numerical Solution Algorithm

Just as frequency domain design usually produced more equations than unknown parameters, matching techniques in dc design instead tend to result in more unknowns than equations (see Equation 6. 40). This implies a

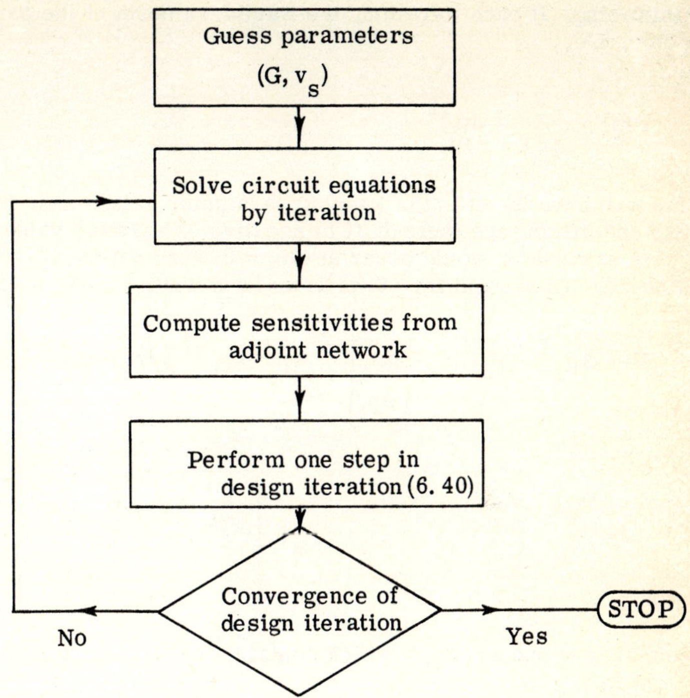

FIG. 6. 9 Flow Chart of DC Design

non-unique solution so that many sets of parameter values are possible. There are a variety of ways of exploiting this freedom. One is to constrain the range of permissable element values using linear programming methods [6. 13]. We will propose another procedure which complements the solution methods of Section 6. 2 and 6. 3.

Let the linearized equation be written as

$$\frac{\partial \underline{f}}{\partial \underline{p}} \Delta \underline{p} = \underset{\sim}{\phi} \Delta \underline{p} = -\underline{f}$$

where the (i, j)th element of $\underset{\sim}{\phi}$ is $\partial f_i / \partial p_j$. We partition $\underset{\sim}{\phi}$ into the form

$$\begin{bmatrix} \underset{\sim}{\phi}_1 & \underset{\sim}{\phi}_2 \end{bmatrix} \begin{bmatrix} \Delta \underline{p}_1 \\ \Delta \underline{p}_2 \end{bmatrix} = -\underline{f} \tag{6. 41}$$

where $\underset{\sim}{\phi}_1$ is square. If $\underset{\sim}{\phi}_1$ can be chosen invertable as well, then we can write

$$\Delta \underline{p}_1 = - (\underset{\sim}{\phi}_1)^{-1} [\, \underline{f} + \underset{\sim}{\phi}_2 \, \Delta \underline{p}_2] \tag{6. 42}$$

Inspecting (6. 16), we observe that these equations can be solved for any choice of $\Delta \underline{p}_2$. We propose to use these r - n degrees of freedom by minimizing, at each iteration, the Euclidean norm of the parameter change vector, i.e.,

$$\tilde{P}(\underline{p}) = \sum_{i=1}^{r} (\Delta p_i)^2 \tag{6.43}$$

This will have the effect of keeping parameter values near their original values, and discourage their drift to negative or extreme values. The minimum of this expression would occur at $\Delta p_i = 0$ were not (6.42) also required to be satisfied. To account for (6.42), we first write (6.43) in the vector form

$$\tilde{P} = [\Delta \underline{p}_1^T \ \Delta \underline{p}_2^T] \begin{bmatrix} \Delta \underline{p}_1 \\ \Delta \underline{p}_2 \end{bmatrix} = [\{-\underset{\sim}{\phi}_1^{-1}(\underline{f}+\underset{\sim}{\phi}_2 \Delta \underline{p}_2)\}^T \quad \Delta \underline{p}_2^T] \times \begin{bmatrix} -\underset{\sim}{\phi}_1^{-1}(\underline{f}+\underset{\sim}{\phi}_2 \Delta \underline{p}_2) \\ \Delta \underline{p}_2 \end{bmatrix} \tag{6.44}$$

We may now choose $\Delta \underline{p}_2$, which contains the free parameters, to minimize $\tilde{P}$ by setting

$$\left(\frac{\partial P}{\partial \Delta \underline{p}_2}\right)^T = 2\{(\underset{\sim}{\phi}_1^{-1} \underset{\sim}{\phi}_2)^T \underset{\sim}{\phi}^{-1} (\underline{f}+\underset{\sim}{\phi}_2 \Delta \underline{p}_2) + \Delta \underline{p}_2\} = 0 \tag{6.45}$$

Solving for $\Delta \underline{p}_2$, we get

$$\{(\underset{\sim}{\phi}_1^{-1} \underset{\sim}{\phi}_2)^T \underset{\sim}{\phi}_1^{-1} \underset{\sim}{\phi}_2 + \underset{\sim}{I}\} \Delta \underline{p}_2 = -(\underset{\sim}{\phi}_1^{-1} \underset{\sim}{\phi}_2) \tag{6.46}$$

This equation, together with (6.42) for obtaining $\Delta \underline{p}_2$, defines the iteration algorithm. When r = n, $\Delta \underline{p}_2$ does not exist and (6.42) again degenerates to a familiar Newton formula.

The significance of minimizing $\tilde{P}$ of (6.43) is indicated by the following simple mathematical problem.

Example 6.7. Consider the plane described by

$$x + y + z = 1 \tag{6.47}$$

We require the point of closest approach of this plane to the origin $x = y = z = 0$. Equivalently, we propose to minimize

$$\tilde{P} = x^2 + y^2 + z^2 \tag{6.48}$$

subject to (6.47)

We first represent (6.47) by

$$[1 \vdots 1\ 1]\begin{bmatrix} x \\ \hdashline y \\ z \end{bmatrix} = 1 \tag{6.49}$$

and identify

$$\underset{\sim}{\phi}_1 = [1] \qquad \underset{\sim}{\phi}_2 = [1\ \ 1] \qquad \underline{f} = -[1]$$

$$\left\{\begin{bmatrix} 1 \\ 1 \end{bmatrix}[1\ \ 1] + \begin{bmatrix} 1 & 0 \\ 0 & 1 \end{bmatrix}\right\}\begin{bmatrix} y \\ z \end{bmatrix} = \begin{bmatrix} 1 \\ 1 \end{bmatrix}[1]$$

or

$$\begin{bmatrix} 2 & 1 \\ 1 & 2 \end{bmatrix}\begin{bmatrix} y \\ z \end{bmatrix} = \begin{bmatrix} 1 \\ 1 \end{bmatrix} \tag{6.50}$$

Solution of (6.50) gives $y = z\ 1/3$; substitution of this result into (6.49) gives $x = 1/3$. This result, where $x = y = z$ at the minimum of (6.48), is quite in keeping with the symmetry of the problem.

6.4.3 A General DC Design Problem (DCOP)

Unline s-plane and frequency domain automated design, the proposed dc iteration is unlikely to be implemented as a special purpose program; the effort to program both the solution and design iterations is of the same order of magnitude for special - and general - purpose applications.

In Appendix Table 6.1, program modifications to DSAP of Chapter V are presented to achieve a dc design capability. Certain characteristics of the program are worthy of note.

(1) Input format. See Appendix C for details. Parameters are tagged with an alphanumeric character in column 20. This includes resistors, both independent current and voltage sources, and both types of controlled current sources. Voltage specifications with identifying node numbers are read as an "O" branch. Specifications can be stepped; this implies that specifications are given at more than one dc operating point (see Example 6.8).

(2) Sensitivity calculation. Subroutine SENN computes sensitivities; these are printed if no specifications are given, so that the program can be used in lieu of DSAP.

(3) Iteration. Subroutine ITRATE uses either least squares ($r \leq n$), or the algorithm of Section 6.4.2 ($r \geq n$). Parameters are prohibited from changing sign during iteration by the method of Section 6.5.1.

<u>Example 6.8.</u> The emitter-coupled logic gate [6.12] of Figure 6.10a should have the property of preserving the dc operating levels of the input signal $v_s(t)$. To achieve this specification (Figure 6.10c) we are prepared to vary R_2, R_3, v_1, and v_2 from the nominal design values shown. The transistor model of Figure 6.10b is to be used.

With the parameter values shown, the output levels are

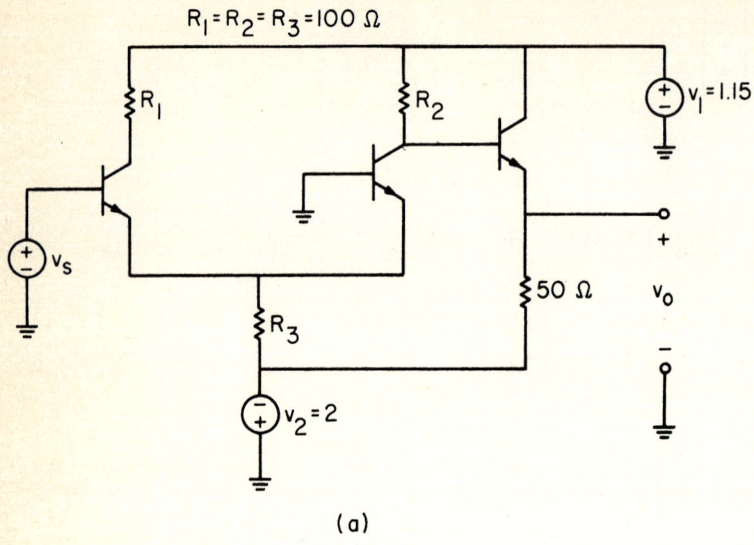

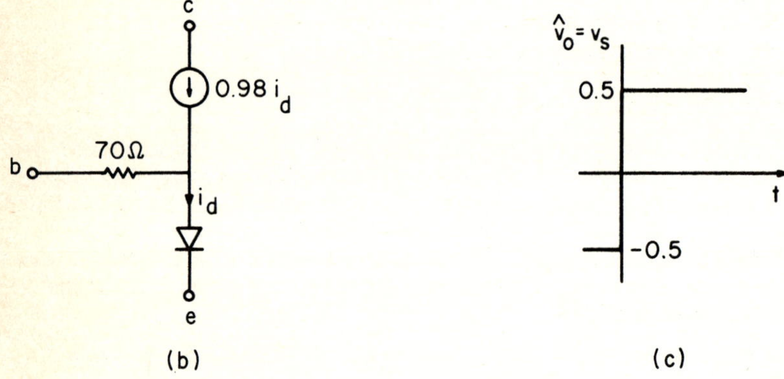

FIG. 6.10 ECL DC Design Example

$$v_o = \begin{cases} -1.02 & v_s = -.5 \\ .395 & v_s = .5 \end{cases}$$

which are far removed from the required values.

Four experiments are run with different combinations of parameters. (Table 6.4). With all the elements as parameters in (a), each parameter changes approximately 10%. This is the result of minimizing $\underset{\sim}{P}$ of (6.43), which attempts to prevent large changes in any one element. However, when the number of parameters is decreased, larger changes in each element value are required. The sensitivities of certain parameters may then change significantly during iteration; for example, in experiment (b) R_3 changes in the opposite direction from the change in (a).

TABLE 6.4. Iteration Results for ECL Gage (v_o realized in each case)

Parameters	Element Values			
	R_2	R_3	v_1	v_2
(a) $R_2, R_3 v_1, v_2$	90.7	109.	1.08	1.75
(b) R_2, R_3, v_2	66.1	23.4	–	.833
(c) v_1, v_2	–	–	1.08	1.52
(d) R_2, R_3	no convergence			

6.5 RELATED TOPICS

6.5.1 Parameter Constraints

One of the risks associated with turning over even a part of the design process to an iteration algorithm is the danger of growth of the design parameters to unreasonable values. We have, for example, no guarantee that parameter values will remain positive with any of the automatic design procedures yet proposed. If these values become negative, we are left to wonder whether (1) a different iteration algorithm would have yielded positive values for parameters or (2) the specifications are unrealizeable with positive-valued parameters, or (3) the inclusion of more parameters (i.e., more degrees of freedom) would avoid negative values. We cannot resolve any of these questions conclusively; however, we can at least increase the chances of finding a realizable solution if one exists.

Our approach is to prohibit values from becoming negative by redefining the iteration parameters. In particular, consider the definition

$$y_i \triangleq \ln p_i \tag{6.51}$$

or

$$p_i = e^{y_i} \tag{6.52}$$

where y_i is updated at each iteration and then p_i determined using (6.52). Clearly p_i cannot become negative for any real value of y_i.

Applying this transformation to any of the Newton-based algorithms already cited, we first note that

$$\underline{f}\,(\underline{p}(y)) = 0$$

implies a Newton iteration formula of the form

$$\left(\frac{\partial \underline{f}}{\partial \underline{p}}\right)\left(\frac{\partial \underline{p}}{\partial \underline{y}}\right) \Delta\underline{y} = -\underline{f} \tag{6.53}$$

where

$$\partial \underline{p}/\partial \underline{y} = \text{diag}\,[\,\partial p_1/\partial y_1 \;\; \partial p_2/\partial y_2 \cdots \partial p_r/\partial y_r]$$

$$= \text{diag}\,[\,p_1 \; p_2 \cdots p_r]$$

If we define $\Delta\tilde{p}_i = p_i\,\Delta y_i$ and $\Delta\tilde{\underline{p}} = [\Delta\tilde{p}_1 \ldots\ldots \Delta\tilde{p}_r]^T$, then (6.53) becomes

$$\left(\frac{\partial \underline{f}}{\partial \underline{p}}\right)' \Delta\tilde{\underline{p}} = -\underline{f}$$

This formula is similar to one we have already developed, except that $\underline{p}$ is updated differently. Knowing $\Delta\tilde{p}_i^m$, we form successively

$$\Delta y_i^m = \Delta\tilde{p}_i^m / p_i^m$$

$$p_i^{m+1} = e^{y_i^m + \Delta y_i^m}$$

or simply

$$p_i^{m+1} = p_i^m \, e^{\Delta y_i^m}$$

By constraining parameter values to remain positive, we can in fact accommodate the more general problem of constraining values to any interval $[p_{min}, p_{max}]$. The most obvious example of this is in the iteration of resistance values, where, by using series and parallel resistances, we can limit a resistance value to any range of positive values.

6.5.2 Preservation of Linear Network Relationships

Nowhere is the value of enlightened problem formulation more important than in automatic design. The form of the dependence of $\underline{f}$ or P on the design parameters can be critical to the efficiency of the iteration process. As we are about to show, certain linear relationships exist between parameters in linear networks. The preservation of this linearity in the iteration process can only enhance the rate and assurance of convergence.

Consider the problem of finding the G_2 in Figure 6.11 such that v_{o2} (i.e., the transfer function v_{o2}/v_s) is 2/33. If we define

$$f(G_3) = v_{o_2} - 2/33 = \frac{G_1 G_3 v_s}{(G_1 + G_2)(G_3 + G_4) + G_3 G_4} - 2/33 \qquad (6.54)$$

then we must solve a nonlinear equation in G_3 with $G_3^0 = 1$. After one Newton iteration, we would have $G_3^1 = 18/11$. However, if we define

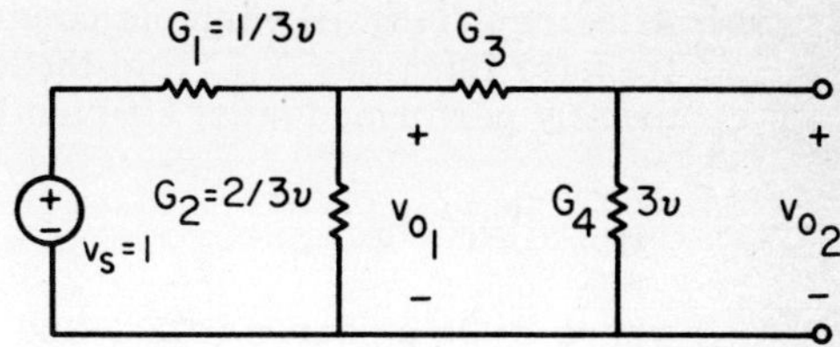

FIG. 6.11 Single Parameter Design Example

$$\tilde{f}(G_3) = G_1G_3 - ((G_1 + G_2)(G_3 + G_4) + G_3 G_4)(2/33) = 0 \tag{6.55}$$

then $\tilde{f}$ is a linear function of G_3 and Newton's method will converge to the correct value $G_3 = 2$ in one iteration!

The above procedure has limited value, however, because it cannot be extended to the case of more than one parameter. For example, if we were to attempt to specify v_{o_1} and v_{o_2} with G_1 and G_4 as parameters, we would find that the cross products involving G_1 and G_4 in (6.55) would again render the problem nonlinear.

Indeed the equations relating linear network resistances and conductances to outputs <u>are</u> nonlinear. Consider the following, however.

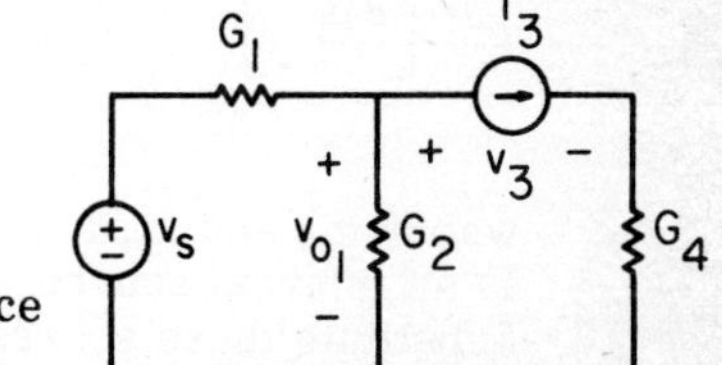

FIG. 6.12 Replacement of Resistance by Source Parameter

Returning to the single parameter case when G_3 is a parameter, we replace G_3 with a current* source i_3 (Figure 6.12). We can then solve by any means available

$$v_{o_1} = T_1(G_1, G_2, G_4)\, i_3 + K_1 \tag{6.56}$$

$$v_3 = T_2(G_1, G_2, G_4)\, i_3 + K_2 \tag{6.57}$$

$$G_3 = v_3/i_3 \tag{6.58}$$

where T_1 and T_2 are transfer functions between the outputs and parameters with external sources (v_s) removed. If we regard i_3, v_3, and G_3 as unknowns, note that only (6.58) represents a nonlinear relationship! Therefore, as suggested by the work of Murray-Lasso [6.13], we can achieve linearity by regarding the element <u>currents and voltages</u> as unknown rather than the element value itself!

The three equations (6.56-6.58) are indicative of three steps to be followed in obtaining a noniterative solution.

---

*A voltage source could also be used.

(1) After replacing all variable components by sources, compute the transfer functions T_{ij} from these sources to all outputs. This is usually performed most efficiently by recognizing from (6.56) that

$$T_{ij} = \partial v_{o_i} / \partial p_j \tag{6.59}$$

where p_j is a source value, and then using the methods of Chapter V. For example, if $p = i_3$ in Figure 6.12, then

$$T_1 = \partial v_o / \partial i_3 = \hat{v}_3$$

where $\hat{v}_3$ is the voltage across $i_3(=0)$ in the adjoint network $\hat{N}$. In any case, with n outputs, only n sets of forward and back substitutions need be performed

(2) Solve for all r unknown sources using one of the Newton-based methods and the sensitivities computed in (6.59). The equations would be similar to (6.56) and of the general form

$$\underline{K} + \left[\frac{\partial v_{o_i}}{\partial p_j}\right]_{n \times r} \underline{z}_s = \underline{z}_o$$

where $\underline{z}_s$ and $\underline{z}_o$ are source and output vectors respectively and $\underline{K}$ is an internal source vector.

(3) Substitute these sources into the original network N and solve for the response across or through the sources. This requires one pair of forward and back substitutions. Then, solve for the branch parameter values necessary to achieve these branch i - v values, similar to (6.58).

6.6 CONCLUSIONS

It is of interest to compare the "automatic" approach to design with the "bench" practice of adjusting a single element at a time and re-determining the network performance. This latter method is known numerically as a relaxation procedure; it typically converges slowly if many elements are involved. This method may be natural to the human – for whom simultaneous parameter adjustment is all but impossible – but it represents a gross inefficiency when a computer is performing the adjustment. As we develop refined parameter adjustment procedures in Chapter XI, we will observe an even wider gulf between human and computer design capabilities for those design problems that can be phrased in proper mathematical format.

Problems

6.1 Show that

$$\left.\frac{d}{d\alpha} P(\underline{p}^m + \alpha \Delta \underline{p}^m)\right|_{\alpha = 0} = -2(e^m)^T (\underline{\phi}^{m\,T} \underline{\phi}^m)^{-1} \underline{e}^m$$

where $\underline{e}^m = \tilde{\phi}^{m\,T} \underline{f}^m$. This derivative being negative assures that an $\alpha > 0$ can be found that decreases P while moving in the direction Δp^m.

6.2 Show that $\tilde{\phi}^{m\,T} \tilde{\phi}^m$ in the above formula is positive definite if $r \leq n$ and $\tilde{\phi}$ is of full rank. Relate this property to the negativity required in Problem 6.1.

6.3 The "least p-th" algorithm [6.9] [6.12] is an extension of the least squares method where

$$P = \sum_{i=1}^{n} f_i^p \qquad p \text{ even integer}$$

Minimization of this criterion function for p large has the effect of minimizing the maximum value of $|f_i|$, i = 1, 2, ... n. Derive the iteration formula for the least p-th algorithm and show, as in Problems 6.1-2, that Δp^m defines a downhill direction for P.

6.4 Generation of Coefficient Derivatives. If the sensitivity of the network determinant $\Delta(s)$ due to a two terminal element is to be found, we first create a port at the element as shown in Figure P6.1. The immittances seen by the sources are then

$$Z = \frac{1}{Y_L(s) + Y_N(s)} \qquad\qquad Y = \frac{1}{Z_L(s) + Z_N(s)} \qquad (P6.1)$$

In either case, the network determinant, since it contains the network natural frequencies, is related to the denominator by a gain constant.

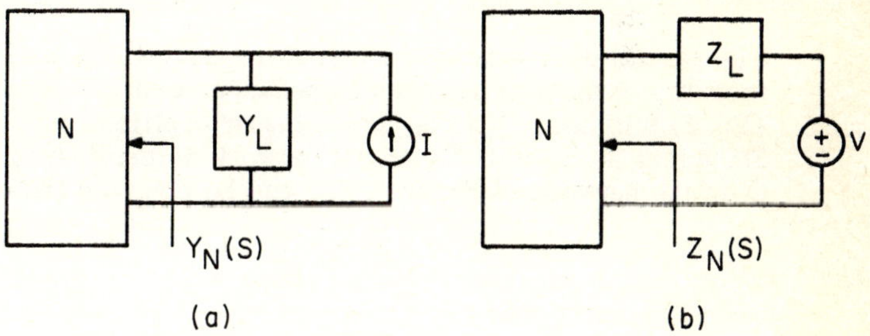

FIG. P6.1 Driving Point Immitances

For example, Equation (P6.1) may be written

$$\frac{N(s)}{\Delta(s)} = \frac{1}{Y_L(s) + \frac{n(s)}{d(s)}} = \frac{d(s)}{Y_L(s)d(s) + n(s)}$$

or

$$\frac{N(s)}{\Delta(s)} = \frac{s^r \sum_{i=0}^{n} d_i s^i}{y_L \sum_{i=0}^{n} d_i s^i + \sum_{b=0}^{n} c_i s^i} \qquad (P6.2)$$

where c_i and/or d_i may be zero, and where $y_L = G, C,$ or $1/L$.

(a) What is r for inductors, capacitors, resistors?
So that the same $\Delta(s)$ is obtained regardless of the location of y_L in the network, normalize $\Delta(s)$ by dividing by its leading coefficient, yielding

$$\Delta(s) = \sum_{i=0}^{n} \frac{y_L d_i + c_i s^i}{(y_L d_n + c_n)}$$

$$N(s) = s^r \sum_{i=0}^{n} \frac{d_i s^i}{(y_L d_n + c_n)}$$

For the purposes of iteration, let $y_L d_n + c_n$ be regarded as a constant during each iteration.*
Then define

$$f_i = \frac{y_L d_i + c_i}{y_L d_n + c_n} - b_o a_i = \underline{0}$$

(b) Show that the partial derivative matrix is given a column at a time, by the coefficients of the polynomial

$$\frac{1}{y_L} \sum_{i=0}^{n} \frac{y_L d_i + c_i}{y_L d_n + c_n} s^i = \sum_{i=0}^{n} \frac{d_i}{y_L b_n + a_n} s^i = s^r N(s)$$

The partial derivatives are therefore trivially obtained from the numerator of the driving point impedance normalized as in Equation (P6. 2) with unity leading denominator coefficient. =

(c) Show that the sensitivity of the network determinant coefficients to C in the network of Figure P6. 2 can be obtained from

$$\frac{V_2}{I_2}(s) = \frac{\frac{s}{C} + \frac{\Gamma}{G_1 C}}{s^2 + \frac{G_1 G_2 + \Gamma C}{G_1 C} s + \frac{G_1 + G_2}{G_1 C}}$$

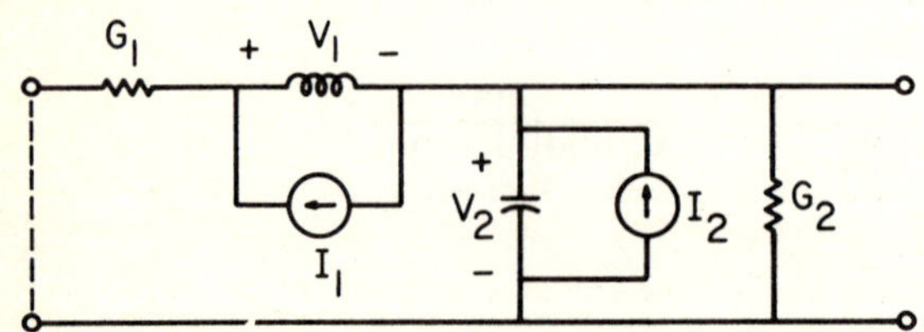

FIG. P6.2 Driving Point Impedance Measurements

*If otherwise, the coefficients to be matched are bilinear rather than linear functions of each element. Range and speed of convergence are then severely affected.

(d) Repeat (c) for the parameter value $1/L$.
(e) Show that the sensitivity matrix analogous to (6.9) is of the form

$$\begin{bmatrix} 0 & \frac{1}{C} & -1. \\ \frac{1}{G_1} & \frac{\Gamma}{G_1 C} & -\sqrt{2} \\ \frac{G_1 + G_2}{G_1 C} & 0 & -1. \end{bmatrix} \begin{bmatrix} \Delta\Gamma \\ \Delta C \\ \Delta b'_o \end{bmatrix} = \begin{bmatrix} -1 + b'_o \\ \frac{-G_1 G_2 + \Gamma C}{G_1 C} + \sqrt{2}\, b'_o \\ \frac{-G_1 + G_2}{G_1 C} + b'_o \end{bmatrix}$$

where $b'_o = b^o/G_1C$.
(f) Show that the number of parameters in the coefficient matching scheme must be greater than or equal to the number of dominant poles described.

6.5 In Problem 6.4, a method was given for the calculation of the derivatives of the coefficients of the polynomial $\Delta(s)$, which would in general form the denominator of any network function. Develop a similar procedure foı calculation of coefficient derivatives of the numerator polynomials assuming a port measurement similar to P6.2.

6.6 A method closely related to coefficient matching is the root-matching procedure. If r_i^o is a desired (complex) root, define

$$f_i = r_i - r_i^o \qquad i = 1, 2, \ldots, m$$

so that $f_i = 0$ is the required solution. If n is the order of the corresponding polynomial, relate the two methods when (a) $m = n$, (b) $m < n$ (the dominant root case).

6.7 It is possible to view the frequency domain matching problem as the minimization of

$$P = \int_{\omega_1}^{\omega_2} \left\{ |T(j\omega)| - |T_s(j\omega)| \right\}^2 d\omega$$

Apply a numerical quadrature formula to the evaluation of P (such as the trapezoidal rule) and compare the results with Equation (6.4).

6.8 The work of Murray-Lasso [6.13] suggests that the specification equation (6.38) can be iterated <u>simultaneously</u> with the circuit equation(s) (6.37). The specification is then viewed as "imbedded" in the network. Develop the corresponding Newton iteration formula for the network of Figure 6.8.

REFERENCES

6.1 Linvill, J. G., "Network Alighment Techniques," Proc. IRE vol. 41, pp. 290-293; February, 1953.

6.1 Aaron, M. R., "The Use of Least Squares in Systems Design," IRE Trans. on Circuit Theory, vol. CT-3, pp. 224-231; December, 1956.

6.3 Desoer, C. A., and S. K. Mitra, "Design of Lossy Ladder Filters by Digital Computer," IRE Trans. on Circuit Theory, vol CT-8, pp. 192-201; September, 1961.

6.4 Calahan, D. A., "Computer Solution of the Network Realization Problem," Proc. Second Allerton Conference, University of Illinois, pp. 175-200; 1964

6.5 Lasdon, L. S., and A. D. Waren, "Optimal Design of Filters with Bounded, Lossy Elements," Trans. IEEE, vol. Ct-13, pp. 175-187; June, 1966.

6.6 Athanassopoulos, I. A., J. D. Schoeffler, and A. D. Waren, "Time-Domain Synthesis by Nonlinear Programming," Fourth Allerton Conference, pp. 766-775; 1966.

6.7 Temes, G. C., and D. A. Calahan, "Computer Aided Network Optimization—The State of the Art," Proc. IEEE; November 1967.

6.8 Gaash, A. A., R. S. Pepper, and D. O. Pederson, "Design of Integrable Desensitized Frequency Selective Amplifiers," Jour. of Solid State Cir., no. 1, pp. 29-34; December, 1966.

6.9 Temes, G. C., and D. Y. F. Zai, "Least p th Approximation," Trans. IEEE, vol. Ct-16, pp. 235-237; May, 1969.

6.10 Director, S. W., "Survey of Circuit-Oriented Optimization Techniques," Trans. IEEE, vol. CT-18, no. 1, pp. 3-9; January, 1971.

6.11 Broderson, A. J., S. W. Director, and W. A. Bristol, "Simultaneous Automated AC and DC Design of Linear Integrated Amplifiers," Trans. IEEE, vol. CT-18, no. 1, pp. 50-57; January, 1971.

6.12 Bandler, J. W., and C. Charalambove, "Generalized least p th Objectives for Networks and Systems," Proc. Fourteenth Midwest Symposium on Circuit Theory, pp. 10.5-1, 10.5-7; University of Denver; May, 1971.

6.13 Murray-Lasso, M. A., and F. B. Kozemchak, "Foundations of Computer-Aided Design by Singular Imbedding," Trans. IEEE, vol. CT-17, no. 1, pp. 105-111; February, 1970.

6.14 Bandler, J. W., "Optimization Methods for Computer Aided Design," Trans. IEEE, vol. MTT-17, pp. 533-552; August, 1969.

7

Tolerance Analysis

7.1 INTRODUCTION

The final step in most circuit design is a tolerance analysis, i.e., a study of the effect of parameter variation on circuit behavior. There are two distinctive applications where such information is vital:

(1) in the design of complicated systems where the electronic circuits must perform flawlessly;

(2) in the economical mass production of circuits, where 100% is not required, but an estimate of yield based on individual component behavior assists in production decisions.

From a computational viewpoint, the principal distinction to be made is whether the element changes may be considered incremental - so that sensitivity information as computed by methods of Chapter V adequately predicts circuit behavior when components vary - or whether the circuit response must be recalculated to simulate the variation. If large variations are required, then the difficulty in re-computation depends greatly on the type of network being analyzed. If the network is linear resistive or a frequency domain model, then a simple linear relation may be written between element values and output quantities (Section 6.5.2); if an element value changes this relation is merely updated and the output re-evaluated. This process can be made extremely efficient, with the result that 50-100 analyses per second are entirely feasible for moderate-sized filters and 2-3 transistor linearized biasing circuits (see Chapter X). Nonlinear dc and especially dynamic networks require significantly more effort for re-solution. For this reason, preliminary studies of these networks are usually based on derivative information; however, it is customary to preform a large-variation tolerance analysis before final acceptance.

Since most of the computational questions concerning tolerance analysis are answerable in terms of the material of Chapters V or X, we will give limited attention to algorithms and focus on formulation of tolerance problems.*

7.2 STATEMENT OF THE PROBLEM

Let a set of network output variables z_i (currents or voltages) be considered a function of parameters p_j that have nominal values p_j^o:

$$\underline{z} = \underline{z}(\underline{p})$$

where

$$\underline{p} = [p_1 \ldots p_r]$$
$$\underline{z} = [z_1 \ldots z_n]$$

Further, let the tolerance on p_j and z_i be defined by

$$\underline{p}_j \leq p_j \leq \bar{p}_j \qquad j = 1, 2, \ldots r \tag{7.1}$$

$$\underline{z}_i \leq z_i \leq \bar{z}_i \qquad i = 1, 2, \ldots n \tag{7.2}$$

Then we can define the following tolerance problems.

(1) Analysis Problem. Given a set of parameter tolerances $(\underline{p}_j, \bar{p}_j)$ for all j, find the set of output extremals $(\underline{z}_i, \bar{z}_i)$ for all i that correspond to the largest changes in z_i for $\underline{p}_j \leq p_j \leq \bar{p}_j$.

(2) Design Problem. Given a set of admissable output tolerances $(\underline{z}_i, \bar{z}_i)$ for all i, find the class of parameter tolerances that guarantee (7.2) to be satisfied.

Contrary to appearances, these problems are not inverses of one another.

7.3 TOLERANCE ANALYSIS BY THE SENSITIVITY APPROACH

7.3.1 Worst-Case Analysis

Let us define, in addition to $\bar{p}_j$ and $\underline{p}_j$ of (7.1), parameter values p_j^o and the corresponding outputs z_i^o, the nominal values of p_j and z_i around which variations are to be considered. Then we can write for the changes in z_i due to changes in p_j

*The serious reader will find in BSTJ, April, 1971 excellent discussions of this topic.

$$\Delta z_i \approx \sum_{j=1}^{r} \Delta p_j \left. \frac{\partial z_i}{\partial p_j} \right|_{p_j = p_j^o} \tag{7.3}$$

where $\Delta p_j = p_j - p_j^o$. Now consider the problem of finding the largest change in z_i when $\underline{p}_j \leq p_j \leq \bar{p}_j$. We can write

$$\Delta z_i^{max} \approx \sum_{j=1}^{r} (p_j - p_j^o) \left(\frac{\partial z_i}{\partial p_j} \right)_{p_j = p_j^o} \tag{7.4}$$

where the value of p_j is chosen either $\bar{p}_j$ or $\underline{p}_j$ depending on the sign of the derivative so that the contribution of each term of the summation is as large a positive number as possible. A similar formula can be used to determine Δz_i^{min}.

In practice, (7.4) is not used to determine Δz_i^{max}; once the choice of $\underline{p}_j$ or $\bar{p}_j$ is made, it is common simply to re-analyze the network with these extremal parameter values.

Example 7.1. The extremal behavior of the dc amplifier model of Figure 7.1 is to be studied, given that

all resistors can vary ± 2.5%

$$20 \leq \beta_i \leq 200$$

$$9.5 \leq v_c \leq 10.5 \text{ volts}$$

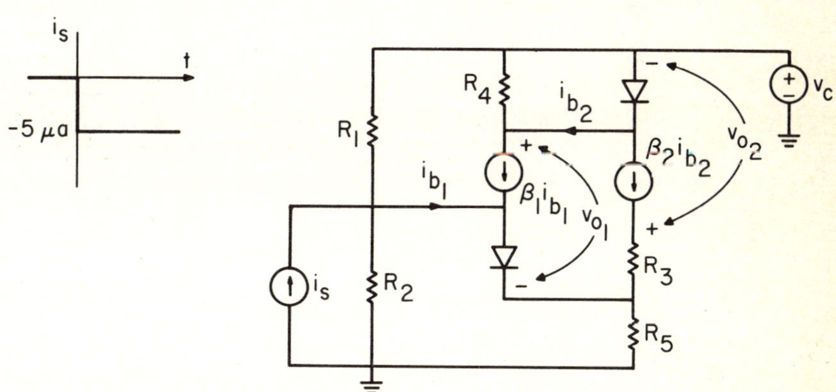

FIG. 7.1 Tolerance Analysis Example

In particular, it is found that with $i_s = 0$, the collector-emitter voltage (v_{o_2}) of the output transistor is -2.12 volts for the nominal parameter values of Table 7.1. This voltage should not become greater than -.5 volts to avoid transistor saturation.

The sensitivities of $\partial v_{o_2}/\partial p_j$ are given in Table 7.1. From their signs, we determine the appropriate tolerance limit to maximize v_{o_2} using (7.4). The approximate sensitivity expression then becomes

$$
\begin{aligned}
\Delta v_{o_2}^{max} &\approx (-2500)(-5.23\times 10^{-5}) + (2500)(4.37\times 10^{-5}) \\
&+ (25)(3.76\times 10^{-3}) + (25)(1.12\times 10^{-3}) \\
&+ (-25)(-3.89\times 10^{-3}) + (150)(1.76\times 10^{-2}) \\
&+ (150)(4.49\times 10^{-3}) + (-.5)(-3.95\times 10^{-2}) \\
&\approx 3.79
\end{aligned}
$$

$$
\bar{v}_{o_2} \approx v_{o_2}^{0} + \Delta v_{o_2}^{max} = 1.27 \text{ volts}
$$

This estimate turns out to be seriously in error; with the extremal parameter values indicated in Table 7.1, the calculated $\bar{v}_{o_2}$ is -.779 volts.

TABLE 7.1 Sensitivity Analysis

	Parameter Value		Sensitivity $\partial v_{o_2}/\partial p_j$ at	
	Nominal	Extremal	Nominal	Extremal
R_1	100 K	97.5 K	-5.23×10^{-5}	-5.05×10^{-5}
R_2	100 K	102.5 K	4.37×10^{-5}	4.61×10^{-5}
R_3	1000	1025	3.76×10^{-3}	4.16×10^{-3}
R_4	1000	1025	1.12×10^{-3}	5.89×10^{-4}
R_5	1000	975	-3.89×10^{-3}	-4.71×10^{-3}
β_1	50	200	1.76×10^{-2}	1.01×10^{-3}
β_2	50	200	4.49×10^{-3}	1.73×10^{-4}
v_c	10	9.5(10.5)	-3.95×10^{-2}	4.19×10^{-2}
v_{o_1}	5.40	4.67(5.15)		
v_{o_2}	-2.02	-.779(-.737)		

7.3.2 Refinement of the Approximation

There is of course no absolute assurance that the sensitivity signs point to correct parameter extrema.* For example, the output v_o of the bridge circuit of Figure 7.2 has zero sensitivity to the resistance R_x, which has zero voltage across it at bridge balance. However, by choosing a nominal value of R_1 slightly above or below its value in (7.5), $\partial v_o/\partial R_x$ can be made either positive or negative. One of these sensitivities must point in the wrong direction for use in (7.4), if tolerance limits are assigned

*Indeed, the true "worst-case" behavior may not occur at the tolerance limitations.

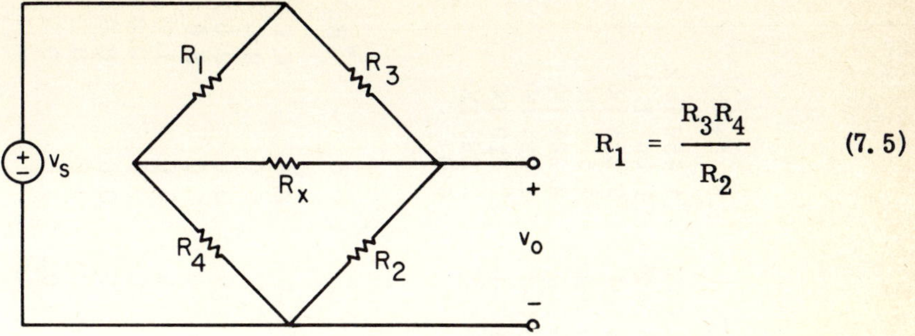

$$R_1 = \frac{R_3 R_4}{R_2} \qquad (7.5)$$

FIG. 7.2 Bridge Circuit Counterexample

to both R_1 and R_x. In other words, the sensitivity is a calculation of local behavior due to one parameter and cannot predict large-change effects or the coupling effects of the variation of many elements.

One common procedure that significantly improves the confidence of a choice of correct element extrema is simply to recalculate the parameter sensitivities at the extremal values. If these have the same signs calculated with nominal parameter values, then the extrema are accepted. If any sensitivity sign has changed, the alternative extrema is assigned. For example, in Table 7.1, the sign of $\partial v_{o_2}/\partial v_c$ changes when the extremal value $v_c = 9.5$ volts is used; setting $v_c = 10.5$ volts then gives a larger value of v_{o_2} (-.737 volts) as shown.

7.3.3 Determination of Absolute Worst Case

For linear resistive networks, the tolerance analysis problem can be phrased in a form that can be solved using linear programming techniques [7.13]. This procedure guarantees a solution which truly minimizes (maximizes) any voltage or current output, accounting for both large-parameter changes and coupling between variations that caused the tolerance analysis of the bridge circuit of Figure 7.2 to fail.

A number of formulations of this solution process are possible. All exploit the property that currents and voltages in a linear network are linearly related, as pointed out on Section 6.5.2. We will adopt a procedure that can be implemented with any node-analysis-based program such as RCAP of Chapter II.

Let each element be assigned that tolerance limit which corresponds to minimal current flow through the element. For example, resistances are given their upper limit $\bar{R}_i$ and controlled current sources (referenced so that nonnegative current flows) are given their minimum value.

If independent current sources are now placed in parallel with each such element, we can write linear equations that relate currents through the sources and voltages across the sources. Table 7.2 shows a set of such relationships for the network of Figure 7.3. If we not view these sources as representing parallel components, we can require that the current through the sources should not exceed that corresponding to the tolerance limitations on these components. For example, if we define in Figure 7.3

$$i_{R_1} = \tilde{R}_1 v_{R_1}$$

TABLE 7.2 Linear Programming Model of Tolerance Analysis Problem

Minimize $(-v_{o_2})$

subject to

v_c	i_{R_1}	i_{R_2}	i_{R_3}	i_{R_5}	i_{c_1}
-1.035 D-3	-1.061 D-1	1.061 D-1	0.0	-2.123 D-3	4.458 D-2
2.071 D-2	2.123 D0	-2.123 D0	0.0	4.246 D-2	1.082 D-1
-5.530 D-1	4.580 D1	-4.580 D1	0.0	-1.088 D-1	2.285 D0
-1.284 D-1	8.933 D1	-8.933 D1	-1.024 D0	7.616 D-1	4.504 D0
-5.530 D-1	4.580 D1	-4.580 D1	0.0	-1.088 D-1	2.285 D0
4.469 D-1	4.580 D1	-4.580 D1	0.0	-1.088 D-1	2.285 D0
4.246 D-1	4.352 D1	-4.352 D1	-1.024 D0	8.704 D-1	2.219 D0
4.469 D-1	4.580 D1	-4.580 D1	0.0	-1.088 D-1	2.285 D0
0.0	0.0	0.0	0.0	0.0	0.0

i_{c_2}	i_s	i_{R_4}	i_{b_1}	i_{b_2}	v_{o_1}	$-v_{o_2}$	v_{R_2}
2.123 D-3	1.061 D-1	-4.246 D-2	1.0	0.0	0.0	0.0	0.0
-4.246 D-2	-2.123 D0	-1.507 D-1	0.0	-1.0	0.0	0.0	0.0
1.088 D-1	-4.580 D1	-2.176 D0	0.0	0.0	1.0	0.0	0.0
2.633 D-1	-8.933 D1	-5.266 D0	0.0	0.0	0.0	1.0	0.0
1.088 D-1	-4.580 D1	-2.176 D0	0.0	0.0	0.0	0.0	1.0
1.088 D-1	-4.580 D1	-2.176 D0	0.0	0.0	0.0	0.0	0.0
1.545 D-1	-4.352 D1	-3.090 D0	0.0	0.0	0.0	0.0	0.0
1.088 D-1	-4.580 D1	-2.176 D0	0.0	0.0	0.0	0.0	0.0
0.0	0.0	0.0	0.0	0.0	0.0	0.0	0.0

TABLE 7.2 (Continued)

$$
\begin{array}{cccc}
v_{R_2} & v_{R_3} & v_{R_5} & v_{R_8} \\
\end{array}
$$

$$
\begin{bmatrix}
0.0 & 0.0 & 0.0 & 0.0 \\
0.0 & 0.0 & 0.0 & 0.0 \\
0.0 & 0.0 & 0.0 & 0.0 \\
0.0 & 0.0 & 0.0 & 0.0 \\
0.0 & 0.0 & 0.0 & 0.0 \\
-1.0 & 0.0 & 0.0 & 0.0 \\
0.0 & -1.0 & 0.0 & 0.0 \\
0.0 & 0.0 & -1.0 & 0.0 \\
0.0 & 0.0 & 0.0 & -1.0
\end{bmatrix}
\begin{bmatrix} \underline{X} \end{bmatrix}
=
\begin{bmatrix}
1.502\ \mathrm{D}{-2} \\
6.924\ \mathrm{D}{-2} \\
7.170\ \mathrm{D}{-1} \\
2.515\ \mathrm{D}0 \\
7.700\ \mathrm{D}{-1} \\
7.700\ \mathrm{D}{-1} \\
1.419\ \mathrm{D}0 \\
1.096\ \mathrm{D}0 \\
3.789\ \mathrm{D}{-1}
\end{bmatrix}
$$

$$-v_c \le -9.5 \qquad \Delta R_2 i_{R_2} \le v_{R_2} \qquad i_{c_1} \le \Delta\beta_1 i_{b_1}$$

$$v_c \le 10.5 \qquad \Delta R_3 i_{R_3} \le v_{R_3} \qquad i_{c_2} \le \Delta\beta_2 i_{b_2}$$

$$\Delta R_1 i_{R_1} \le v_{R_1} \qquad \Delta R_4 i_{R_4} \le v_{R_4} \qquad i_s \le .05$$

$$\Delta R_5 i_{R_5} \le v_{R_5}$$

then the parallel resistances R_1 and $\tilde{R}_1$ will not exceed the lower tolerance $\underline{R}_1$ so long as

$$\frac{R_1 \tilde{R}_1}{R_1 + \tilde{R}_1} \geq \underline{R}_1$$

It is readily shown that this is equivalent to requiring

$$\Delta R_1 i_{R_1} \leq v_{R_1}$$

where $\Delta R_1 = \bar{R}_1 \underline{R}_1 / (\bar{R}_1 - \underline{R}_1)$. Similarly, the parallel current source i_{c_1} is limited by

$$i_{c_1} \leq \Delta \beta_1 i_{b_1}$$

where $\Delta\beta_1 = \bar{\beta}_1 - \underline{\beta}_1$. The complete set of such inequalities is shown in Table 7.2.

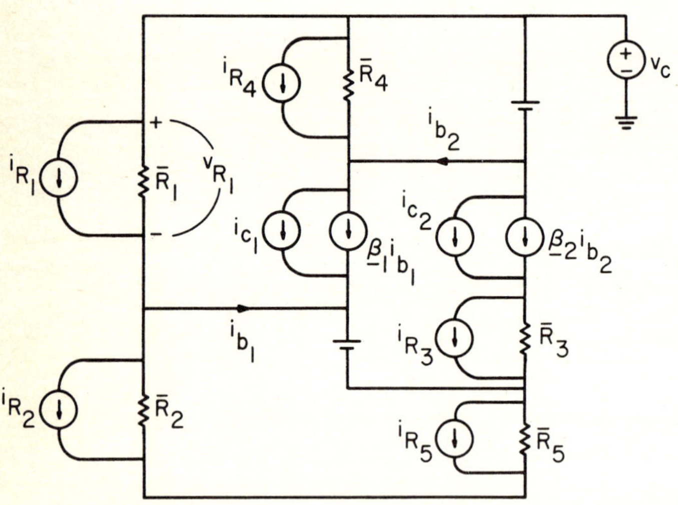

FIG. 7.3 Augmented Network

The problem now is to maximize v_{o_2} (or minimize $-v_{o_2}$) subject to the constraints of Table 7.2. Such a mathematical model can be solved by linear programming, which yields the absolute (usually unique) maximum of v_{o_2}. In this case, the result is within .1% of that obtained in Section 7.3.2. The discrepancy is probably caused by the linearization of the network model in Figure 7.1. While there is assurance of obtaining true worst case behavior, this method is seemingly worth the programming and computational effort only when the network is almost linear, such as for an amplifier model. Also, the method is not directly extendable to frequency domain problems if magnitude specifications are involved.

7.4 THE TOLERANCE DESIGN PROBLEM

The essential difference between the tolerance analysis and design problems is that the latter usually has no unique solution. Typically, we can relax tolerances on one parameter at a cost of closer tolerances on other parameters. This precludes a direct solution procedure such as linear programming affords for the analysis problem. The best we can hope to achieve is to make the designer aware of his options. Since we cannot display all possible choices in the space of r parameters, we limit our discussion to two parameters to gain some insight into a reasonable design procedure [7.10].

Consider initially a linear resistive network with outputs z_1 and z_2 and parameters p_1 and p_2 corresponding to element values. It can be shown that z_i can be written as a multilinear function of p_1 and p_2 of the form, viz,

$$z_1 = \frac{a_0 + a_1p_1 + a_2p_2 + a_3p_1p_2}{1 + b_1p_1 + b_2p_2 + b_3p_1p_2} \qquad (7.6)$$

where p_i appears at most to the first power.

The admissable regions for parameters can now be conveniently displayed as follows. Letting z_1 be constant, we can write

$$(z_1 - a_0) + (b_1z_1 - a_1)\,p_1 + (b_2z_1 - a_2)\,p_2 + (b_3z_1 - a_3)p_1p_2 = 0 \qquad (7.7)$$

This equation in general has the form of a hyperbola in the $p_1 - p_2$ plane with asymptotes readily calculated from (7.7). Given $\bar{z}_1$ and $\underline{z}_1$, two hyperbolas can then be constructed that must contain all allowable parameter sets (p_1, p_2). Such a construction is shown in Figure 7.4a. For z_2, a similar region can be determined. The resultant region of parameter admissability must then be carefully studied to select parameter tolerances. For example, in Figure 7.4b selection of $\bar{p}_2$ and $\underline{p}_2$ determines $\bar{p}_1$ and $\underline{p}_1$ as shown. Note that the admissable region does not include the entire region bounded by the four hyperbolas, but is instead a parallelopiped.

Design from parameter plane intersections can accommodate any number of outputs z_i. However, higher dimensional parameter studies must be handled by combinatorial analysis [7.10].

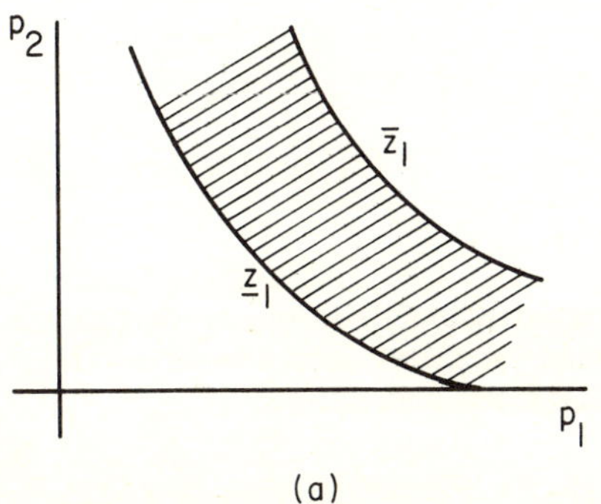

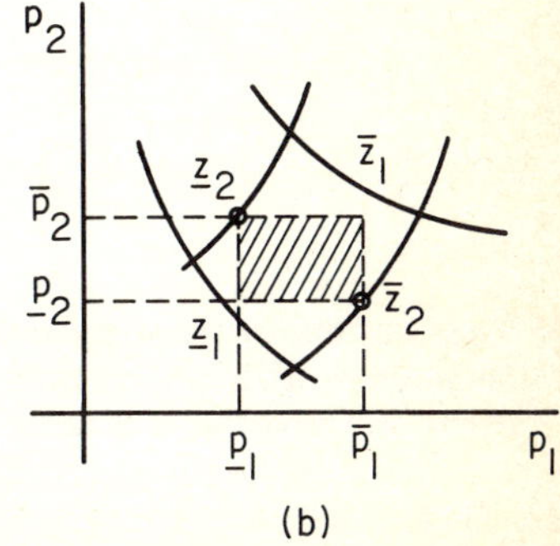

FIG. 7.4 Admissable Regions in Parameter Space

Where frequency domain magnitude specifications are involved, (7.6) would have the form

$$|z_i| = \frac{|a_0 + a_1 p_1 + a_2 p_2 + a_3 p_1 p_2|}{|1 + b_1 p_1 + b_2 p_2 + b_3 p_1 p_2|}$$

where the a_i and b_i are complex constants. The equation equivalent to (7.7) is now quite complicated and the tolerance contours similar to those of Figure 7.4 have no useful properties. Indeed the nature of the cancellations involved in the behavior of most frequency domain circuit models suggests that the contours are likely to be quite irregular and even disjoint. Similar comments apply for the tolerance analysis of nonlinear dc networks.

Despite the limited applicability of (7.6), the pairwise display of parameter contours as in Figure 7.4 is presently the best method of communicating large-change sensitivity data to the designer who must decide parameter tolerance tradeoffs.

7.5 TOLERANCE ANALYSIS FROM STATISTICAL DATA

The simplified assumptions on element variations and their first order effects on network behavior make worst-case design extremely attractive to a designer with a minimum of information on actual element changes. However, when probability distributions associated with element variation are available, more representative information can be obtained on response variations.

7.5.1 Review of Statistical Analysis

Let $[x_1, x_2]$ be an interval in the domain of the variable x. Let us further define a function h(x) over the interval such that the probability pr $(0 \leq pr \leq 1)$ of occurrence of the value x' $(x \leq x' \leq x + dx)$ is $h(x)dx$. Then h(x) is called a distribution function and x a random variable. The distribution function has the following properties:

1. $h(x) \geq 0$

2. $\int_a^b h(x)dx = pr\{a \leq x \leq b\} \qquad x_1 \leq a \leq b \leq x_2$

3. $\int_{x_2}^{x_1} h(x)dx = 1$

That is, the probability that x lies between a and b is the area under h between a and b, and this total area must be unity. For example, the uniform distribution of Figure 7.5a must have the properties

1. $c > 0$

2. $pr\{a \leq x \leq b\} = c(b - a) \qquad x_1 \leq a \leq b \leq x_2$

3. $c(x_2 - x_1) = 1$

For the case of a random variable assuming only a discrete set of values, a similar set of conditions can be given.

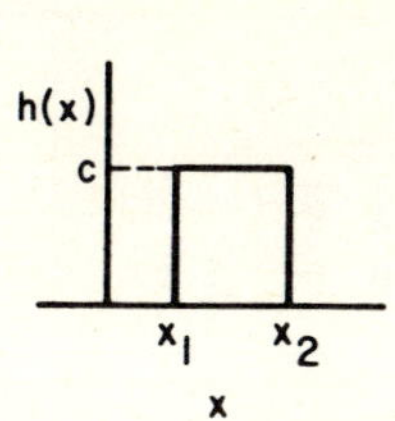

(a) UNIFORM DISTRIBUTION

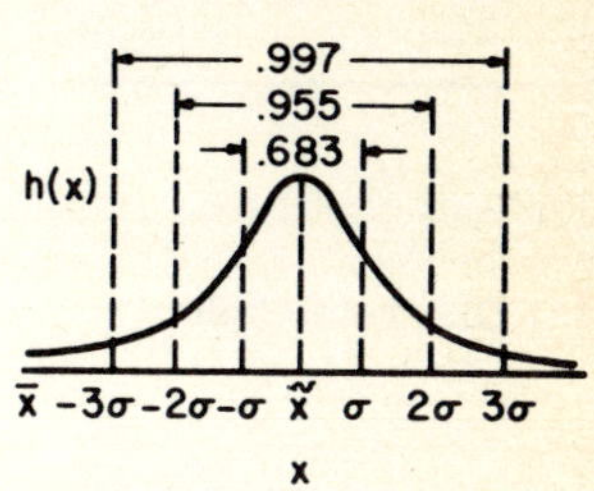

(b) GAUSSIAN DISTRIBUTION AND PROBABILITIES

FIG. 7.5 Common Distributions

A frequently encountered distribution is the Gaussian (normal) distribution given by

$$h(x) = \frac{1}{\sigma\sqrt{2\pi}} e^{-(x-\tilde{x})^2/2\sigma^2}$$

which has the form of Figure 7.5b. As above, the area within any interval is the probability that x has a value in that interval; these probabilities are indicated in the figure.

Closely related to the frequency distribution is the histogram. Consider, for example, the results of drawing resistors from a bin, grouping the values, and displaying the number of resistors in each group as in Figure 7.6a. The histogram differs from a plot of a distribution function in that the area need not be unity. In the case of equal intervals it is convenient to make the heights of each group equal to the number of samples in that group (Figure 7.6b).

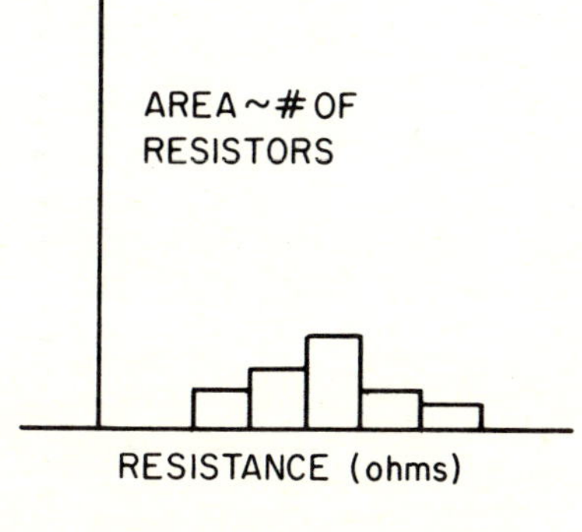

(a) HISTOGRAM

NO. OF RESISTORS
23
17
5
4

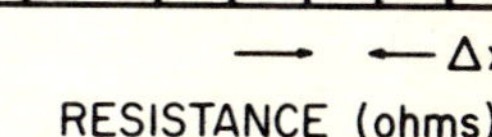

(b) EQUAL INTERVAL HISTOGRAM

FIG. 7.6 Histograms

TABLE 7. 3 Mean and Variance Summarized

Definition:	Mean ($\tilde{x}$)	Standard Deviation (σ) Variance (σ^2)
Discrete Case	$\tilde{x} = \frac{1}{n} \sum_{i=1}^{n} x_i h_i$	$\sigma^2 = \frac{1}{n} \sum_{i=1}^{n} (x_i - \tilde{x})^2 h_i$ $= \frac{\sum_{i=1}^{n} x_i^2 h_i}{n} - x^2$
Continuous Case	$\tilde{x} = \int_{x_1}^{x_2} xh(x)dx$	$\sigma^2 = \int_{x_1}^{x_2} (x-\tilde{x})^2 h(x)dx$
Unclassified Data	Same as discrete, with $h_i = 1$	
Uniform Distribution ($x_2 = 1$, $x_1 = 0$)	$\tilde{x} = \frac{1}{2}$	$\sigma^2 = \frac{1}{12}$
Gaussian Distribution	See Figure 7. 5	

The shape of a distribution function is characterized in part by its mean (expectation) and variance, as defined in Table 7. 3. The former is a measure of the location of the center of $h(x)$; the latter is an indication of the deviation of $h(x)$ from this center. Sample calculations of mean and variance are given in Table 7. 4.

When two or more random variables are defined, not only their own variances are important, but coupling between variables is also important. The correlation coefficient is defined, very similarly to the coupling coefficient in circuit theory, as relating variables in pairs. The result is the correlation matrix $\underset{\sim}{R}$ (Table 7. 5). An example calculation is given in Table 7. 4.

TABLE 7.4 Calculation of Statistical and Sensitivity Data

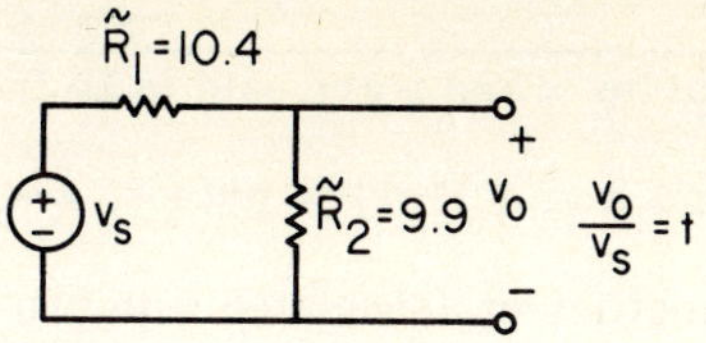

$\tilde{R}_1 = 10.4$, $\tilde{R}_2 = 9.9$, v_s, v_o, $\frac{v_o}{v_s} = t$

R_1	R_2	$R_1-\tilde{R}_1$	$R_2-\tilde{R}_2$	$(R_1-\tilde{R}_1)^2$	$(R_2-\tilde{R}_2)^2$	$(R_1-\tilde{R}_1)(R_2-\tilde{R}_2)$
10.	10.	-0.4	0.1	0.16	0.01	-0.04
10.5	9.	0.1	-0.9	0.01	0.81	-0.09
11.	9.	0.6	-0.9	0.36	0.81	-0.54
10.	11.	-0.4	1.1	0.16	1.21	-0.44
9.5	10.	-0.9	0.1	0.81	0.01	-0.09
9.5	11.	-0.9	1.1	0.81	1.21	-0.99
11.	10.	0.6	0.1	0.36	0.01	0.06
11.	9.5	0.6	-0.4	0.36	0.16	-0.24
10.	10.5	-0.4	0.6	0.16	0.36	-0.24
11.5	9.	1.1	-0.9	1.21	0.81	-0.99
104.0	99.0			4.40	5.40	-3.60

Statistical Data:

$$\tilde{R}_1 = 10.4 \qquad \tilde{R}_2 = 9.9 \qquad \sigma_{R_2} = \frac{1}{10}(5.4) = 0.734$$

$$\sigma_{R_1} = \frac{1}{10}(4.4) = 0.663 \qquad r = \frac{-3.60}{10(0.663)(0.734)} = -0.738$$

Circuit Data:

$$\frac{\partial t}{\partial R_1} = \frac{-\tilde{R}_2}{(\tilde{R}_1 + \tilde{R}_2)^2} = -0.024 \qquad \frac{\partial t}{\partial R_2} = \frac{\tilde{R}_1}{(\tilde{R}_1 + \tilde{R}_2)^2} = 0.0253$$

TABLE 7.5 Independence and Correlation Summarized

1. Two random variables x and y are said to be independent if and only if

$$\mathrm{pr}\{a \le x \le b \text{ and } c \le y \le d\} = \mathrm{pr}\{a \le x \le b\}\,\mathrm{pr}\{c \le y \le d\}.$$

2. The correlation coefficient associated with two random variables x' and x'' is defined (for unclassified data)

$$r = \frac{\sum (x_i' - \tilde{x}')(x_i'' - \tilde{x}'')}{n\sigma'\sigma''}$$

where $\tilde{x}'$, $\tilde{x}''$, σ', and σ'' are the means and standard deviations associated with x' and x'' respectively.

3. Independent variables are uncorrelated, but uncorrelated variables need not be independent.

4. If the variance of h(x) is σ^2, the variance of h(ax) is $a^2\sigma^2$. This shows the effect of scaling.

5. If random variables $x_1, x_2 \ldots x_n$ with variances $\sigma_1^2, \sigma_2^2, \cdots \sigma_n^2$ are described by the correlation matrix.

$$\underset{\sim}{R} = \begin{bmatrix} 1 & r_{12} & \cdot\ \cdot & r_{1n} \\ r_{21} & 1 & \cdot\ \cdot & r_{2n} \\ \cdot & & & \cdot \\ \cdot & & & \cdot \\ r_{n1} & r_{n2} & \cdot\ \cdot & 1 \end{bmatrix}$$

then the variance σ_x^2 of the random variable

$$x = \sum_{i=1}^{n} x_i$$

is given by

$$\sigma_x^2 = \underline{\sigma}\,\underset{\sim}{R}\,\underline{\sigma}^T$$

where $\underline{\sigma} = [\sigma_1 \ldots \sigma_n]^T$. If $\underset{\sim}{R} = \underset{\sim}{I}$ so that the variables are uncorrelated, this equation becomes

$$\sigma_x^2 = \sum_{i=1}^{n} \sigma_i^2$$

7.5.2 The Method of Moments

Returning to the equation

$$\Delta z_i \approx \sum_{j=1}^{r} \Delta p_j \frac{\partial z_i}{\partial p_j} \tag{7.8}$$

suppose we are given the variances σ_j^2 of the distributions associated with Δp_j and are interested in the variance σ_j^2 associated with the variable z_i. If the partial derivatives are real numbers, they may be considered as scaling factors for the distributions of Δp_j. From Table 7.5, we may define the scaling factor $a_j = \partial z_i/\partial p_j$, so that $\Delta p'_j = a_j p_j$ and the standard deviation σ_j of $\Delta p'_j$ is

$$\sigma_j' = a_j \sigma_j$$

But now from (7.8), Δz_i is the sum of random variables $\Delta p'_j$. From Table 7.5, therefore,

$$\sigma_z^2 = \begin{bmatrix} \frac{\partial z_i}{\partial p_1} \sigma_1 \cdots \frac{\partial z_i}{\partial p_r} \sigma_r \end{bmatrix} \underset{\sim}{R} \begin{bmatrix} \frac{\partial z_i}{\partial p_1} \sigma_1 \\ \vdots \\ \frac{\partial z_i}{\partial p_r} \sigma_r \end{bmatrix}$$

Thus, the variance of the response may be estimated from the variance of element variations if the sensitivities are given. This is called the method of moments [7.4].

A sample calculation of the statistical variances and network sensitivities is carried out in Table 7.4 for a voltage divider. The variance in the voltage transfer function is given by

$$\sigma_t^2 = \begin{bmatrix} \sigma_{R_1} \frac{\partial t}{\partial R_1} & \sigma_{R_2} \frac{\partial t}{\partial R_2} \end{bmatrix} \begin{bmatrix} 1. & r \\ r & 1. \end{bmatrix} \begin{bmatrix} R_1 \frac{\partial t}{\partial R_1} \\ R_2 \frac{\partial t}{\partial R_2} \end{bmatrix}$$

$$= [-0.016 \quad 0.0185] \begin{bmatrix} 1.0 & -0.738 \\ -0.738 & 1.0 \end{bmatrix} \begin{bmatrix} -0.016 \\ 0.0185 \end{bmatrix}$$

$$= 0.001035$$

and

$$\sigma_t = 0.0321$$

7.5.3 Comparison of Derivative Methods

A tolerance analysis problem involving random failure to start either (1) from a set of manufacturer's component specifications or (2) from measurements on a sample of components manufactured under local control. The preferred method of analysis depends principally on the origin of this tolerance data.

In Case (1), the actual statistical distribution of the component values is likely to resemble Figure 7.7a, because the manufacturer has selectively withdrawn element values having closer tolerance. Thus, components are quite likely to assume one of the two limiting tolerances. This fact makes worst-case design quite reasonable, since the probability is 1/2 that a component will assume a value near the limit maximizing the response variation.

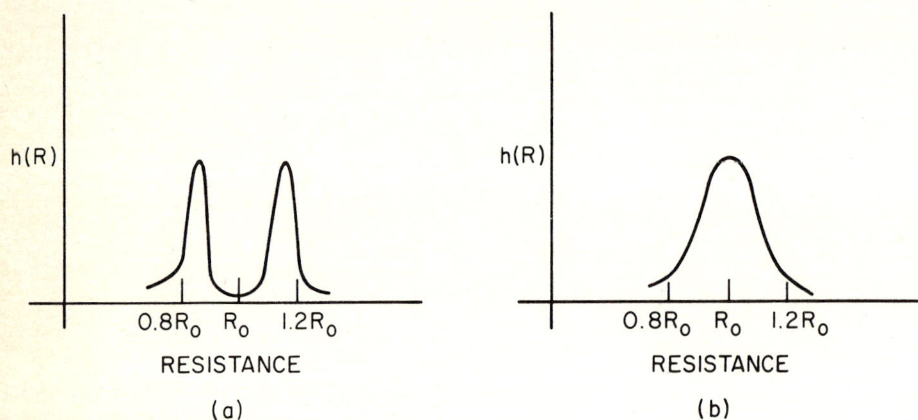

FIG. 7.7 Distributions of 20% Resistor

For Case (2), the distribution is more likely to assume a form resembling Figure 7.7. Clearly, use of worst-case criterion here would result in an overly pessimistic design except in the most critical circumstances. In fact the case for statistical design becomes even stronger if the correlation matrix can be calculated as in Table 7.4 and has off diagonal terms with large magnitudes. This latter case occurs frequently in integrated circuit manufacture, where all elements of the same type tend to be in error by the same percent.

It would appear that both worst-case and statistical design yield only a single number representing the spread of f. However, in the special (but common) case when all elements have a Gaussian distribution, it may be shown that z is also Gaussian distributed – assuming (7.8) is valid. The distribution of z is then completely determined!

7.5.4 Monte Carlo Analysis

When large statistical variations must be considered because the partial derivatives do not adequately represent variations in network behavior, a brute force approach known as Monte Carlo analysis is used. The procedure is simply to re-analyze the circuit a large number of times, each time letting the varying elements assume values according to prescribed distributions. An accurate distribution for the circuit response is thus evolved, since the procedure simulates the choosing of circuits from a physical sample. This process requires the generation of random numbers having a prescribed distribution.

Subroutines to generate uniformly - and Gaussian - distributed random numbers are available on most digital computers. Fortunately, the uniform distribution can be used to generate nearly all other distributions. The procedure is the following. Given a desired probability density function h(x) defined over an interval $[x_1, x_2]$ and strictly positive over this interval:

a. integrate h(x) to yield

$$g(x) = \int_{x_1}^{x} h(x')dx'$$

as in Figure 7.8b; g(x) is then necessarily strictly increasing and achieves a maximum value of unity;

b. choose a random number from a distribution uniformly distributed over the interval [0, 1]; let the associated random variable be y;

c. set y = g(x) and find x as in Figure 7.8b.

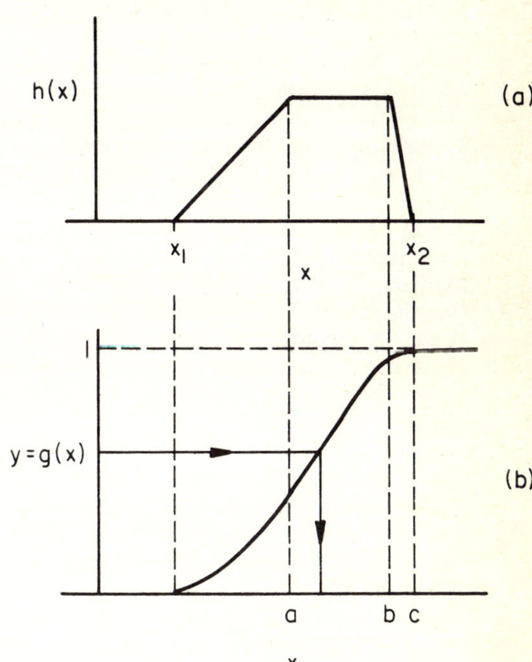

FIG. 7.8 Generation of Arbitrary Distributions

The random variable x now has the desired distribution function h(x). Note, even without justification of the above procedure (Problem 7.3), it appears reasonable that more numbers will be chosen in the interval [a, b] than in [b, c] if y is uniformly distributed.

Problems

7.1 Suggest a method for augmenting the node equations of a network to yield the constant-output contours of Figure 7.4.

7.2 Formulate a linear programming problem which would locate one or more of the eight intersections shown in Figure 7.4.

7.3 Show the method given in the text for generating nonuniform distributions from uniform distributions is valid. You will need the relationship

$$h(y)dy = h[\, g(x)] \left(\frac{dg(x)}{dx}\right) dx$$

REFERENCES

7.1 Clunies-Ross, C., and S. S. Husson, "Statistical Techniques in Circuit Optimization," Proc. National Electronics Conference, pp. 325-334; 1962.

7.2 Klapp, S. T., "Empirical Parameter Variation Analysis for Electronic Circuits," IEEE International Convention Record; 1963.

7.3 Mark, D. G., "A New Design Tool: The Matched Characteristic Method of Nonlinear Analysis," Trans. IEEE on Aerospace, vol. 2, no. 2, p. 312; April, 1964.

7.4 Mark, D. G., "Choosing the Best Method for a Variability Analysis," Electronic Design; November 8, 1963.

7.5 Mark, D. G., and L. H. Stember, "Variability Analysis," Electro-Technology, pp. 37-48; July, 1965.

7.6 Hald, A., Statistical Theory with Engineering Applications, Wiley, Chapman and Hall Ltd.

7.7 Mark, D. G., "Tolerance Analysis of Nonlinear Circuits," Proc. NEC; November, 1967.

7.8 Monroe, M. E., Theory of Probability, McGraw-Hill; 1951.

7.9 Chambers, R. P., "Random Number Generation on Digital Computers," IEEE Spectrum, vol. 4, pp. 48-56; February, 1967.

7.10 Karafin, B. J., "Optimum Assignment of Component Tolerances for Electrical Networks," BSTJ, vol. 50, no. 4, pp. 1225-1242; April, 1971.

7.11 Butler, E. M., "Large Change Sensitivities for Statistical Design," BSTJ, vol. 50, no. 4, pp. 1209-1224; April, 1971.

7.12 Goddard, P. J., P. A. Villalaz, and R. Spence, "Method for the Efficient Computation of the Large-Change Sensitivity of Linear Nonreciprocal Networks," Electronics Letters, vol. 7, no. 4, pp. 112-113; February 25, 1971.

7.13 Dantzig, G. B., Linear Programming and Extensions, Princeton University Press; 1963.

8

Introduction to Advanced Techniques of Equation Formulation

8.1 INTRODUCTION

The node analysis formulation procedure used thus far in the text was adopted because of its familiarity and ease of programming. The primary disadvantages it poses are awkwardness and apparent inefficiency in modeling common electrical components such as voltage sources, inductors, etc. Rather than adding new voltage variables to expose branch currents - as a gyrator does - we should look for formulation methods that allow branch currents as natural solution variables.

In later chapters, we will find occasion to sharpen this quest for efficiency to the point of inquiring which variables - fromthe set of all branch currents, voltages, and combinations thereof - result in the most rapid total response calculation. To attempt an answer of any generality will require a familiarity with the structural or topological properties of a network. Toward this end, in this chapter we undertake a formal study of the (1) extensions of Kirchhoff's simple current and voltage laws, (2) relationships between these laws, and (3) methods of combining structural and element value information to yield complete circuit descriptions.

8.2 ELEMENTARY GRAPH THEORY

8.2.1 Introduction

If we are to write mixed sets of voltage and current equations for a network we must assure that the equations are both independent and complete, the latter requiring that all branch currents and voltages can be obtained from solution of the equations. These issues can be settled without regard to the type of elements involved - resistor, capacitor, etc., - by considering only the network structure. Graph theory is the formal study of this network analysis.

The amount of graph theory one needs to understand network analysis beyond node analysis depends on the type of solution (analytical or numerical) and the depth of understanding required. The least demanding is a "cookbook knowledge" necessary to formulate general forms of equations for computer analysis from a branch level description (similar to RCAP, DCAP, and TCAP of previous chapters). For this purpose, Sections 8.2 and 8.4 offer sufficient background for the discussion of Section 8.7. The graph theoretical justification of the "recipe" is considered in Section 8.3, which also serves as a basis for procedures favoring an analytical form of solution in Section 8.5.

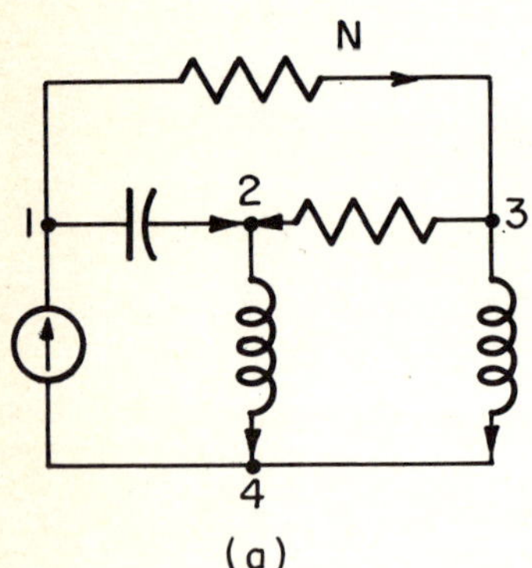

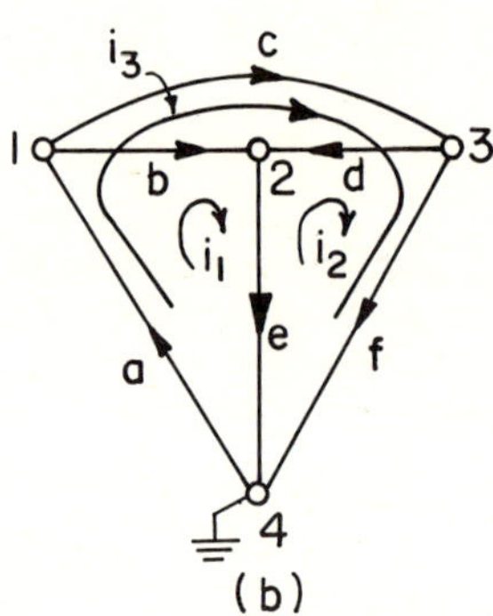

FIG. 8.1 Example Network and Its Graph

8.2.2 Basic Concepts

Consider the electrical network of Figure 8.1a. The structure of the network N may be represented as in Figure 8.1b; this is known as the graph of N, and is composed of n_e edges (representing branches) connecting n vertices (representing nodes). If we assign reference directions for branch current flow in Figure 8.2a, then carrying these over to the graph yields a directed graph, as shown. Adding letter designations to all edges then permits the following description of the network structure (topology).

We can associate with the graph a set of Kirchhoff Current Law (KCL) equations, one equation for each node of N:

$$-i_a + i_b + i_c = 0$$

$$\begin{aligned} -i_c + i_d + i_f &= 0 \\ i_a - i_e - i_f &= 0 \end{aligned} \tag{8.1}$$

In matrix form this becomes

$$\begin{array}{c} \text{Node} \\ \\ 1 \\ 2 \\ 3 \\ 4 \end{array} \quad \begin{array}{c} \text{Branch} \\ \begin{array}{cccccc} a & b & c & d & e & f \end{array} \\ \begin{bmatrix} -1 & 1 & 1 & 0 & 0 & 0 \\ 0 & -1 & 0 & -1 & 1 & 0 \\ 0 & 0 & -1 & 1 & 0 & 1 \\ 1 & 0 & 0 & 0 & -1 & -1 \end{bmatrix} \end{array} \begin{bmatrix} i_a \\ i_b \\ i_c \\ i_d \\ i_e \\ i_f \end{bmatrix} = \begin{bmatrix} 0 \\ 0 \\ 0 \\ 0 \end{bmatrix} \tag{8.2}$$

or, more succinctly,

$$\underset{\sim}{A}_a \underline{i}_b = \underline{0}$$

$\underset{\sim}{A}_a$ is termed an augmented incidence matrix, and has the following characteristics:

(1) all branches of N are represented whether they describe two-terminal elements such as inductors or independent sources, or multi-terminal elements such as controlled sources;

(2) the sum of any column is identically zero, since each column contains $\underline{+1}$ and a $\underline{-1}$. For this reason, the network can in fact be described by eliminating (grounding) any node and deleting the resultant node equations, viz,

$$\begin{bmatrix} -1 & 1 & 1 & 0 & 0 & 0 \\ 0 & -1 & 0 & -1 & 1 & 0 \\ 0 & 0 & -1 & 1 & 0 & 1 \end{bmatrix} \begin{bmatrix} i_a \\ i_b \\ i_c \\ i_d \\ i_e \\ i_f \end{bmatrix} = \begin{bmatrix} 0 \\ 0 \\ 0 \end{bmatrix} \tag{8.3}$$

yielding

$$\underset{\sim}{A}_i \underline{i}_b = \underline{0}$$

$\underset{\sim}{A}_i$ is called simply an <u>incidence matrix</u>. The missing row can be uniquely restored.

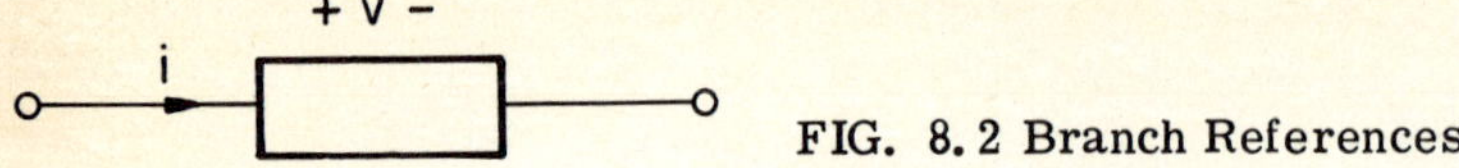

FIG. 8.2 Branch References

Rather surprisingly, the incidence matrix may also be used to relate voltage variables in a network, provided a consistent sign convention is followed relating branch currents and voltages (Figure 8.2). Let $\underline{v}_n$ be a vector of node voltages with respect to the same ground node deleted from the augmented incidence matrix. Then we can write

$$\underline{v}_b = \underset{\sim}{A}_i^T \underline{v}_n \tag{8.5}$$

where $\underline{v}_b$ is a vector of branch voltages ordered the same as $\underline{i}_b$. For example, in Figure 8.2, we find by transposing the $\underset{\sim}{A}_i$ of (8.3)

$$\begin{bmatrix} v_a \\ v_b \\ v_c \\ v_d \\ v_e \\ v_f \end{bmatrix} = \begin{bmatrix} -1 & 0 & 0 \\ 1 & -1 & 0 \\ 1 & 0 & -1 \\ 0 & -1 & 1 \\ 0 & 1 & 0 \\ 0 & 0 & 1 \end{bmatrix} \begin{bmatrix} v_1 \\ v_2 \\ v_3 \end{bmatrix}$$

Inspection of this example demonstrates the generality of the result. Each row of $\underset{\sim}{A}_i^T$ has a +1 and -1 in the column corresponding to the connection nodes, unless the column is associated with the ground node.

We now state the following with further discussion deferred to Section 8.3.

"Equations (8.4) and (8.5) offer both a complete and independent description of the structural equations of a network."

Of more significance for programming purposes is that only the incidence matrix $\underset{\sim}{A}_i$ is involved; this matrix can be trivially found from a scan of input data containing node connections.

8.3 GRAPH THEORETICAL RELATIONSHIPS

8.3.1 The Circuit Matrix

Despite the completeness of (8.6), it involves node voltages, which are not fundamental (basis) network variables in a graph theoretical sense. To formalize our discussion, we first examine equations that involve only the branch vectors $\underline{v}_b$ and $\underline{i}_b$.

We can obtain equations involving solely $\underline{v}_b$ by using KVL around closed paths in the network. For example, we can write for the network of Figure 8.1,

$$\begin{array}{c} \text{Loop} \\ \left[\begin{array}{c} 1 \\ 2 \\ 3 \end{array}\right. \end{array} \begin{array}{c} \begin{array}{cccccc} a & b & c & d & e & f \end{array} \\ \left.\begin{array}{cccccc} 1 & 1 & 0 & 0 & 1 & 0 \\ 0 & 0 & 0 & -1 & -1 & 1 \\ 1 & 0 & 1 & 0 & 0 & 1 \end{array}\right] \end{array} \begin{bmatrix} v_a \\ v_b \\ v_c \\ v_d \\ v_e \\ v_f \end{bmatrix} = \begin{bmatrix} 0 \\ 0 \\ 0 \end{bmatrix} \qquad (8.6)$$

or

$$\underset{\sim}{B}v_b = \underline{0} \qquad (8.7)$$

where $\underset{\sim}{B}$ is known as a circuit matrix. $\underset{\sim}{B}$ in general has entries of the form

b_{ij} = 1 if branch j is in loop i and the branch reference agrees with the loop reference.
b_{ij} = -1 if branch j is in loop i and the branch reference does not agree with the loop reference.
b_{ij} = 0 if branch j is not in loop i.

If the matrices $\underset{\sim}{A}_i$ and $\underset{\sim}{B}$ are arranged in the same column order, then

$$\underset{\sim}{A}_i\underset{\sim}{B}^T = \underset{\sim}{0} \qquad (8.8)$$

This may be checked from (8. 4) and (8. 6) and can be shown in general— regardless of the choice of loops. The argument hinges on the observation that the product $a_{ij}b_{kj}$ is nonzero if (1) a node i is connected to branch j, and (2) loop k passes through branch j. From Figure 8. 3, there must be precisely two such nonzero products associated with each node (row of $\underset{\sim}{A}_i$) and each passage of a loop current past node i. These products always sum to zero, regardless of the reference directions chosen, since reversing the reference current direction ($a_{ij} \to -a_{ij}$) also reverses the reference voltage direction ($b_{kj} \to -b_{kj}$).

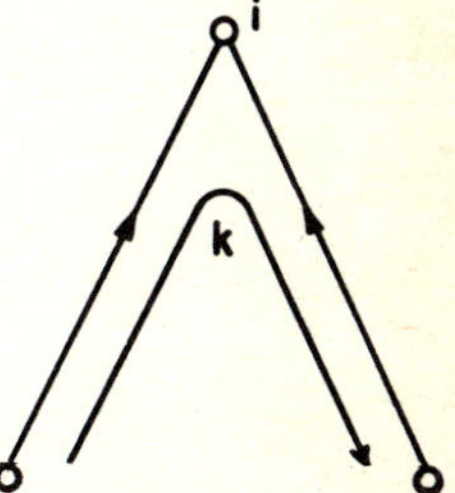

FIG. 8. 3 Node-Loop Intersection

8. 3. 2 Voltage-Current Relationships

In node (mesh) analysis of circuits, the equations are related by branches connecting the various nodes (meshes). It is, of course, simple to write equations for this form of coupling. When solution variables include both voltages and currents, however, the coupling is more complicated and a complete set of circuit equations are seemingly more difficult to formulate. This difficulty is more apparent than real; by writing the branch current-voltage relationships in a special form, a set of clearly independent KCL equations similar to (8.3) can be simply related to a set of KVL equations similar to (8.6).

The key to relating B and A_i further is the concept of a tree. A tree is a connected* subgraph which contains all the vertices of the graph, but no circuits (closed paths). Examples of trees of the graph of Figure 8.1b are shown in Figure 8.4.

A tree has the following useful characteristics.

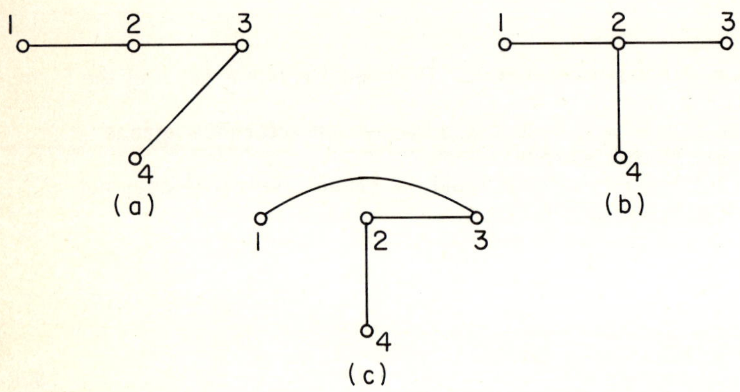

FIG. 8.4 Examples of Trees (see Figure 8.2b)

Tree Property 1. "A tree contains exactly n-1 edges (branches)." Consider a graph consisting of any branch b of a tree. If we add to G a new tree branch connected to b, this must result in the addition of a new vertex or complete a circuit. Since the latter is not allowed, a new vertex must have been added. Adding another branch and repeating the argument, we can conclude that there is a one-to-one correspondence between new tree branches and new vertices, resulting in exactly n-1 branches.

Tree Property 2. "Any subgraph G of n-1 edges having no circuits must be a tree." Assume G is not connected and contains no circuits. Then it must consist of a number (p) of subgraphs G_i, each of which is its own tree (Figure 8.6) with v_i vertices v_i -1 branches. The total number of branches is then

$$\sum_{i=1}^{p} (v_i - 1) = \sum_{i=1}^{p} v_i - p$$

$$= n - p$$

*A path exists between any two vertices.

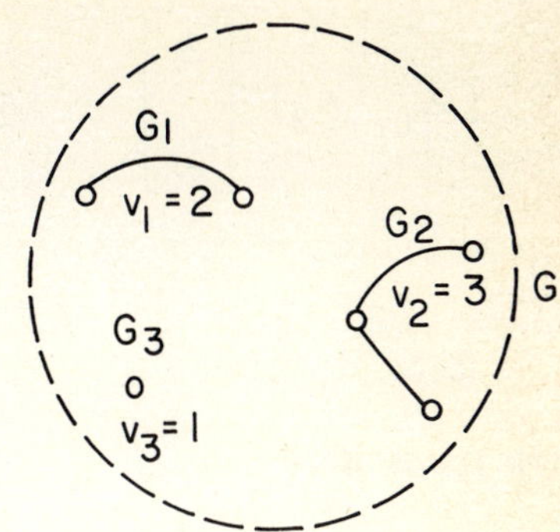

FIG. 8.5 Subgraph Being Tested As Tree

But the number of branches is given as n-1, so p = 1. Consequently, the graph G is connected and is its own tree.

The elements of the graph not in a particular tree are known as the <u>links</u> of the tree. A branch may be part of one tree and yet a link of another tree.

From a knowledge of the properties of a tree, we now show that equations can be obtained from $\underset{\sim}{A}_i$ of the form

$$[\underset{\sim}{I} \;\; \underset{\sim}{F}]\begin{bmatrix} \underline{i}_t \\ \\ \underline{i}_\ell \end{bmatrix} = \underline{0} \qquad [-\underset{\sim}{F}^T \;\; \underset{\sim}{I}]\begin{bmatrix} \underline{v}_t \\ \\ \underline{v}_\ell \end{bmatrix} = \underline{0} \tag{8.9}$$

where "t" and "ℓ" represent tree and link variables. The branch voltage and current equations are thus relatable by the single matrix $\underset{\sim}{F}$. Further, in contrast to (8.4) and (8.5), the presence of the identity matrix guarantees the independence of these equations.

The algorithm for determination of $\underset{\sim}{F}$ proceeds as follows.

1. Partition $\underset{\sim}{A}_i$ into the form

$$\begin{array}{ccc} & \text{branch sets} & \\ & \{b_1\} \quad \{b_2\} & \\ \underset{\sim}{A}_i & = & [\underset{\sim}{A}_1 \quad \underset{\sim}{A}_2] \end{array} \tag{8.10}$$

where $\{b_1\}$ is a tree branch set and $\{b_2\}$ the corresponding link branch set. For the network of Figure 8.1a we could choose

$$\underset{\sim}{A}_i = \begin{array}{c} \\ 1 \\ 2 \\ 3 \end{array}\begin{array}{c} \begin{array}{cccccc} c & d & e & a & b & f \end{array} \\ \left[\begin{array}{ccc|ccc} 1 & 0 & 0 & -1 & 1 & 0 \\ 1 & -1 & 1 & 0 & -1 & 0 \\ -1 & 1 & 0 & 0 & 0 & 1 \end{array}\right] \end{array} \tag{8.11}$$

2. Solve the equations

$$[\mathbf{A}_1 \quad \mathbf{A}_2]\begin{bmatrix}\underline{i}_t \\ \underline{i}_\ell\end{bmatrix} = \mathbf{A}_1\underline{i}_t + \mathbf{A}_2\underline{i}_\ell = \underline{0} \tag{8.12}$$

or

$$\mathbf{A}_1\underline{i}_t = -\mathbf{A}_2\underline{i}_\ell \tag{8.13}$$

for $\underline{i}_t$, yielding

$$\underline{i}_t = -\mathbf{A}_1^{-1}\mathbf{A}_2 i_\ell \tag{8.14}$$

Equation (8.14) in matrix form is then

$$[\mathbf{I} \quad \mathbf{A}_1^{-1}\mathbf{A}_2]\begin{bmatrix}\underline{i}_t \\ \underline{i}_\ell\end{bmatrix} = \underline{0} \tag{8.15}$$

Comparing (8.15) and (8.9) suggests choosing

$$\mathbf{F} = \mathbf{A}_1^{-1}\mathbf{A}_2 \tag{8.16}$$

The calculation of $\mathbf{F}$ therefore takes the form of reducing the tree branch columns of $\mathbf{A}_i$ to a unit matrix. The remaining columns then yield $\mathbf{F}$. For example, the reduction of the $\mathbf{A}_i$ of (8.11) a column at a time is given below.

$$\begin{bmatrix}1 & 0 & 0 & -1 & 1 & 0 \\ 0 & 1 & 1 & 0 & -1 & 0 \\ -1 & 1 & 0 & 0 & 0 & 1\end{bmatrix} \longrightarrow \begin{bmatrix}1 & 0 & 0 & -1 & 1 & 0 \\ 0 & -1 & 1 & 0 & -1 & 0 \\ 0 & 1 & 0 & -1 & 1 & 1\end{bmatrix}$$

$$\downarrow \tag{8.17}$$

$$\begin{bmatrix}1 & 0 & 0 & -1 & 1 & 0 \\ 0 & 1 & 0 & -1 & 1 & 1 \\ 0 & 0 & 1 & -1 & 0 & 1\end{bmatrix} \longleftarrow \begin{bmatrix}1 & 0 & 0 & -1 & 1 & 0 \\ 0 & 1 & -1 & 0 & 1 & 0 \\ 0 & 0 & 1 & -1 & 0 & 1\end{bmatrix}$$

To satisfy ourselves that the resulting current and voltage equations do indeed describe the network in some sense, we may note the following.

(1) The current equation

$$[-\mathbf{F}^T \quad \mathbf{I}]\begin{bmatrix}\underline{i}_t \\ \underline{i}_\ell\end{bmatrix} = \begin{bmatrix}1 & 0 & 0 & -1 & 1 & 0 \\ 0 & 1 & 0 & -1 & 1 & 1 \\ 0 & 0 & 1 & -1 & 0 & 1\end{bmatrix}\begin{bmatrix}i_c \\ i_d \\ i_e \\ i_a \\ i_b \\ i_f\end{bmatrix} = \begin{bmatrix}0 \\ 0 \\ 0\end{bmatrix}$$

$$= \underline{0} \tag{8.18}$$

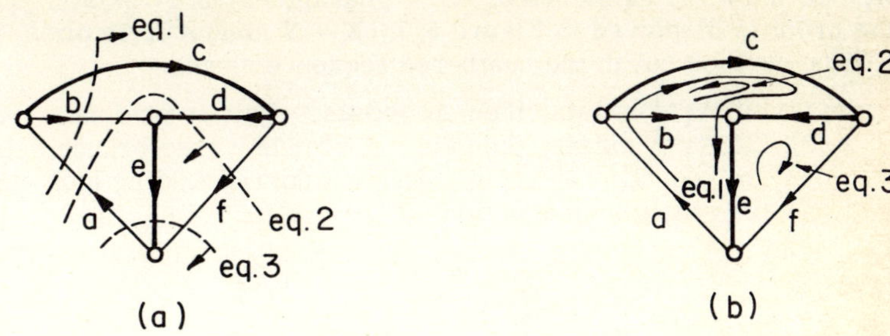

FIG. 8.6 Fundamental Cut Sets and Loops

describes the current flow between parts of the network (Figure 8.6a) in the directions shown. These equations represent, then, extensions of Kirchhoff's Current Law at a node, and are called fundamental cut set equations. The branches involved in a cut set equation (such as a-b-c) form a cut set.

(2) The voltage equations are represented by

$$[-\underset{\sim}{F}^T \; \underset{\sim}{I}]\begin{bmatrix} \underline{v}_t \\ \underline{v}_\ell \end{bmatrix} = \begin{bmatrix} 1 & 1 & 1 & 1 & 0 & 0 \\ -1 & -1 & 0 & 0 & 1 & 0 \\ 0 & -1 & -1 & 0 & 0 & 1 \end{bmatrix} \begin{bmatrix} v_c \\ v_d \\ v_e \\ v_a \\ v_b \\ v_f \end{bmatrix} = \underline{0} \qquad (8.19)$$

$$= \underline{0}$$

are simply the KVL equations for the loops of Figure 8.6b. More explicitly, KCL and KVL equations of the form (8.9) define fundamental cut sets and fundamental loops, in contrast to equations of more general form (8.4 and 8.6).

The above procedure evidently depends on the existence of $\underset{\sim}{A}_1^{-1}$. In the next section, we show that choosing a set of tree branches for $\{b_1\}$ guarantees $\underset{\sim}{A}_1^{-1}$ exists and results in equations of the form of (8.9). We will also show the converse, namely, the existence of $\underset{\sim}{A}_1^{-1}$ implies $\{b_1\}$ is a tree branch set and that (8.9) exists. The latter is more important computationally, since, using the procedure demonstrated by (8.17), we can identify trees quite simply.

8.3.3 Proof of the Algorithm

Given $\underset{\sim}{A}_i$ of order $(n-1) \times n_e$ partitioned in the form

$$\underset{\sim}{A}_i = [\underset{\sim}{A}_1 \quad \underset{\sim}{A}_2] \qquad \text{with columns } \{b_1\} \; \{b_2\}$$

we now show the existence of $\tilde{A}_1^{-1}$ guarantees (8.9) exists. The outline of the proof is displayed in Figure 8.7 ($X \rightarrow Y$ means X implies Y). The proof will be carried out in the numbered sequence shown.

(1) "If $\tilde{A}_1^{-1}$ exists, then $\tilde{A}_f$ exists." This is obvious.
(2) "If $\tilde{A}_1^{-1}$ exists, then the set $\{b_1\}$ is a tree branch set." If the set $\{b_1\}$ contains a circuit, define a corresponding loop current to yield the single loop equation

$$[\;(\underline{B}_1)^T \quad (\underline{B}_2{}^T)\;] \begin{bmatrix} \underline{v}_{b_1} \\ \underline{v}_{b_2} \end{bmatrix} = \underline{0} \tag{8.20}$$

where $\underline{B}_1$ and $\underline{B}_2$ are column vectors and where the partition is again made on $\{b_1\}$. Note that $\underline{B}_2 = \underline{0}$, since only branches of $\{b_1\}$ appear in the loop.

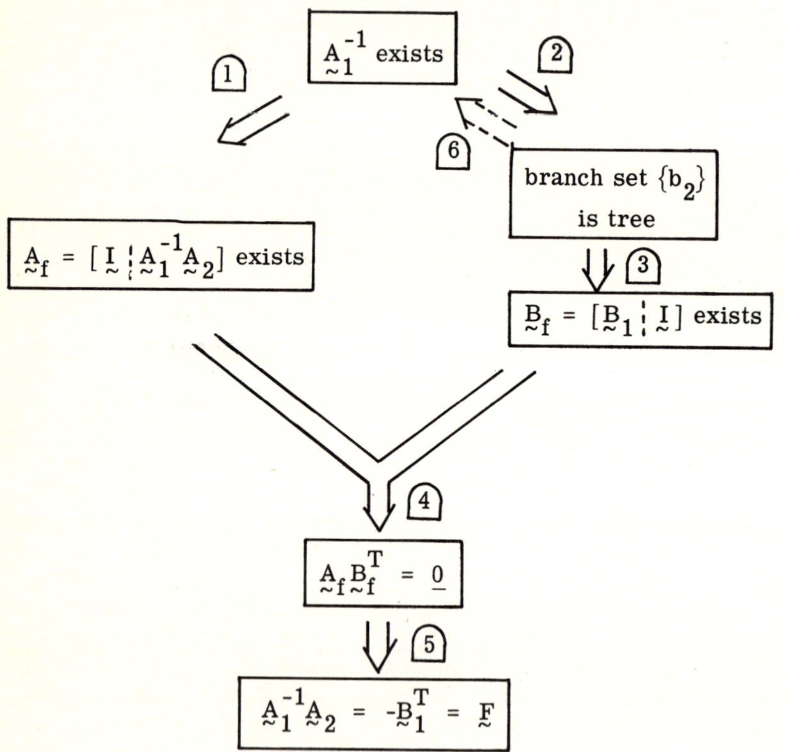

FIG. 8.7 Sequenced Outline of Proof

From (8.8) we can write

$$\tilde{A}_i \tilde{B}^T = \tilde{0}$$

for <u>any</u> circuit matrix $\underset{\sim}{B}$, so

$$\left[\underset{\sim}{A}_1 \quad \underset{\sim}{A}_2\right] \begin{bmatrix} \underline{B}_1 \\ \\ \underline{B}_2 \end{bmatrix} = \underset{\sim}{A}_1 \underline{B}_1 = \underline{0} \tag{8.22}$$

Now if $\underset{\sim}{A}_1^{-1}$ exists, we would conclude

$$\underline{B}_1 = \underset{\sim}{A}_1^{-1}\, \underline{0} = \underline{0} \tag{8.23}$$

which is assuredly wrong. Consequently, if $\underset{\sim}{A}_1^{-1}$ exists, the set of n-1 branches comprising $\{b_1\}$ cannot contain a circuit. From Tree Property 2, the set must be a tree.

(3) "If the set $\{b_1\}$ is a tree, then loop equations of the form

$$\begin{matrix} \{b_1\} & \{b_2\} \end{matrix}$$
$$\left[\underset{\sim}{B}_1 \quad \underset{\sim}{I}\right] \begin{bmatrix} \underline{v}_{b_1} \\ \\ \underline{v}_{b_2} \end{bmatrix} = \underline{0} \tag{8.24}$$

exist." Define a set of loops by successively adding links to the tree branch set (for example, in Figure 8.6b, the links a-b-f have been added to define three loops). Since each link appears in only one loop equation, the associated column in $\underset{\sim}{B}$ contains only one entry—which can be brought to the diagonal position. An equation can be developed for each link, so the unit matrix of (8.24) results.

(4) "If $\underset{\sim}{A}_f$ and $\underset{\sim}{B}_f$ exist, then $\underset{\sim}{A}_f \underset{\sim}{B}_f = \underset{\sim}{0}$." We know $\underset{\sim}{A}_i \underset{\sim}{B}_f^T = \underset{\sim}{0}$, since $\underset{\sim}{B}_f$ is a circuit matrix. Now consider any matrix $\underset{\sim}{A}'$ with rows which are linear combinations of the rows of $\underset{\sim}{A}_i$, i.e.,

$$\underset{\sim}{A}_i = \begin{bmatrix} \underline{A}_1^T \\ \underline{A}_2^T \\ \vdots \\ \underline{A}_{n-1}^T \end{bmatrix} \qquad \underset{\sim}{A}' = \begin{bmatrix} \sum_{k=1}^{n-1} \alpha_{ik} \underline{A}_k^T \\ \vdots \\ \sum_{k=1}^{n-1} \alpha_{mk} \underline{A}_k^T \end{bmatrix} \tag{8.25}$$

Now let $\underset{\sim}{B}_f^T$ be written in the form of column vectors, as

$$\underset{\sim}{B}_f^T = \left[\underline{B}_1\ \underline{B}_2 \cdots \underline{B}_r\right] \tag{8.26}$$

The condition $\underset{\sim}{A}_i \underset{\sim}{B}_f^T = 0$ then requires $\underline{A}_k^T \underline{B}_j = 0$ for all j and k. But the product $\underset{\sim}{A}' \underset{\sim}{B}_f$ has an (i, j) entry of the form

$$\left(\sum_{k=1}^{n-1} \alpha_{ik}\underset{\sim}{A}_k^T\right)\underset{\sim}{B}_j = \sum_{k=1}^{n-1} \alpha_{ik}\underset{\sim}{A}_k^T\underset{\sim}{B}_j \tag{8.27}$$

$$= 0 \tag{8.28}$$

Therefore $\underset{\sim}{A}'\underset{\sim}{B}_f = \underset{\sim}{0}$. Now

$$\underset{\sim}{A}_f = [\underset{\sim}{I} \,\vdots\, \underset{\sim}{A}_1^{-1}\underset{\sim}{A}_2] \tag{8.29}$$

$$= \underset{\sim}{A}_1^{-1}[\underset{\sim}{A}_1 \,\vdots\, \underset{\sim}{A}_2] \tag{8.30}$$

$$= \underset{\sim}{A}_1^{-1}\underset{\sim}{A}_i \tag{8.31}$$

Each row of $\underset{\sim}{A}_f$ is then a linear combination of the rows of $\underset{\sim}{A}_i$, so $\underset{\sim}{A}_f\underset{\sim}{B}_f = \underset{\sim}{0}$.

(5) "If $\underset{\sim}{A}_f\underset{\sim}{B}_f^T = \underset{\sim}{0}$, then $\underset{\sim}{A}_1{}^{-1}\underset{\sim}{A}_2 = \underset{\sim}{B}_1^T$." This follows directly from the form of $\underset{\sim}{A}_f$ and $\underset{\sim}{B}_f$, viz.

$$[\underset{\sim}{I} \,\vdots\, \underset{\sim}{A}_1^{-1}\underset{\sim}{A}_2] \begin{bmatrix} \underset{\sim}{B}_1^T \\ \\ \underset{\sim}{I} \end{bmatrix} = \underset{\sim}{B}_1^T + \underset{\sim}{A}_1^{-1}\underset{\sim}{A}_2$$

$$= \underset{\sim}{0} \tag{8.32}$$

We have now completed our five-part proof. Beginning with a partition of $\underset{\sim}{A}_i$, we need determine only if $\underset{\sim}{A}^{-1}\underset{\sim}{A}_2 = \underset{\sim}{F}$ exists, which is a trivial computational exercise. Unfortunately, this procedure gives no insight into the class of networks that can be so partitioned and thus described by (8.9). We now show that all connected networks are amenable to such analysis.

Since every connected network contains at least one tree, we establish the generality of the method with the following theorem (Figure 8.6).

(6) "If the branch set $\{b_1\}$ is a tree branch set, then $\underset{\sim}{A}_1^{-1}$ exists." We prove this by showing that if $\{b_1\}$ is a tree branch set, then the rows and columns of $\underset{\sim}{A}_i$ can be arranged such that $\underset{\sim}{A}_1$ is of the upper triangular form

$$\underset{\sim}{A}_1 = \begin{bmatrix} \pm 1 & a_{12} & \cdot\,\cdot & a_{1,n-1} \\ 0 & \pm 1 & \cdot\,\cdot & a_{2,n-1} \\ \cdot & \cdot & \cdot\,\cdot & \cdot \\ \cdot & \cdot & \cdot\,\cdot & \cdot \\ 0 & 0 & \cdot\,\cdot & \pm 1 \end{bmatrix} \tag{8.33}$$

The inverse of $\underset{\sim}{A}_1$ then exists (by analogy to Gauss reduction).

Equation (8. 34) evidently describes a special kind of branch connection. Examining the columns from left to right, the first branch is connected to the first node and to the ground node; the second element is between the second node and either the first or the ground nodes; etc. In other words, if we begin with a subgraph of only one branch (the first), then adding the next branch results in the addition of exactly one new node. Evidently, a tree branch set can be arranged to have this property by adding only branches connected to a subgraph of the tree already formed. Since a tree is connected, a branch can be found which adds at most one new node; since a tree contains no circuits, every remaining branch must add at least one new node (see Tree Property 1).

An example of the formation of a tree to yield an upper triangular $\underset{\sim}{A}_1$ is shown in Figure 8. 8; the incidence matrix is

$$
\begin{array}{cc}
 & \text{Branch} \\
\text{Node} & \begin{array}{cccccc} d & c & e & a & b & f \end{array}
\end{array}
$$

$$
\underset{\sim}{A}_i = \begin{array}{c} 2 \\ 1 \\ 4 \end{array}
\begin{bmatrix}
-1 & 0 & 1 & 0 & -1 & 0 \\
0 & 1 & 0 & -1 & 1 & 0 \\
0 & 0 & -1 & 1 & 0 & -1
\end{bmatrix}
$$

which has the desired upper triangular form for $\underset{\sim}{A}_1$.

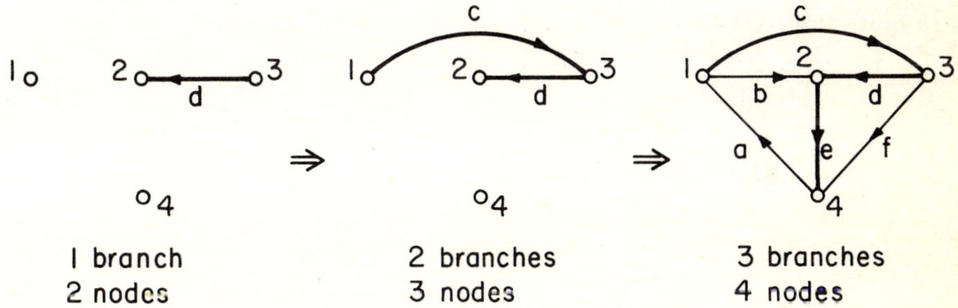

FIG. 8. 8 Growth of Tree

8. 3. 4 Summary

We have now achieved the following. Given the incidence matrix $\underset{\sim}{A}_i$ for any connected graph, we can write an independent set of n-1 KCL equations in either of the forms

$$
\underset{\sim}{A}_i \underline{i}_b = \underline{0} \qquad \text{or} \qquad [\underset{\sim}{I} \quad \underset{\sim}{F}] \begin{bmatrix} \underline{i}_t \\ \underline{i}_\ell \end{bmatrix} = \underline{0}
$$

Similarly, we can write an independent set of voltage equations in the form

$$
[-\underset{\sim}{F}^T \quad \underset{\sim}{I}] \begin{bmatrix} \underline{v}_t \\ \underline{v}_\ell \end{bmatrix} = \underline{0} \tag{8.34}
$$

In no case have we shown that these are equations we complete, i.e., that every other current or voltage relationships are linear combinations of those equations. These proofs of completeness are significantly more complicated than the independence proofs we have presented [8.5].

The node voltage relationship

$$\underline{v}_b = \underset{\sim}{A}_i^T \underline{v}_n \tag{8.35}$$

can always be used in place of (8.34). To see this, with proper branch ordering we can write (8.35) in the form

$$\begin{bmatrix} -\underset{\sim}{A}_1^T & \underset{\sim}{I} & \underset{\sim}{0} \\ -\underset{\sim}{A}_2^T & \underset{\sim}{0} & \underset{\sim}{I} \end{bmatrix} \begin{bmatrix} \underline{v}_n \\ \underline{v}_t \\ \underline{v}_\ell \end{bmatrix} = \underline{0}$$

Using ordinary Gaussian elimination techniques, we can write

$$\begin{bmatrix} \underset{\sim}{I} & -(\underset{\sim}{A}_1^T)^{-1} & \underset{\sim}{0} \\ \underset{\sim}{0} & -\underset{\sim}{A}_2^T(\underset{\sim}{A}_1^T)^{-1} & \underset{\sim}{I} \end{bmatrix} \begin{bmatrix} \underline{v}_n \\ \underline{v}_t \\ \underline{v}_\ell \end{bmatrix} = \underline{0}$$

or in more familiar terms

$$\underline{v}_n = (A_1^T)^{-1} \underline{v}_t$$

$$[-\underset{\sim}{F}^T \quad \underset{\sim}{I}] \begin{bmatrix} \underline{v}_t \\ \underline{v}_\ell \end{bmatrix} = 0$$

Thus, (8.5) contains the fundamental loop equations and in addition provides a way of determining $\underline{v}_n$.

8.4 CONSTITUTIVE RELATIONSHIPS

The fundamental loop and fundamental cut set equations are only a rearrangement of KCL and KVL equations and so provide structural information only. A complete network solution requires equations that yield element value value information as well. Such information is generalized by the concept of constitutive relationships (CR). These can be of many forms, including

(1) branch relationships, such as

$$i_r = Gv_r, \quad i_c = \frac{Cdv_c}{dt},$$

$$i_s = I_s \quad i_2 = g_m v_1, \quad i_d = I_s(e^{\lambda v_d} - 1)$$

(2) port relationships, such as

transformer:
$$\begin{bmatrix} i_1 \\ v_2 \end{bmatrix} = \begin{bmatrix} 0 & n \\ n & 0 \end{bmatrix} \begin{bmatrix} v_1 \\ i_2 \end{bmatrix}$$

transistor:
$$i_e = I_{es}(e^{\lambda v_{eb}} - 1) - \alpha_r I_{cs}(e^{\lambda v_{cb}} - 1)$$

$$i_c = -\alpha_f I_{es}(e^{\lambda v_{eb}} - 1) + I_{cs}(e^{\lambda v_{cb}} - 1)$$

The most common class of constitutive relationships is of course the linear one-port branch set such as

$$\begin{bmatrix} i_1 \\ i_2 \\ \vdots \\ i_n \end{bmatrix} = \begin{bmatrix} G_1 & 0 & \dots & 0 \\ 0 & G_2 & \dots & 0 \\ \vdots & \vdots & & \vdots \\ 0 & 0 & \dots & G_n \end{bmatrix} \begin{bmatrix} v_1 \\ v_2 \\ \vdots \\ v_n \end{bmatrix} \tag{8.36}$$

or

$$\underline{i} = \underset{\sim}{G}\underline{v}$$

and

$$\begin{bmatrix} v_1 \\ v_2 \\ \vdots \\ v_n \end{bmatrix} = \begin{bmatrix} L_1 & 0 & \dots & 0 \\ 0 & L_2 & \dots & 0 \\ \vdots & \vdots & & \vdots \\ 0 & 0 & \dots & L_n \end{bmatrix} \frac{d}{dt} \begin{bmatrix} i_1 \\ i_2 \\ \vdots \\ i_n \end{bmatrix} \tag{8.37}$$

or

$$\underline{v} = \underset{\sim}{L}\frac{d\underline{i}}{dt}$$

The large variety of CR's (linear and nonlinear; differential and algebraic; single- and multi-port) would appear to complicate the formulation problem for general forms of networks. There are two solutions to this problem, depending on whether an <u>analytic</u> solution or a <u>numerical</u> solution is sought. In the latter case, we can apply the linearization and discretization methods of Chapter III and IV to reduce Equations (8.36-8.37) to either (a) linear resistive elements, or (2) sources. To develop insight for and appreciate the limitations of the numerical approach, however, we will now study analytic solutions of special forms of networks; namely, networks of linear RLC elements with independent sources.

8.5 FORMULATION FOR ANALYTIC SOLUTION

8.5.1 The Network Tableau

The combining of topological and constitutive equations to yield a solution can be performed in a variety of ways. Our approach will be similar to ones proposed by Pottle [8.2] and by Hachtel et al., [8.1] in which all the network equations are written in the form of a single, partitioned matrix equations.* To illustrate, we now consider at length the description and solution of an arbitrary network of resistors only.

We begin by performing a tree and link sort so that we can write all the network equations in the form

$$\begin{matrix} \text{KVL} \\ \text{KCL} \\ \text{CR} \\ \\ \end{matrix} \begin{bmatrix} \underset{\sim}{I} & \underset{\sim}{0} & \underset{\sim}{0} & -\underset{\sim}{F}^T \\ \underset{\sim}{0} & \underset{\sim}{I} & \underset{\sim}{F} & \underset{\sim}{0} \\ -\underset{\sim}{G} & \underset{\sim}{0} & \underset{\sim}{I} & \underset{\sim}{0} \\ \underset{\sim}{0} & -\underset{\sim}{R} & \underset{\sim}{0} & \underset{\sim}{I} \end{bmatrix} \begin{bmatrix} \underline{v}_g \\ \underline{i}_r \\ \underline{I}_g \\ \underline{V}_r \end{bmatrix} = \begin{bmatrix} \underline{0} \\ \underline{0} \\ \underline{0} \\ \underline{0} \end{bmatrix} \tag{8.38}$$

Here, upper case notation for i and v is used to designate link currents and tree voltages, since knowledge of these variables allows immediate calculation of tree currents and link voltages (respectively) from (8.9). $\underset{\sim}{R}$ and $\underset{\sim}{G}$ are matrices containing tree resistances and link conductances of the form

$$\underset{\sim}{R} = \begin{bmatrix} R_1 & 0 & \cdots & 0 \\ 0 & R_2 & & \vdots \\ \vdots & & \ddots & \vdots \\ 0 & \cdots & \cdots & R_p \end{bmatrix} \qquad \underset{\sim}{G} = \begin{bmatrix} G_{p+1} & 0 & \cdots & 0 \\ 0 & G_{p+2} & & \vdots \\ \vdots & & \ddots & \vdots \\ 0 & \cdots & \cdots & G_{p+q} \end{bmatrix} \tag{8.39}$$

The problem of solving a matrix equation in which the matrix elements are themselves matrices is considered in Problem 8.1. The first step would involve a partition of the form

$$\begin{bmatrix} \underset{\sim}{A}_{11} & \underset{\sim}{A}_{12} \\ \underset{\sim}{A}_{21} & \underset{\sim}{A}_{22} \end{bmatrix} = \left[\begin{array}{c|ccc} \underset{\sim}{I} & \underset{\sim}{0} & \underset{\sim}{0} & -\underset{\sim}{F}^T \\ \hline \underset{\sim}{0} & \underset{\sim}{I} & \underset{\sim}{F} & \underset{\sim}{0} \\ -\underset{\sim}{G} & \underset{\sim}{0} & \underset{\sim}{I} & \underset{\sim}{0} \\ \underset{\sim}{0} & -\underset{\sim}{R} & \underset{\sim}{0} & \underset{\sim}{I} \end{array}\right] \tag{8.40}$$

*The particular tableau proposed in [8.1] is intended for numerical solution (see Section 8.7).

The result of applying the reduction formula $\underset{\sim}{A}_{22} - \underset{\sim}{A}_{21}\underset{\sim}{A}_{11}^{-1}\underset{\sim}{A}_{12}$ is the matrix given below. Results of subsequent reduction steps are also shown.

$$\begin{bmatrix} \underset{\sim}{I} & \underset{\sim}{F} & \underset{\sim}{0} \\ \underset{\sim}{0} & \underset{\sim}{I} & -\underset{\sim}{G}\,\underset{\sim}{F}^T \\ -\underset{\sim}{R} & \underset{\sim}{0} & \underset{\sim}{I} \end{bmatrix} \Rightarrow \begin{bmatrix} \underset{\sim}{I} & -\underset{\sim}{G}\,\underset{\sim}{F}^T \\ \underset{\sim}{R}\,\underset{\sim}{F} & \underset{\sim}{I} \end{bmatrix} \Rightarrow [\underset{\sim}{I} + \underset{\sim}{R}\,\underset{\sim}{F}\,\underset{\sim}{G}\,\underset{\sim}{F}^T] \tag{8.41}$$

The network equations in terms of the voltages are then

$$[\underset{\sim}{I} + \underset{\sim}{R}\,\underset{\sim}{F}\,\underset{\sim}{G}\,\underset{\sim}{F}^T]\ \underline{V}_r = \underline{0}$$

Of course we come to the trivial conclusion that $\underline{V}_r = \underline{0}$, since the network contains no sources. However, the above reduction sequence from the network tableau is an important procedure which we will apply to increasingly broader classes of networks.

8.5.2 Tree Generation

Whenever more than one type of element is present, some care must be exercised in choosing a tree. Certain kinds of elements are preferred for tree branches while others are preferred for link branches. For example, independent voltage sources are <u>always</u> forced to be tree branches whereas independent current sources are <u>always</u> forced to be link branches. If either cannot be accommodated, then a loop of independent voltage sources or a cut-set of current sources must exist. The network is therefore inconsistent, since the sources cannot be truly independent.

An algorithm to perform a preferred sorting of tree and link branches can be made part of the tree selection process. For example, let us arrange the columns of the incidence matrix to correspond to the tree branch priority of Table 8.1. Adopting the procedure of left to right column reduction as (8.17) we observe the procedure fails only if the diagonal element of the column being reduced is zero. For example, the partial column reduction for two branch arrangements of the network of Figure 8.2 are

$$\begin{array}{c} \\ 1 \\ 3 \\ 4 \end{array}\begin{array}{c} \begin{array}{ccccccc} c & e & d & a & b & f & g \end{array} \\ \begin{bmatrix} 1 & 0 & 0 & -1 & 1 & 0 & 1 \\ -1 & 0 & 1 & 0 & 0 & 1 & -1 \\ 0 & -1 & 1 & 1 & 0 & -1 & 0 \end{bmatrix} \end{array} \qquad \begin{array}{c} \\ 1 \\ 3 \\ 4 \end{array}\begin{array}{c} \begin{array}{ccccccc} c & g & e & d & a & b & f \end{array} \\ \begin{bmatrix} 1 & 1 & 0 & 0 & -1 & 1 & 0 \\ -1 & -1 & 0 & 1 & 0 & 0 & 1 \\ 0 & 0 & -1 & 0 & 1 & 0 & -1 \end{bmatrix} \end{array}$$

$$\downarrow \qquad\qquad\qquad\qquad \downarrow$$

$$\begin{bmatrix} 1 & 0 & 0 & -1 & 1 & 0 & 1 \\ 0 & 0 & 1 & -1 & 1 & 1 & 0 \\ 0 & -1 & 0 & 1 & 0 & -1 & 0 \end{bmatrix} \qquad \begin{bmatrix} 1 & 1 & 0 & 0 & -1 & 1 & 0 \\ 0 & 0 & 0 & 1 & -1 & 1 & 1 \\ 0 & 0 & -1 & 0 & 1 & 0 & -1 \end{bmatrix}$$

TABLE 8.1 Branch Priority

1. voltage sources
2. capacitors
3. resistors
4. inductors
5. current sources

A zero is encountered in the diagonal position of the second column in both cases; however, in the first case, the column being reduced contains nonzero entries below the diagonal, so an interchange of rows avoids this apparent degeneracy. In the second case, the only nonzero elements are above the diagonal; therefore, no linear combination of the rows below the diagonal can produce a nonzero entry in the offending column. The branch cannot then form a tree with the branches of the reduced columns, so this branch set must form a circuit. Since the branch is of lower priority than those of the reduced columns to its left, it is ignored in further reduction, viz,

$$\begin{array}{c} \\ 1 \\ 3 \\ 4 \end{array} \begin{array}{c} \begin{array}{ccccccc} c & e & d & a & b & f & g \end{array} \\ \begin{bmatrix} 1 & 0 & 0 & -1 & 1 & 0 & 1 \\ 0 & 0 & 1 & -1 & 1 & 1 & 0 \\ 0 & -1 & 0 & 1 & 0 & -1 & 0 \end{bmatrix} \end{array}$$

The result of any such tree-link sort based on the priorities of Table 8.1 will be arrangement of branch voltages and currents of the form

$$\underline{i}_b = \begin{bmatrix} \underline{i}_t \\ \underline{I}_\ell \end{bmatrix} = \begin{bmatrix} \underline{i}_e \\ \underline{i}_c \\ \underline{i}_r \\ \underline{i}_\ell \\ \underline{I}_c \\ \underline{I}_g \\ \underline{I}_\ell \\ \underline{I}_j \end{bmatrix} \begin{array}{l} \text{(Independent Voltage Sources)} \\ \text{(Tree Capacitors)} \\ \text{(Tree Resistors)} \\ \text{(Tree Inductors)} \\ \text{(Link Capacitors} \\ \text{(Link Resistors)} \\ \text{(Link Inductors)} \\ \text{(Independent Current Sources)} \end{array} \begin{bmatrix} \underline{V}_e \\ \underline{V}_c \\ \underline{V}_r \\ \underline{V}_\ell \\ \underline{v}_c \\ \underline{v}_g \\ \underline{v}_\ell \\ \underline{v}_j \end{bmatrix} = \begin{bmatrix} \underline{V}_t \\ \underline{v}_\ell \end{bmatrix} = \underline{v}_b \qquad (8.42)$$

where again upper case variables indicate tree voltages and currents, the fundamental cut set and loop equations can be written as

$$\underline{i}_t = -\mathbf{F}\,\underline{I}_\ell \quad \text{and} \quad \underline{v}_\ell = \mathbf{F}^T \underline{V}_t$$

or in partitioned form as

$$\begin{bmatrix} \underline{i}_e \\ \underline{i}_c \\ \underline{i}_r \\ \underline{i}_\ell \end{bmatrix} = - \begin{bmatrix} \mathbf{F}_{ec} & \mathbf{F}_{eg} & \mathbf{F}_{e\ell} & \mathbf{F}_{ej} \\ \mathbf{F}_{cc} & \mathbf{F}_{cg} & \mathbf{F}_{c\ell} & \mathbf{F}_{cj} \\ \mathbf{0} & \mathbf{F}_{rg} & \mathbf{F}_{r\ell} & \mathbf{F}_{rj} \\ \mathbf{0} & \mathbf{0} & \mathbf{F}_{\ell\ell} & \mathbf{F}_{\ell j} \end{bmatrix} \begin{bmatrix} \underline{I}_c \\ \underline{I}_g \\ \underline{I}_\ell \\ \underline{I}_j \end{bmatrix} \tag{8.43}$$

$$\begin{bmatrix} \underline{v}_c \\ \underline{v}_g \\ \underline{v}_\ell \\ \underline{v}_j \end{bmatrix} = \begin{bmatrix} \mathbf{F}_{ec}^T & \mathbf{F}_{cc}^T & \mathbf{0} & \mathbf{0} \\ \mathbf{F}_{eg}^T & \mathbf{F}_{cg}^T & \mathbf{F}_{rg}^T & \mathbf{0} \\ \mathbf{F}_{e\ell}^T & \mathbf{F}_{c\ell}^T & \mathbf{F}_{r\ell}^T & \mathbf{F}_{\ell\ell}^T \\ \mathbf{F}_{ej}^T & \mathbf{F}_{cj}^T & \mathbf{F}_{rj}^T & \mathbf{F}_{\ell j}^T \end{bmatrix} \begin{bmatrix} \underline{V}_e \\ \underline{V}_c \\ \underline{V}_r \\ \underline{V}_\ell \end{bmatrix} \tag{8.44}$$

Although this may seem to be a complicated representation, the four partitions may be regarded as a natural result of the goal of identifying voltage (tree)-type and current (link)-type elements as well as dynamic- and resistive-type elements. As we will see, each of these sixteen submatrices will be handled separately as our equation formulation proceeds. The ultimate result will be a set of equations sufficiently complicated to suggest the value of (8.43) and (8.44) is more conceptional than algorithmic.

8.5.3 A Tableau for Source-resistor Networks

As a prelude to the admittedly complicated formulation when dynamic elements are allowed, we now examine the analysis of networks that contain resistors and independent sources. In contrast to (8.41), we can now expect to obtain a non trivial solution for branch currents and voltages.

Based on the partitions of (8.43) and (8.44), the tableau of (8.38) can be expanded to the following form:

$$\left[\begin{array}{cccc|cccc} \mathbf{I} & & & & & \mathbf{F}_{ej} & & \mathbf{F}_{eg} \\ & \mathbf{I} & & & -\mathbf{F}_{ej}^T & & -\mathbf{F}_{rj}^T & \\ & & \mathbf{I} & & & \mathbf{F}_{rj} & & \mathbf{F}_{rg} \\ & & & \mathbf{I} & -\mathbf{F}_{eg}^T & & -\mathbf{F}_{rg}^T & \\ \hline & & & & \mathbf{I} & & & \\ & & & & & \mathbf{I} & & \\ & & -\mathbf{R} & & & & \mathbf{I} & \\ & & & -\mathbf{G} & & & & \mathbf{I} \end{array}\right] \left[\begin{array}{c} \underline{i}_e \\ \underline{v}_j \\ \underline{i}_r \\ \underline{v}_g \\ \hline \underline{V}_e \\ \underline{I}_j \\ \underline{V}_r \\ \underline{I}_g \end{array}\right] = \left[\begin{array}{c} \underline{0} \\ \underline{0} \\ \underline{0} \\ \underline{0} \\ \hline \underline{E} \\ \underline{J} \\ \underline{0} \\ \underline{0} \end{array}\right] \tag{8.45}$$

The first reduction step, based on the dotted partition shown above, yields

$$\left[\begin{array}{cc|cc} \mathbf{I} & \mathbf{0} & \mathbf{0} & \mathbf{0} \\ \mathbf{0} & \mathbf{I} & \mathbf{0} & \mathbf{0} \\ \hline \mathbf{0} & \mathbf{R}\,\mathbf{F}_{rj} & \mathbf{I} & \mathbf{R}\,\mathbf{F}_{rg} \\ -\mathbf{G}\,\mathbf{F}_{eg}^T & \mathbf{0} & & \mathbf{I} \end{array}\right] \left[\begin{array}{c} \underline{V}_e \\ \underline{I}_j \\ \hline \underline{V}_r \\ \underline{I}_g \end{array}\right] = \left[\begin{array}{c} \underline{E} \\ \underline{J} \\ \hline \underline{0} \\ \underline{0} \end{array}\right] \qquad (8.46)$$

The right-hand side, being nonzero, must be considered in the next reduction step. For the above partition, the formulae of Problem 8.1 yield

$$\left[\begin{array}{c|c} \mathbf{I} & \mathbf{R}\,\mathbf{F}_{rg} \\ \hline -\mathbf{G}\,\mathbf{F}_{rg}^T & \mathbf{I} \end{array}\right] \left[\begin{array}{c} \underline{V}_r \\ \hline \underline{I}_g \end{array}\right] = \left[\begin{array}{c} -\mathbf{R}\,\mathbf{F}_{rj}\underline{J} \\ \hline \mathbf{G}\,\mathbf{F}_{eg}^T\,\underline{E} \end{array}\right] \qquad (8.47)$$

This equation is then solved to yield

$$[\,\mathbf{I} + \mathbf{G}\,\mathbf{F}_{rg}^T\,\mathbf{R}\,\mathbf{F}_{rg}\,]\,\underline{I}_g = [\,-\mathbf{G}\,\mathbf{F}_{rg}^T\,\mathbf{R}\,\mathbf{F}_{rj}\,\underline{J} + \mathbf{G}\,\mathbf{F}_{eg}^T\,\underline{E}\,] \qquad (8.48)$$

Example 8.1

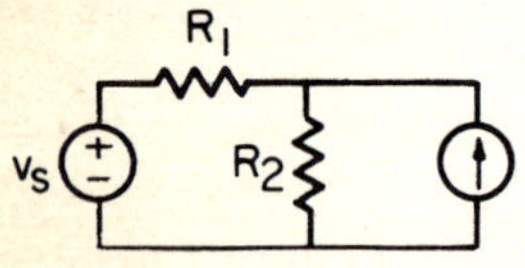

FIG. 8.9

Incident Matrix:

$$\mathbf{A}_i = \begin{array}{c} 1 \\ 2 \end{array} \overset{\begin{array}{cccc} v_s & R_1 & R_2 & i_s \end{array}}{\begin{bmatrix} 1 & 1 & 0 & 0 \\ 0 & -1 & 1 & -1 \end{bmatrix}}$$

Reduced Matrix:

$$[\mathbf{I}\ \ \mathbf{F}] = \left[\begin{array}{cc|cc} 1 & 0 & 1 & -1 \\ 0 & 1 & -1 & 1 \end{array}\right]$$

Partitioned Cut-Set Matrix:

$$\underset{\sim}{F} = \left[\begin{array}{c|c} \underset{\sim}{F}_{eg} & \underset{\sim}{F}_{ej} \\ \hline \underset{\sim}{F}_{rg} & \underset{\sim}{F}_{rj} \end{array}\right] = \begin{array}{c} \begin{array}{cc} R_2 & i_s \end{array} \\ \left[\begin{array}{c|c} 1 & -1 \\ \hline -1 & 1 \end{array}\right] \end{array}$$

Constitutive Equations:

$$\underset{\sim}{R} = [R_1] \qquad \underset{\sim}{G} = [G_2]$$

Network Equations:

$$[1 + G_2(-1)(R_1)(-1)]I_{G_2} = -(G_2)(-1)(R_1)(1)(i_s) + (G_2)(1)(v_s)$$

or

$$(1 + G_2 R_1) I_{G_2} = G_2 R_1 i_s + G_2 v_s$$

This result can be checked by viewing the network alternately as a current and voltage divider.

8.6 THE NETWORK STATE EQUATIONS

8.6.1 State Variable Analysis

An important mathematical concept applicable to most forms of discrete and continuous systems is that of state. Knowledge of the state of a system at a time t_o and knowledge of the inputs for $t > t_o$ is the necessary and sufficient information to determine the behavior of a system for $t > t_o$. It is not surprising that the "state" variables which carry this important information for electrical networks are the capacitor voltages and the inductor currents, since knowledge of the energy stored in the network can be determined solely from these variables.

We intend to show in this section that first order differential equations of the form

$$\dot{\underline{x}} = \underset{\sim}{A}\,\underline{x} + \underset{\sim}{B}\,\underline{u} \tag{8.49}$$

can be written for most dynamic networks, where

$\underline{X}$ is state vector

$$= [v_{C_1} \; . \; . \; . \; v_{C_n} \quad i_{L_1} \; . \; . \; . \; i_{L_m}]^T$$

$\underline{u}$ is a vector of independent source inputs

$$= [u_1 \; . \; . \; . \; u_s]$$

$\underset{\sim}{A}, \underset{\sim}{B}$ are constant matrices

Moreover, since the state variables may not include the output variables of interest, we may have to write an output equation

$$\underline{y} = \underset{\sim}{C}\,\underline{x} + \underset{\sim}{D}\,\underline{u} \tag{8.50}$$

where $\underline{y}$ is a vector of outputs. For example, the circuit of Figure 8.10 is described by the state and output equations

$$\frac{d}{dt}\begin{bmatrix} v_{C_1} \\ i_{L_1} \end{bmatrix} = \begin{bmatrix} -G_2/C_1 & 1/C_1 \\ -1/L_1 & -R_1/L_1 \end{bmatrix}\begin{bmatrix} v_{C_1} \\ i_{L_1} \end{bmatrix} + \begin{bmatrix} 0 \\ 1/L_1 \end{bmatrix} v_s$$

$$v_o = [1 \quad 0]\begin{bmatrix} v_{C_1} \\ i_{L_1} \end{bmatrix} + [0]\, v_s$$

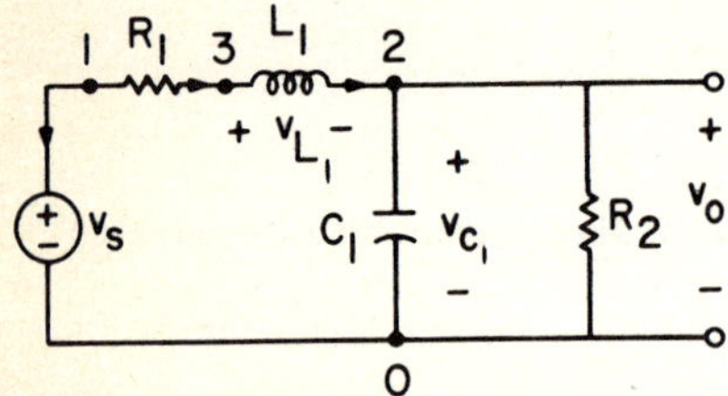

FIG. 8.10 Dynamic Network

8.6.2 State Equations for RCL Networks

The Proper Tree. The tree which is used as the basis for finding $\underset{\sim}{F}$ must have special properties for state variable purposes. In particular it must contain

1. as many voltage sources and capacitors as possible, in that order of importance,
2. as few current sources and inductros as possible, respectively.

Such a tree is a tree is termed a proper tree.

The tree branch priority of Table 8.1, which yields tree-link variables in the order of (8.42), is clearly matched to the above requirements. The corresponding partion of $\underset{\sim}{F}$ as given in (8.43) and (8.44) will therefore turn out to be a natural one for the state variable analysis techniques.

Constitutive Relationships. For each of the eight categories represented by (8.42), we must define a branch voltage-current relationship. Accordingly, we represent the constitutive relationships as:

Independent Voltage Sources: $\underline{V}_e = \underline{E}$

Tree Capacitors: $\underline{i}_c = \underset{\sim}{C}\,\dfrac{d\underline{V}_c}{dt}$

Tree Resistors: $\underline{V}_r = \underset{\sim}{R}\,\underline{i}_r$

Tree Inductors: $\underline{V}_\ell = \underset{\sim}{L}_2\,\dfrac{d\underline{i}_\ell}{dt}$

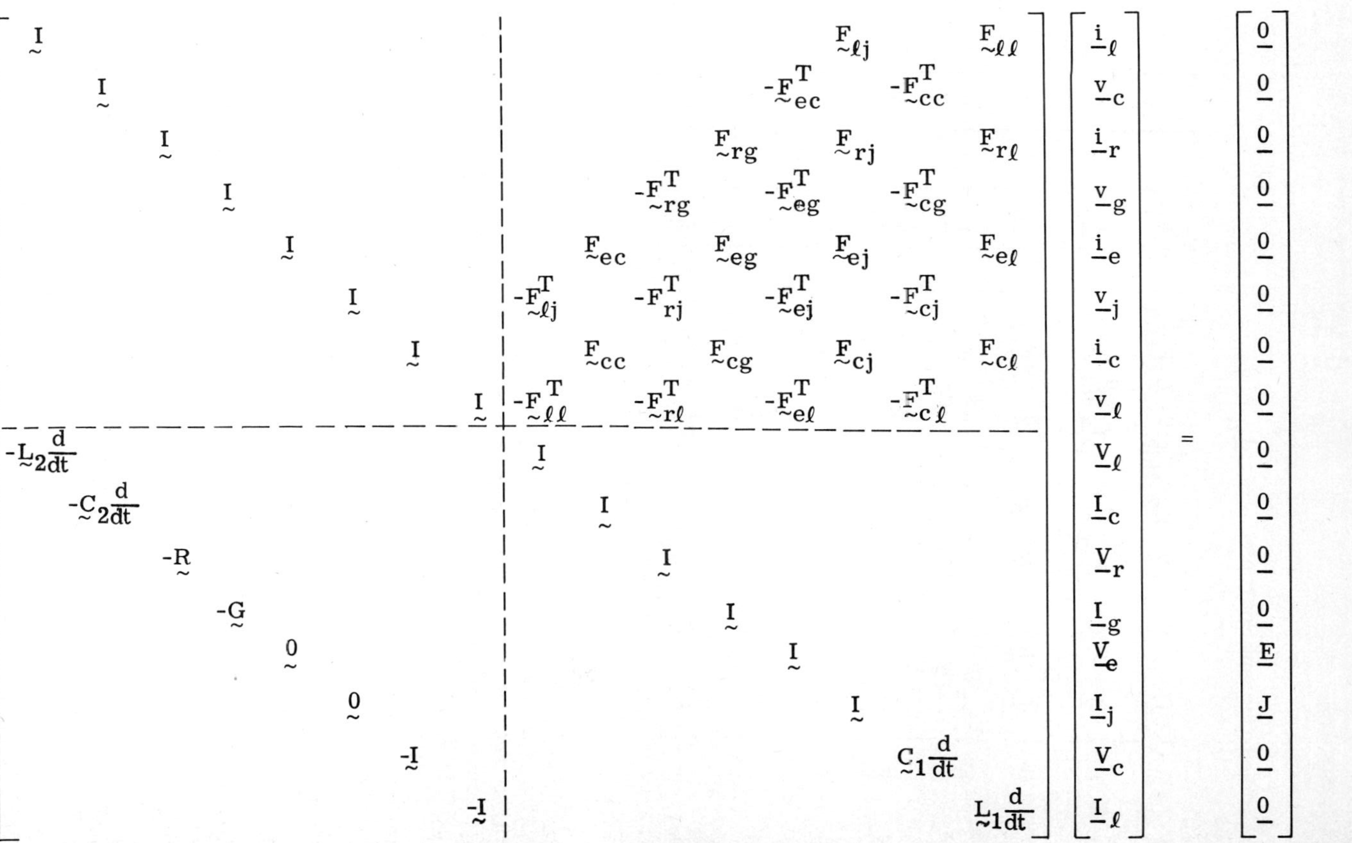

$$\left[\begin{array}{cccccccc|cccccccc}
\mathbf{I} & & & & & & & & & & & & & \mathbf{F}_{\ell j} & & \mathbf{F}_{\ell\ell} \\
& \mathbf{I} & & & & & & & & & & & -\mathbf{F}^T_{ec} & & -\mathbf{F}^T_{cc} & \\
& & \mathbf{I} & & & & & & & & & \mathbf{F}_{rg} & & \mathbf{F}_{rj} & & \mathbf{F}_{r\ell} \\
& & & \mathbf{I} & & & & & & & -\mathbf{F}^T_{rg} & & -\mathbf{F}^T_{eg} & & -\mathbf{F}^T_{cg} & \\
& & & & \mathbf{I} & & & & & \mathbf{F}_{ec} & & \mathbf{F}_{eg} & & \mathbf{F}_{ej} & & \mathbf{F}_{e\ell} \\
& & & & & \mathbf{I} & & & -\mathbf{F}^T_{\ell j} & & -\mathbf{F}^T_{rj} & & -\mathbf{F}^T_{ej} & & -\mathbf{F}^T_{cj} & \\
& & & & & & \mathbf{I} & & & \mathbf{F}_{cc} & & \mathbf{F}_{cg} & & \mathbf{F}_{cj} & & \mathbf{F}_{c\ell} \\
& & & & & & & \mathbf{I} & -\mathbf{F}^T_{\ell\ell} & & -\mathbf{F}^T_{r\ell} & & -\mathbf{F}^T_{e\ell} & & -\mathbf{F}^T_{c\ell} & \\
\hline
-\mathbf{L}_2\frac{d}{dt} & & & & & & & & \mathbf{I} & & & & & & & \\
& -\mathbf{C}_2\frac{d}{dt} & & & & & & & & \mathbf{I} & & & & & & \\
& & -\mathbf{R} & & & & & & & & \mathbf{I} & & & & & \\
& & & -\mathbf{G} & & & & & & & & \mathbf{I} & & & & \\
& & & & \mathbf{0} & & & & & & & & \mathbf{I} & & & \\
& & & & & \mathbf{0} & & & & & & & & \mathbf{I} & & \\
& & & & & & -\mathbf{I} & & & & & & & & \mathbf{C}_1\frac{d}{dt} & \\
& & & & & & & -\mathbf{I} & & & & & & & & \mathbf{L}_1\frac{d}{dt}
\end{array}\right]
\left[\begin{array}{c}
\underline{i}_\ell \\ \underline{v}_c \\ \underline{i}_r \\ \underline{v}_g \\ \underline{i}_e \\ \underline{v}_j \\ \underline{i}_c \\ \underline{v}_\ell \\ \underline{V}_\ell \\ \underline{I}_c \\ \underline{V}_r \\ \underline{I}_g \\ \underline{V}_e \\ \underline{I}_j \\ \underline{V}_c \\ \underline{I}_\ell
\end{array}\right]
=
\left[\begin{array}{c}
\underline{0} \\ \underline{0} \\ \underline{0} \\ \underline{0} \\ \underline{0} \\ \underline{0} \\ \underline{0} \\ \underline{0} \\ \underline{0} \\ \underline{0} \\ \underline{0} \\ \underline{0} \\ \underline{E} \\ \underline{J} \\ \underline{0} \\ \underline{0}
\end{array}\right] \tag{8.51}$$

Link Capacitors: $\underline{I}_c = \mathbf{C}_2 \dfrac{dv_c}{dt}$

Link Resistors: $\underline{I}_g = \mathbf{G}\,\underline{V}_g$

Independent Current Sources: $\underline{I}_j = \underline{J}$

$$\left[\begin{array}{cc|cccccc}
\mathbf{I} & \mathbf{0} & \mathbf{0} & \mathbf{0} & \mathbf{0} & \mathbf{L}_2\mathbf{F}_{\ell j}\frac{d}{dt} & 0 & \mathbf{L}_2\mathbf{F}_{\ell\ell}\frac{d}{dt} \\
\mathbf{0} & \mathbf{I} & \mathbf{0} & \mathbf{0} & -\mathbf{C}_2\mathbf{F}_{ec}^T\frac{d}{dt} & \mathbf{0} & -\mathbf{C}_2\mathbf{F}_{cc}^T\frac{d}{dt} & \mathbf{0} \\
\hline
\mathbf{0} & \mathbf{0} & \mathbf{I} & \mathbf{R}\mathbf{F}_{rg} & \mathbf{0} & \mathbf{R}\mathbf{F}_{rj} & \mathbf{0} & \mathbf{R}\mathbf{F}_{r\ell} \\
\mathbf{0} & \mathbf{0} & -\mathbf{G}\mathbf{F}_{rg}^T & \mathbf{I} & -\mathbf{G}\mathbf{F}_{eg}^T & \mathbf{0} & -\mathbf{G}\mathbf{F}_{cg}^T & \mathbf{0} \\
\mathbf{0} & \mathbf{0} & \mathbf{0} & \mathbf{0} & \mathbf{I} & \mathbf{0} & \mathbf{0} & \mathbf{0} \\
\mathbf{0} & \mathbf{0} & \mathbf{0} & \mathbf{0} & \mathbf{0} & \mathbf{I} & \mathbf{0} & \mathbf{0} \\
\mathbf{0} & \mathbf{F}_{cc} & \mathbf{0} & \mathbf{F}_{cg} & \mathbf{0} & \mathbf{F}_{cj} & \mathbf{C}_1\frac{d}{dt} & \mathbf{F}_{c\ell} \\
-\mathbf{F}_{\ell\ell}^T & \mathbf{0} & -\mathbf{F}_{r\ell}^T & \mathbf{0} & -\mathbf{F}_{e\ell}^T & \mathbf{0} & -\mathbf{F}_{c\ell}^T & \mathbf{L}_1\frac{d}{dt}
\end{array}\right]
\begin{bmatrix} \underline{V}_1 \\ \underline{I}_c \\ \underline{V}_r \\ \underline{I}_g \\ \underline{V}_e \\ \underline{I}_j \\ \underline{V}_c \\ \underline{I}_\ell \end{bmatrix}
=
\begin{bmatrix} \underline{0} \\ \underline{0} \\ \underline{0} \\ \underline{0} \\ \underline{E} \\ \underline{J} \\ \underline{0} \\ \underline{0} \end{bmatrix}
\tag{8.52}$$

The Complete Tableau (T_a). Combining the eight topological and eight constitutive relationships into a single matrix equation yields the tableau of (8.51). The variables in this case have been specifically arranged to yield state equations of the form of (8.49) as the tableau is reduced in the manner of Section 8.3.4.

The first reduction step using the partition shown in (8.51) removes all tree branch currents and link branch voltages as variables and yields a reduced tableau of mixed topological and constitutive submatrices. This tableau is shown in (8.52).

The results of successive reduction steps are displayed in (8.53) and (8.54) using the partition shown in the previous tableau. It is particularly important to note that the derivatives of variables associated with excess elements vanish in the reduction due to the absence of certain topological submatrices in (8.43) and (8.44).

Equation (8.54) can be finally reduced to the following state form:

$$\begin{bmatrix} \mathbf{C}_{12} & \mathbf{0} \\ \mathbf{0} & \mathbf{L}_{12} \end{bmatrix}
\begin{bmatrix} \dot{\underline{V}}_c \\ \dot{\underline{I}}_\ell \end{bmatrix}
=
\begin{bmatrix} \mathbf{Y}_a & \boldsymbol{\alpha}_a \\ \boldsymbol{\mu}_a & \mathbf{Z}_a \end{bmatrix}
\begin{bmatrix} \underline{V}_c \\ \underline{I}_\ell \end{bmatrix}
+
\begin{bmatrix} \mathbf{Y}_b & \boldsymbol{\alpha}_b \\ \boldsymbol{\mu}_b & \mathbf{Z}_b \end{bmatrix}
\begin{bmatrix} \underline{E} \\ \underline{J} \end{bmatrix}
+
\begin{bmatrix} \mathbf{Y}'_b & \mathbf{0} \\ \mathbf{0} & \mathbf{Z}'_b \end{bmatrix}
\begin{bmatrix} \dot{\underline{E}} \\ \dot{\underline{J}} \end{bmatrix}$$

$$\left[\begin{array}{cc|cccc} \underset{\sim}{I} & \underset{\sim}{R}\underset{\sim}{F}_{rg} & \underset{\sim}{0} & \underset{\sim}{R}\underset{\sim}{F}_{rj} & \underset{\sim}{0} & \underset{\sim}{R}\underset{\sim}{F}_{r\ell} \\ -\underset{\sim}{G}\underset{\sim}{F}_{rg}^{T} & \underset{\sim}{I} & -\underset{\sim}{G}\underset{\sim}{F}_{eg}^{T} & \underset{\sim}{0} & -\underset{\sim}{G}\underset{\sim}{F}_{cg}^{T} & \underset{\sim}{0} \\ \hline \underset{\sim}{0} & \underset{\sim}{0} & \underset{\sim}{I} & \underset{\sim}{0} & \underset{\sim}{0} & \underset{\sim}{0} \\ \underset{\sim}{0} & \underset{\sim}{0} & \underset{\sim}{0} & \underset{\sim}{I} & \underset{\sim}{0} & \underset{\sim}{0} \\ \underset{\sim}{0} & \underset{\sim}{F}_{cg} & \underset{\sim}{F}_{cc}\underset{\sim}{C}_{2}\underset{\sim}{F}_{ec}^{T}\frac{d}{dt} & \underset{\sim}{F}_{cj} & \underset{\sim}{C}_{1}\frac{d}{dt}+F_{cc}C_{2}F_{cc}^{T}\frac{d}{dt} & \underset{\sim}{F}_{c\ell} \\ -\underset{\sim}{F}_{r\ell}^{T} & \underset{\sim}{0} & -\underset{\sim}{F}_{e\ell}^{T} & \underset{\sim}{F}_{\ell\ell}^{T}\underset{\sim}{L}_{2}\underset{\sim}{F}_{\ell j}\frac{d}{dt} & -\underset{\sim}{F}_{c\ell}^{T} & \underset{\sim}{L}_{1}\frac{d}{dt}+\underset{\sim}{F}_{\ell\ell}^{T}\underset{\sim}{L}_{2}\underset{\sim}{F}_{\ell\ell}\frac{d}{dt} \end{array}\right] \begin{bmatrix} \underline{V}_{r} \\ \underline{I}_{g} \\ \underline{V}_{e} \\ \underline{I}_{j} \\ \underline{V}_{c} \\ \underline{I}_{\ell} \end{bmatrix} = \begin{bmatrix} \underline{0} \\ \underline{0} \\ \underline{E} \\ \underline{J} \\ \underline{0} \\ \underline{0} \end{bmatrix} \qquad (8.53)$$

$$
\left[\begin{array}{c:c:c:c}
I & 0 & 0 & 0 \\
0 & I & 0 & 0 \\
\hdashline
\begin{array}{l} F_{cg}R_1^{-1}F_{eg}^T \\ +F_{cc}C_2F_{ec}^T\frac{d}{dt} \end{array} & F_{cj}-F_{cg}R_1^{-1}F_{rg}^TR-F_{rj} & \begin{array}{l} F_{cg}R_1^{-1}F_{cg}^T \\ +(C_1+F_{cc}C_2F_{cc}^T)\frac{d}{dt} \end{array} & F_{c\ell}-F_{cg}R_1^{-1}F_{rg}^TRF_{r\ell} \\
-F_{e\ell}^T+F_{r\ell}^TRF_{rg}R_1^{-1}F_{eg}^T & \begin{array}{l} F_{r\ell}^TG_1^{-1}F_{rj} \\ +F_{\ell\ell}^TL_2F_{\ell j}\frac{d}{dt} \end{array} & -F_{c\ell}^T+F_{r\ell}^TRF_{rg}R_1^{-1}F_{cg}^T & \begin{array}{l} F_{r\ell}^TG_1^{-1}F_{r\ell} \\ +(L_1+F_{\ell\ell}^TL_2F_{\ell\ell})\frac{d}{dt} \end{array}
\end{array}\right]
\begin{bmatrix} \underline{V}_e \\ \underline{I}_j \\ \underline{V}_c \\ \underline{I}_\ell \end{bmatrix}
=
\begin{bmatrix} \underline{E} \\ \underline{J} \\ \underline{0} \\ \underline{0} \end{bmatrix}
$$

where $R_1 = G^{-1} + F_{rg}^T R F_{rg}$

$$G_1 = R^{-1} + F_{rg} G F_{rg}^T \tag{8.54}$$

where

$$\underset{\sim}{C}_{12} = \underset{\sim}{C}_1 + \underset{\sim}{F}_{cc}\underset{\sim}{C}_2\underset{\sim}{F}_{cc}^T$$

$$\underset{\sim}{L}_{12} = \underset{\sim}{L}_1 + \underset{\sim}{F}_{\ell\ell}^T\underset{\sim}{L}_2\underset{\sim}{F}_{\ell\ell}$$

$$\underset{\sim}{Y}_a = -\underset{\sim}{F}_{cg}\underset{\sim}{R}_1^{-1}\underset{\sim}{F}_{cg}^T$$

$$\underset{\sim}{\alpha}_a = -\underset{\sim}{F}_{c\ell} + \underset{\sim}{F}_{cg}\underset{\sim}{R}_1^{-1}\underset{\sim}{F}_{rg}^T\underset{\sim}{R}\,\underset{\sim}{F}_{r\ell} = -\underset{\sim}{\mu}_a^T$$

$$\underset{\sim}{Z}_a = -\underset{\sim}{F}_{r\ell}^T\underset{\sim}{G}_1^{-1}\underset{\sim}{F}_{r\ell}$$

$$\underset{\sim}{Y}_b = -\underset{\sim}{F}_{cg}\,R_1^{-1}\underset{\sim}{F}_{eg}^T$$

$$\underset{\sim}{\alpha}_b = -\underset{\sim}{F}_{cj} + \underset{\sim}{F}_{cg}\underset{\sim}{R}_1^{-1}\underset{\sim}{F}_{eg}^T$$

$$\underset{\sim}{\mu}_b = \underset{\sim}{F}_{e\ell}^T - \underset{\sim}{F}_{r\ell}^T\underset{\sim}{R}\,\underset{\sim}{F}_{rg}\underset{\sim}{R}_1^{-1}\underset{\sim}{F}_{eg}^T$$

$$\underset{\sim}{Z}_b = -\underset{\sim}{F}_{r\ell}^T\underset{\sim}{G}_1^{-1}\underset{\sim}{F}_{rj}$$

$$\underset{\sim}{Y}_b' = \underset{\sim}{F}_{cc}\underset{\sim}{C}_2\underset{\sim}{F}_{ec}^T$$

$$\underset{\sim}{Z}_b' = -\underset{\sim}{F}_{\ell\ell}^T\underset{\sim}{L}_2\underset{\sim}{F}_{\ell j}$$

The equation is remarkable in the sense that it is not the desired form since derivatives of the sources $\underline{J}$ and $\underline{E}$ are present. If we insist on choosing capacitor voltages and inductor currents as state variables, then we cannot in general avoid such terms. For example, two possible state equations for the network of Figure 8.11 are

$$\frac{dv_{C_2}}{dt} = \left(\frac{C_1}{C_1 + C_2}\right)\frac{dv_s}{dt}, \qquad \frac{dv_{C_1}}{dt} = \left(\frac{C_2}{C_1 + C_2}\right)\frac{dv_s}{dt}$$

if a single capacitor voltage is to be chosen for a state variable.

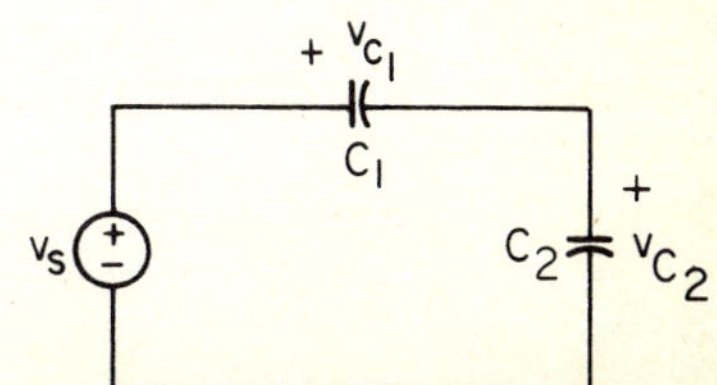

FIG. 8.11 Network Exhibiting Input Derivatives

Note, however, that the existence of the matrices Z'_b and Y'_b which account for these derivatives both require tree inductors and link capacitors be present so that $\underset{\sim}{F}_{cc}$ and $\underset{\sim}{F}^T_{\ell\ell}$ are nonzero.

<u>Example 8.2</u>

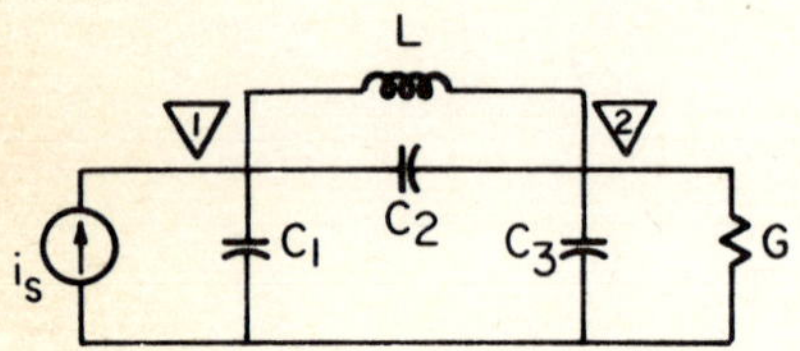

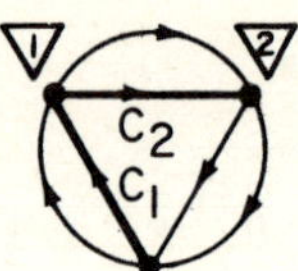

FIG. 8.12 Example Network and Graph

Incident Matrix:

$$\underset{\sim}{A}_i = \begin{array}{c} \\ 1 \\ 2 \end{array}\begin{array}{c} \begin{array}{cccccc} C_1 & C_2 & C_3 & G & L & i_s \end{array} \\ \begin{bmatrix} 1 & 1 & 0 & 0 & 1 & -1 \\ 0 & -1 & 1 & 1 & -1 & 0 \end{bmatrix} \end{array}$$

Reduced Matrix:

$$\left[\underset{\sim}{I} \;\middle|\; \underset{\sim}{F}\right] = \begin{array}{c} \begin{array}{cccc} C_3 & G & L & i_s \end{array} \\ \left[\begin{array}{cc|cccc} 1 & 0 & 1 & 1 & 0 & -1 \\ 0 & 1 & -1 & -1 & 1 & 0 \end{array}\right] \end{array}$$

Partitioned Cut-Set Matrix:

$$\underset{\sim}{F} = \left[\begin{array}{c|c|c|c} 1 & 1 & 0 & -1 \\ -1 & -1 & 1 & 0 \end{array}\right] = \left[\begin{array}{c|c|c|c} \underset{\sim}{F}_{cc} & \underset{\sim}{F}_{cg} & \underset{\sim}{F}_{c\ell} & \underset{\sim}{F}_{cj} \end{array}\right]$$

Constitutive Equations:

$$\underset{\sim}{C}_1 = \begin{bmatrix} C_1 & 0 \\ 0 & C_2 \end{bmatrix} \qquad \underset{\sim}{C}_2 = [\underset{\sim}{C}_3] \qquad \underset{\sim}{L}_1 = [L]$$

$$\underset{\sim}{G} = [G] \qquad \underline{J} = [i_s] \qquad R_1^{-1} = [G^{-1}]^{-1} = [G]$$

Component Matrices:

$$\underset{\sim}{C}_{12} = \begin{bmatrix} C_1 & 0 \\ 0 & C_2 \end{bmatrix} + \begin{bmatrix} 1 \\ -1 \end{bmatrix} [C_3][1 \quad -1] = \begin{bmatrix} C_1 + C_3 & -C_3 \\ -C_3 & C_2 + C_3 \end{bmatrix}$$

$$\underset{\sim}{L}_{12} = [L]$$

$$\underset{\sim}{Y}_a = -\begin{bmatrix} 1 \\ -1 \end{bmatrix} [G][1 \quad -1] = \begin{bmatrix} -G & G \\ G & -G \end{bmatrix}$$

$$\underset{\sim}{\alpha}_a = -\begin{bmatrix} 0 \\ 1 \end{bmatrix} \qquad \mu_a = [0 \quad 1]$$

$$\underset{\sim}{\alpha}_b = -\begin{bmatrix} -1 \\ 0 \end{bmatrix}$$

State Equations:

$$\left[\begin{array}{cc|c} C_1 + C_3 & -C_3 & 0 \\ -C_3 & C_2 + C_3 & 0 \\ \hline 0 & 0 & L \end{array}\right] \frac{d}{dt} \left[\begin{array}{c} v_{C_1} \\ v_{C_2} \\ \hline i_\ell \end{array}\right] = \left[\begin{array}{cc|c} -G_1 & G_1 & 0 \\ G_1 & -G_1 & -1 \\ \hline 0 & 1 & 0 \end{array}\right] \left[\begin{array}{c} v_{C_1} \\ v_{C_2} \\ \hline i_\ell \end{array}\right] + \left[\begin{array}{c} 1 \\ 0 \\ \hline 0 \end{array}\right] i_s$$

8.7 FORMULATION FOR NUMERICAL SOLUTION

8.7.1 The Network Tableau (T_n)

Similar to the formulation for analytic solution, we form a tableau of both topological and constitutive equations. If the network is linear, we can write from (8.4) and (8.5)

$$\begin{array}{l} \text{KVL (8.5)} \\ \text{KCL (8.4)} \\ \text{CR} \end{array} \begin{bmatrix} \underset{\sim}{A}_i^T & -\underset{\sim}{I} & \underset{\sim}{0} \\ \underset{\sim}{0} & \underset{\sim}{0} & \underset{\sim}{A}_i \\ \underset{\sim}{0} & \underset{\sim}{Y} & \underset{\sim}{Z} \end{bmatrix} \begin{bmatrix} \underset{\sim}{v}_n \\ \underline{v}_b \\ \underline{i}_b \end{bmatrix} = \begin{bmatrix} \underline{0} \\ \underline{0} \\ \underline{f} \end{bmatrix}$$

where $\underset{\sim}{Y}$, $\underset{\sim}{Z}$, and $\underline{f}$ contain the parameter and operator information necessary to define the constitutive relationships.

Example 8.3. The equations for the network of Figure 8.12 are to be reformulated for numerical solution.

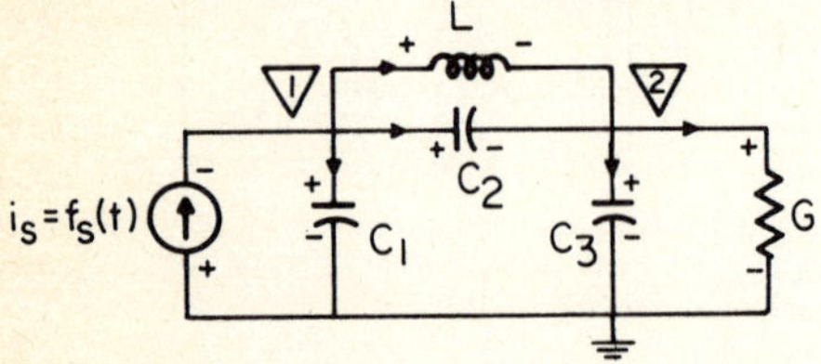

FIG. 8.13 Example 8.2 Re-visited

Incidence Matrix:

$$
\underset{\sim}{A}_i = \begin{array}{c} \\ ① \\ ② \end{array}
\begin{array}{c}
\begin{array}{cccccc} C_1 & L & C_2 & G & C_3 & i_s \end{array} \\
\begin{bmatrix} 1 & 1 & 1 & 0 & 0 & -1 \\ 0 & -1 & -1 & 1 & 1 & 0 \end{bmatrix}
\end{array}
$$

Network Tableau

$$
\begin{array}{l}
\text{KVL} \\ \\ \\ \\ \\ \\ \text{KCL} \\ \\ \text{CR} \\ \\ \\ \\ \\ \\
\end{array}
\left[\begin{array}{cc|cccccc|cccccc}
1 & 0 & -1 & & & & & & & & & & & \\
1 & -1 & & -1 & & & & & & & & & & \\
1 & -1 & & & -1 & & & & & & & & & \\
0 & 1 & & & & -1 & & & & & & & & \\
0 & 1 & & & & & -1 & & & & & & & \\
-1 & 0 & & & & & & -1 & & & & & & \\
\hline
 & & & & & & & & 1 & 1 & 1 & 0 & 0 & -1 \\
 & & & & & & & & 0 & -1 & -1 & 1 & 1 & 0 \\
\hline
 & & C_1 d/dt & & & & & & -1 & & & & & \\
 & & & -1 & & & & & & L d/dt & & & & \\
 & & & & C_2 d/dt & & & & & & -1 & & & \\
 & & & & & G & & & & & & -1 & & \\
 & & & & & & C_3 d/dt & & & & & & -1 & \\
 & & & & & & & 0 & & & & & & 1
\end{array}\right]
\left[\begin{array}{c}
v_1 \\ v_2 \\ \hline v_{c_1} \\ v_\ell \\ v_{c_2} \\ v_g \\ v_{c_3} \\ v_s \\ \hline i_{c_1} \\ i_\ell \\ i_{c_2} \\ i_g \\ i_{c_3} \\ i_s
\end{array}\right]
=
\left[\begin{array}{c}
\underline{0} \\ \hline \\ \\ \underline{0} \\ \\ \\ \hline 0 \\ 0 \\ 0 \\ 0 \\ 0 \\ f_s(t)
\end{array}\right]
$$

Column groups: node voltages | branch voltages | branch currents. The node-voltage block of the KVL rows is $\underset{\sim}{A}_i^T$; the branch-current block of the KCL rows is $\underset{\sim}{A}_i$; the branch-voltage block of the CR rows is $\underset{\sim}{Y}$ and the branch-current block of the CR rows is $\underset{\sim}{Z}$.

Any differential or nonlinear operators in the constitutive equations may be handled by discretization or linearization before the formation of the tableau. For example, a nonlinear element described by i = f(v) would first be represented in its linearized form

$$i^{m+1} = i^m + \left(\frac{\partial i}{\partial v}\right)^m (v^{m+1} - v^m)$$

if Newton iteration were being used for solution. As noted in Chapter III, this relationship describes the linearized branch model of Figure 8.14a;

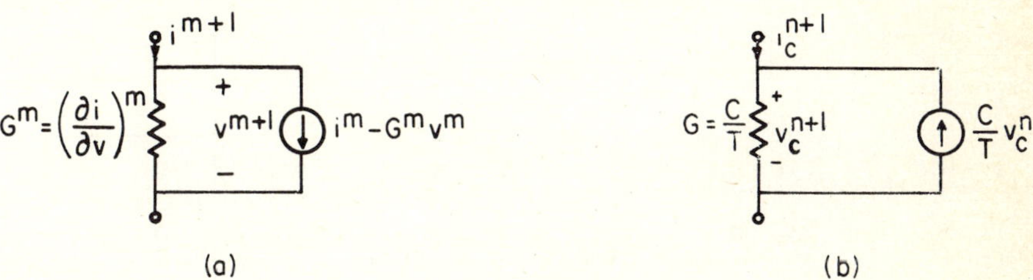

FIG. 8.14 Linearized, Discretized Branch Models

this model would appear in the tableau as two elements (a conductance and a current source) or, more economically, the source would simply appear on the right hand side of the constitutive equation for the nonlinear element. A similar process is followed for the capacitor described by

$$i_c = \frac{C dv_c}{dt}$$

This expression is discretized using the backward Euler approximation (or similar multistep formula of Chapter IX) to yield

$$i_c^{n+1} = \frac{C(v_c^{n+1} - v_c^n)}{T}$$

which is represented in Figure 8.14b.

<u>Example 8.4.</u> The transient response of the network of Figure 8.15 is to be found using the backward Euler formula with Newton iteration.

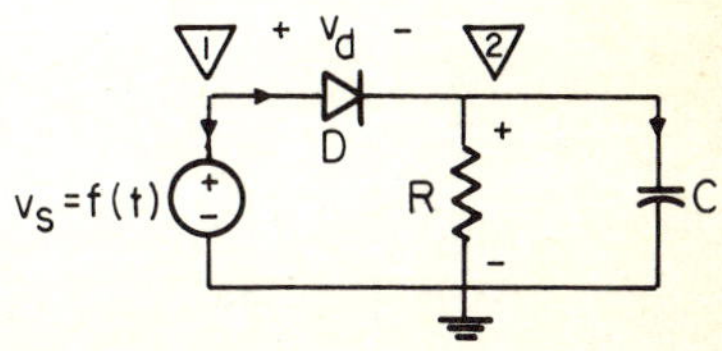

FIG. 8.15 Nonlinear, Dynamic Example

The tableau at the (m+1)st iteration during the (n+1)st time step is to be formed.

Incidence Matrix:

$$
\underset{\sim}{A}_i = \begin{array}{c} \textcircled{1} \\ \textcircled{2} \end{array}
\overset{\begin{array}{cccc} v_s & D & R & C \end{array}}{\begin{bmatrix} 1 & 1 & 1 & 1 \\ 0 & -1 & 0 & 0 \end{bmatrix}}
$$

Network Tableau (T_n)

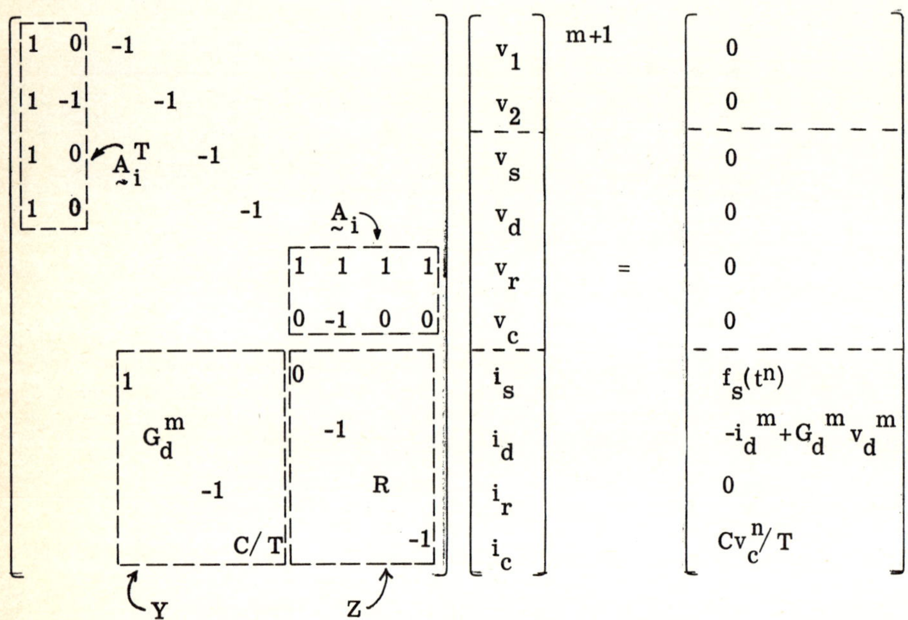

8.7.2 Numerical Solution of T_n

The solution of the network tableau equations is carried out by Gaussian elimination or LU factorization with interchange of rows and columns to avoid zero-valued pivots. Obviously, special processing techniques to exploit the matrix sparsity must be used. These are considered in detail in Chapter X.

8.7.3 Comparison of T_a and T_n

Both tableaus T_a and T_n have strengths and weaknesses deserving of comment.

(a) Both the tree-link sort and the reduction of T_a by elimination of the submatrices of $\underset{\sim}{F}$ are replaced by one Gaussian elimination step in the solution of $\underset{\sim}{T}_n$. A scan of the network branch list permits direct insertion (in any order) of both the topological and constitutive information into T_n.

(b) There are no assurances that T_n does not have a zero-valued determinant. In some instances, this occurs because the network is inconsistent (independent voltage sources in a loop), in which case it might be difficult to determine why the inconsistency occurred and so inform the user. In other cases (see Problem 8.2), this occurs because some uninteresting network variables are in fact indeterminant and, because T_n involves <u>all</u> network variables, the determinant must necessarily be zero-valued.

(c) T_n is an excellent vehicle for studying the relative solution efficiency of different sets of solution variables. For example, by selecting a special pivot order, the node voltages can be made the last n-1 solution variables, thus simulating node analysis. Similarly, another pivot order can be chosen to represent state variable analysis.

Problems

8.1 Let a matrix be partitioned into the form

$$\begin{bmatrix} \underset{\sim}{A}_{11} & \underset{\sim}{A}_{12} \\ \underset{\sim}{A}_{21} & \underset{\sim}{A}_{22} \end{bmatrix} \begin{bmatrix} \underline{x}_1 \\ \underline{x}_2 \end{bmatrix} = \begin{bmatrix} \underline{b}_1 \\ \underline{b}_2 \end{bmatrix}$$

Show that a solution for $\underline{x}_2$ exists in the form

$$[\underset{\sim}{A}_{22} - \underset{\sim}{A}_{21} \underset{\sim}{A}_{11}^{-1} \underset{\sim}{A}_{12}] \underline{x}_2 = \underline{b}_2 - \underset{\sim}{A}_{21} \underset{\sim}{A}_{11}^{-1} \underline{b}_1$$

8.2 It is sometimes necessary in performing an integration to solve the tableau T_n with $T = 0$ (for example, if a signal step is applied as t^{n-} and the response at t^{n+} is desired). What special property could T_n have under these conditions? How could it be solved? Does a similar problem occur when $T = \infty$, i.e., the dc case? You will find the T_a tableau of (8.51) useful in solving this problem.

8.3 It is desired to be able to describe certain pedagogical one-port elements known as the norator and nullator. In the latter, both v and i are fixed ($v = c_1$, $i = c_2$) and in the former both v and i are arbitrary.

(a) How are these elements represented in the network tableau T_n?
(b) What relation must exist between the number of nullators and the number of norators?

8.4 The solution of the tableau T_n of Example 8.4 by Newton iteration can be carried out without completing the back substitution of the topological equations at each iteration. How can this procedure be justified? (For numerical reasons, however, the back substitution may indeed have to be carried out.)

REFERENCES

8.1 Hachtel, G. D., R. R. Brayton, and F. G. Gustavson, "The Sparse Tableau Approach to Network Analysis and Design," Trans. IEEE, vol. CT-18, no. 1, pp. 101-113; January, 1971.
8.2 Pottle, C., "State Space Techniques for General Active Network Analysis," Chapter 3 of System Analysis by Digital Computer (Kuo and Kaiser, editors), Wiley; 1966.
8.3 Chua, L. O., and D. A. Pereault, "Analysis of Nonlinear RLC Networks with Controlled Sources," Digest of 1970 IEEE International Symposium on Circuit Theory; pp. 89-90.
8.4 Branin, F. H., Jr., and K. U. Wang, "A New Hybrid Formulation of the Network Equations," IBM Report TR21.409 (Kingston, N. Y.); December, 1970.
8.5 Seshu, S., and M. B. Reed, Linear Graphs and Electrical Networks, Addison Wesley; 1961.
8.6 Ohtsuki, T., and H. Watanabe, "State Variable Analysis of RLC Networks Containing Nonlinear Coupling Elements," Trans. IEEE, vol. CT-16, no. 1, pp. 26-38; February, 1969.

9

Advanced Numerical Techniques in Transient Analysis

9.1 INTRODUCTION

The limited discussion in Chapter IV of numerical techniques for transient analysis is far removed from reflecting the state of the art. First, the numerical integration methods themselves were not accurate (the trapezoidal rule being second order and the most accurate). Second, the subject of error control was ignored. Finally, stability analysis was limited to single linear differential equations.

The first step in expanding the scope of this analysis is to consider the transient solution of systems of linear differential equations. We will spend considerable time on this topic, since the linear case is easiest to analyze and offers considerable insight into the solution of nonlinear equations.

Next, a very powerful implicit integration algorithm will be discussed in detail, including an associated error-control technique. This will lead us to the subject of simultaneous solution of differential and algebraic equations (as arise, for example, in node analysis) and a study of the stability of the solution of nonlinear equations. In the process, we will come to some rather remarkable conclusions, such as "integration in a capacitor voltage variable rather a capacitor charge variable can result in a form of numerical instability!"

9.2 ANALYTIC SOLUTION OF LINEAR DIFFERENTIAL EQUATIONS

To appreciate the nature of the approximation involved in numerical solution of systems of linear differential equations, knowledge of the form of the exact solution is important. Accordingly, we derive the analytic solution of the vector differential equation

$$\dot{\underline{x}} = \underset{\sim}{A}\underline{x} + \underset{\sim}{B}\underline{u} \tag{9.1}$$

where

$\underline{x}$ is a vector of ℓ solution variables
$\underline{u}$ is an input vector
$\underset{\sim}{A}$ and $\underset{\sim}{B}$ are constant matrices

In the same manner that the exponential function is defined

$$e^{\alpha t} = 1 + \alpha t + \frac{(\alpha t)^2}{2!} + \ldots + \frac{(\alpha t)^k}{k!} + \ldots = \sum_{k=0}^{\infty} \frac{(\alpha t)^k}{k!} \tag{9.2}$$

so too can the matrix exponential $e^{\underset{\sim}{A}t}$ be defined by the series expansion

$$e^{\underset{\sim}{A}t} = \underset{\sim}{I} + \underset{\sim}{A}t + \frac{(\underset{\sim}{A}t)^2}{2!} + \ldots + \frac{(\underset{\sim}{A}t)^k}{k!} + \ldots = \sum_{k=0}^{\infty} \frac{(\underset{\sim}{A}t)^k}{k!} \tag{9.3}$$

Now consider the case $\underline{u} = \underline{0}$; Eq. (9.1) becomes

$$\dot{\underline{x}} = \underset{\sim}{A}\underline{x} \tag{9.4}$$

We now show that the matrix solution of (9.4) is

$$\underline{x}(t) = e^{\underset{\sim}{A}t}\underline{x}(0) \tag{9.5}$$

Substituting the expansion of (9.3) into (9.4) we have

$$\frac{d}{dt}\left[e^{\underset{\sim}{A}t}\underline{x}(0)\right] = \sum_{k=1}^{\infty} \frac{\underset{\sim}{A}^k t^{k-1}}{(k-1)!}\underline{x}(0)$$

$$= \underset{\sim}{A} \sum_{k=0}^{\infty} \frac{\underset{\sim}{A}^k t^k}{k!}\underline{x}(0)$$

$$= \underset{\sim}{A}\, e^{\underset{\sim}{A}t}\underline{x}(0) = \underset{\sim}{A}\,\underline{x}(t)$$

$$= \underset{\sim}{A}\underline{x}(t)$$

which satisfies (9.4).

When $\underline{u} \neq \underline{0}$, then the complete solution may be obtained by assuming a solution of the form

$$\underline{x}(t) = e^{\underset{\sim}{A}t}\underline{f}(t) \tag{9.6}$$

where $\underline{f}(t)$ is to be found. Then

$$\dot{\underline{x}} = \underset{\sim}{A}e^{\underset{\sim}{A}t}\underline{f} + e^{\underset{\sim}{A}t}\dot{\underline{f}}$$

$$= \underset{\sim}{A}e^{\underset{\sim}{A}t}\underline{f} + \underset{\sim}{B}\underline{u}$$

so

$$e^{\underset{\sim}{A}t}\dot{\underline{f}} = \underset{\sim}{B}\underline{u} \tag{9.7}$$

Integrating (9.7) from $-\infty$ to t, we have

$$\underline{f}(t) = \int_{-\infty}^{t} e^{-\underset{\sim}{A}\lambda}\underset{\sim}{B}\underline{u}(\lambda)d\lambda$$

so (9.6) yields

$$\begin{aligned}\underline{x}(t) &= e^{\underset{\sim}{A}t}\int_{-\infty}^{t} e^{-\underset{\sim}{A}\lambda}\underset{\sim}{B}\underline{u}(\lambda)d\lambda \\ &= e^{\underset{\sim}{A}t}\int_{-\infty}^{0} e^{-\underset{\sim}{A}\lambda}\underset{\sim}{B}\underline{u}(\lambda)d\lambda + e^{\underset{\sim}{A}t}\int_{0}^{t} e^{-\underset{\sim}{A}\lambda}\underset{\sim}{B}\underline{u}(\lambda)d\lambda\end{aligned} \tag{9.8}$$

Comparing (9.8) and (9.5) when $\underline{u} = \underline{0}$, we associate

$$\underline{x}(0) = \int_{-\infty}^{0} e^{-\underset{\sim}{A}\lambda}\underset{\sim}{B}\underline{u}(\lambda)d\lambda$$

so the complete solution is

$$\underline{x}(t) = e^{\underset{\sim}{A}t}\underline{x}(0) + e^{\underset{\sim}{A}t}\int_{0}^{t} e^{-\underset{\sim}{A}\lambda}\underset{\sim}{B}\underline{u}(\lambda)d\lambda$$

Since we may regard any time as the initial $t = 0$, the solution at $t=t^{n+1}$ may well be written in terms of the solution at $t=t^{n}$ as

$$\underline{x}[(n+1)T] = e^{\tilde{A}T}\underline{x}(nT) + e^{\tilde{A}T}\int_0^T e^{-\tilde{A}\lambda}\tilde{B}\underline{u}(\lambda+nT)d\lambda \tag{9.9}$$

where $t^n = nT$. This formula will be the basis of the following analyses.

9.3 ANALYSIS OF NUMERICAL STABILITY

9.3.1 Introduction

In Chapter IV, the stability of single equations was analyzed, although we indicated that numerical instability was a problem only for systems of equations with widely separated time constants. (These are sometimes called "stiff equations"). In this section, we extend these previous results by showing that all of the eigenvalues of $\tilde{A}$ of (9.1) must be within the stability regions derived in Chapter IV for single equations.

For a system of equations described by

$$\dot{\underline{x}} = \tilde{A}\underline{x} \tag{9.10}$$

consider an integration rule of the form

$$\underline{x}^{n+1} = \tilde{H}\underline{x}^n$$

where $\tilde{H}$ is a function of $\tilde{A}T$, i.e., $\tilde{H} = \tilde{H}(\tilde{A}T)$. We restrict $\tilde{H}$ to have a full set of eigenvectors. That is, if

$$\tilde{H}\tilde{Y} = \tilde{Y}\tilde{\Lambda}$$

where $\tilde{Y}$ is a matrix of eigenvectors, and $\tilde{\Lambda} = \text{diag}\,[\lambda_1 \ldots . \lambda_\ell]$, then $\tilde{Y}^{-1}$ exists. We can immediately write

$$\tilde{H} = \tilde{Y}\tilde{\Lambda}\tilde{Y}^{-1} \tag{9.11}$$

Given an initial vector $\underline{x}^0$, we then have

$$\begin{aligned} \underline{x}^{n+1} &= \tilde{H}\underline{x}^n \\ &= \tilde{H}\tilde{H}\underline{x}^{n-1} \\ &= \tilde{H}^{(n+1)}\underline{x}^0 \end{aligned} \tag{9.12}$$

where the notation $\tilde{H}^{(n+1)}$ means $\tilde{H}$ raised to the (n+1) st power. Expanding (9.12), we find

$$\begin{aligned} \underline{x}^{n+1} &= (\tilde{Y}\tilde{\Lambda}\tilde{Y}^{-1}) \ldots (\tilde{Y}\tilde{\Lambda}\tilde{Y}^{-1})\underline{x}^0 \\ &= \tilde{Y}\tilde{\Lambda}^{(n+1)}\tilde{Y}^{-1}\underline{x}^0 \\ &= \tilde{Y}\,\text{diag}\,[\lambda_1^{(n+1)}, \lambda_2^{(n+1)}, \ldots \lambda_\ell^{(n+1)}]\,\tilde{Y}^{-1}\underline{x}^0 \end{aligned} \tag{9.13}$$

If any $|\lambda_i| > 1$ then a corresponding component of $\underset{\sim}{A}$ will grow with n. Although this does not guarantee that the components of $\underline{x}$ will increase in size, in practice such growth will nearly always occur. We term such an integration unstable. Therefore, stability requires that all the eigenvalues of $\underset{\sim}{H}$ lie within or on the unit circle in the λ-plane (a necessary condition).

The stability criterion established we must now be able to relate, for a given method, the eigenvalues λ_i of $\underset{\sim}{H}(\underset{\sim}{A}T)$ to the eigenvalues s_i of $\underset{\sim}{A}$. For example, the forward Euler method is of the form

$$\underline{x}^{n+1} = [\underset{\sim}{I} + \underset{\sim}{A}T]\, x^n$$

$$= \underset{\sim}{H}(\underset{\sim}{A}T)\underline{x}^n$$

so that the eigenvalues of $\underset{\sim}{A}$ and $\underset{\sim}{I} + \underset{\sim}{A}T$ must be related. We appeal to a well-known theorem of linear algebra [9.21] which requires that the eigenvalues λ_i of a polynomial function of $\underset{\sim}{A}$ be the same polynomial function of the eigenvalues of $\underset{\sim}{A}$, i.e.,

$$\lambda_i = H(s_i T) \tag{9.14}$$

Thus, numerical stability requires

$$|H(s_i T)| \leq 1 \qquad i = 1, 2, \ldots \ell \tag{9.15}$$

In the forward Euler method, for every eigenvalue s_i of $\underset{\sim}{A}$ there exists a corresponding eigenvalue $\lambda_i = 1 + s_i T$. For stability, then,

$$|1+s_i T| \leq 1 \qquad i = 1, 2, \ldots \ell$$

which agrees with Eq. (4.26) for the single equation case.

A significant computational question which may be addressed with (9.15) is the stability of the power series expansion of $e^{\underset{\sim}{A}T}$. We can approximate the solution of (9.10) by the expansion

$$\underline{x}^{n+1} = (\underset{\sim}{I} + \underset{\sim}{A}T + 1/2\ \underset{\sim}{A}^2 T^2 + \ldots \frac{1}{k!}\underset{\sim}{A}^k T^k)\,\underline{x}^n \tag{9.16}$$

From (9.15), this integration will be stable in regions of the complex plane where

$$|1 + s_i T + 1/2\ (s_i T) + \ldots + \frac{1}{k}(s_i T)^k| \leq 1$$

The stable regions associated with different k's is shown in Figure 9.1. We observe that the stability regions do not grow markedly with k, the negative real axis intercept being given by

$$\sigma^k_{\min} = -\ 2 - ke^{-1}$$

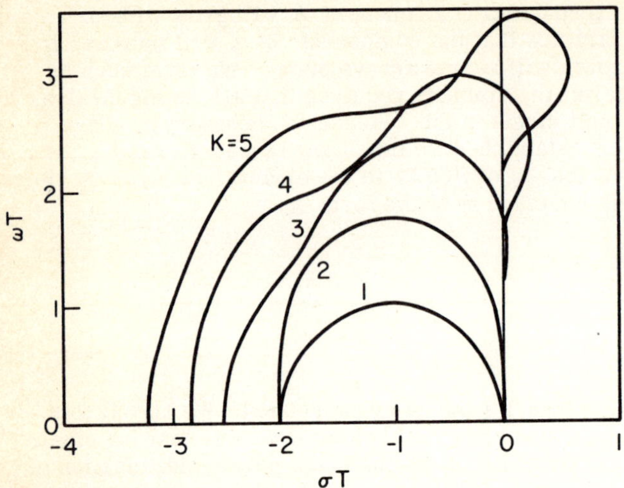

FIG. 9.1 sT-plane Stability Contours for Truncated $e^{\tilde{A}T}$ Expansions

for large k. Thus, a problem with widely split eigenvalues is not practical to integrate from the series expansion of $e^{\tilde{A}T}$ for reasonable values of k.

9.3.2 A-stability

The ideal integration formula yields a stable difference equation approximation to a stable differential equation. Such a method is termed A-stable, i.e., absolutely stable. If s_i and λ_i are the i-th eigenvalues of $\tilde{A}$ and $\tilde{H}$ respectively, A-stability requires

$$|\lambda_i| \leq 1 \quad \text{for} \quad \text{Re } s_i T \leq 0 \tag{9.17}$$

Note that A-stability does not consider the behavior of $|\lambda_i|$ for Re $s_i T > 0$, so that the difference equation approximation to an unstable differential equation may be either stable or unstable. Thus, the question of integrating an unstable differential equation within reasonable limits is not addressed.

The criterion of (9.17) permits A-stability to be phrased in a simple algebraic form which should be recognizable to students of circuit theory. A-stability would demand

$$|H(s_i)| \leq 1 \qquad \text{Re } s_i \leq 0$$

But this condition is precisely the requirement for passivity of a one-port in terms of the scattering coefficient. The test for A-stability immediately follows from classical network theory [9.22], namely,

(a) H(s) contains no poles in Re $s_i \leq 0$

(b) $|H(j\omega)| \leq 1 \quad 0 \leq \omega \leq \infty$

The backward Euler is an example of an A-stable method. In this case $H(s_i) = (1 - s_i T)^{-1}$, which has a single pole at $\sigma_1 = 1/T$; furthermore,

$$\max_{0 \leq \omega \leq \infty} |H(j\omega)| = H(j0) = 1$$

9.4 SINGLE-STEP INTEGRATION

9.4.1 Introduction

The integration methods we discussed in Chapter IV are all examples of single-step integration. This class of integration rules is characterized by the requirement that

$$\underline{x}^{n+1} = \underline{f}(\underline{x}^n, t^n) \tag{9.18}$$

i.e., that $\underline{x}^{n+1}$ does not explicitly depend on $\underline{x}^r$ or $\dot{\underline{x}}^r$ for $r < n$.

Within this class of rules, there are two common classes of solution methods. The first is applicable to linear systems of equations and follows from the view that the principal numerical problem is the efficient approximation of $e^{\underset{\sim}{A}T}$ in (9.9). A second class of methods is applicable to nonlinear differential equations as well and, though difficult to derive for simultaneous equations, includes linear equations as a special case.

9.4.2 Approximations of $e^{\underset{\sim}{A}T}$

From an approximation of $e^{\underset{\sim}{A}T}$, we can develop a number of approximate techniques for finding $\underline{x}^{n+1}$ in (9.9). Two of these are considered in Problem 9.1 and 9.2. In approximating $e^{\underset{\sim}{A}T}$, we can, of course, use the series expansion of (9.16), if we take care to avoid the numerical stability manifested in Figure 9.1. We will concentrate instead on A-stable methods.

Conditions (a) and (b) of Section 9.3.2 are powerful tools for developing new A-stable integration methods. For example, it is well known [9.23] [9.24] that $e^{\underset{\sim}{A}T}$ can be approximated by the so-called Pade approximants listed in Table 9.1. It is easy to show that, for $s = j\omega$, $|H(j\omega)| = 1$ for all ω, thus satisfying (b). It is then a simple matter to factor the denominator polynomials to show that all roots lie in the right half plane, which satisfies (a).

The difficulty in implementing this procedure is the apparent necessity for inverting a matrix which becomes more dense as it is raised to higher powers. If this matrix remains relatively sparse, then the inversion is best replaced by an LU factorization, with a forward and back substitution at each time step. The reader is invited to perform an operation count to take N integration steps if $\underset{\sim}{A}$ is tridiagonal, for different orders of Pade approximants.

9.4.3 The Runge-Kutta Integration

The Runge-Kutta procedure is an example of a class of single-step methods which, requiring no knowledge of $\underline{x}(t)$ for $t < t^n$, involve evaluation of $\underline{x}(t)$ at subintervals of t between $t^{\overline{n}}$ and t^{n+1}. The fourth-order Runge-Kutta will serve as an example. If the circuit equations are written in the general form

$$\dot{\underline{x}} = \underline{f}(\underline{x}, t) \tag{9.19}$$

TABLE 9.1 Summary of Single Step Methods

	$f(\underset{\sim}{A})^*$	Stability Conditions on Natural Frequencies	Stability Region	Accuracy Related to
Exact	$f(\underset{\sim}{A}) = e^{\underset{\sim}{A}T}$	$\left\lvert e^{s_i T} \right\rvert \leq 1$	Re $s_i \leq 0$	precise
Power series	$f(\underset{\sim}{A}) = \sum_{j=0}^{k} \frac{(\underset{\sim}{A}T)^j}{j!}$	$\left\lvert \sum_{j=0}^{k} \frac{(s_i T)^j}{j!} \right\rvert \leq 1$	See Figure 9.1	$(\underset{\sim}{A}T)^{k+1}$
Pade' approximants n = 1 (trapezoidal rule)	$f(\underset{\sim}{A}) = [\underset{\sim}{I} - \frac{1}{2}\underset{\sim}{A}T]^{-1} [\underset{\sim}{I} + \frac{1}{2}\underset{\sim}{A}T]$	$\left\lvert \frac{1 + \frac{1}{2}s_i}{1 - \frac{1}{2}s_i} \right\rvert \leq 1$	Re $s_i \leq 0$	$(\underset{\sim}{A}T)^3$
n = 2	$f(\underset{\sim}{A}) = [\underset{\sim}{I} - \frac{1}{2}\underset{\sim}{A}T + \frac{1}{12}(\underset{\sim}{A}T)^2]^{-1}$ x $[\underset{\sim}{I} + \frac{1}{2}\underset{\sim}{A}T + \frac{1}{12}(\underset{\sim}{A}T)^2]$	$\left\lvert \frac{1 + \frac{1}{2}s_i + \frac{1}{12}s_i^2}{1 - \frac{1}{2}s_i + \frac{1}{12}s_i^2} \right\rvert \leq 1$	Re $s_i \leq 0$	$(\underset{\sim}{A}T)^5$
n = 3	$f(\underset{\sim}{A}) = [\underset{\sim}{I} - \frac{1}{2}\underset{\sim}{A}T + \frac{1}{10}(\underset{\sim}{A}T)^2 - \frac{1}{120}(\underset{\sim}{A}T)^3]^{-1}$ x $[\underset{\sim}{I} + \frac{1}{2}\underset{\sim}{A}T + \frac{1}{10}(\underset{\sim}{A}T)^2 + \frac{1}{120}(\underset{\sim}{A}T)^3]$	$\left\lvert \frac{1 + \frac{1}{2}s_i + \frac{1}{10}s_i^2 + \frac{1}{120}s_i^3}{1 - \frac{1}{2}s_i + \frac{1}{10}s_i^2 - \frac{1}{120}s_i^3} \right\rvert \leq 1$	Re $s_i \leq 0$	$(\underset{\sim}{A}T)^7$

* Higher order Pade' approximants ($n > 3$) of this form are not stable in the half plane.

where $\underline{f} = [f_1, f_2, \ldots, f_n]^T$, then the procedure is to evaluate successively

$$
\begin{aligned}
\dot{\underline{x}}_1^{n+1} &\triangleq \underline{f}(x^n, t^n) \\
\dot{\underline{x}}_2^{n+1} &\triangleq \underline{f}[\underline{x}^n + \frac{T}{2}\dot{\underline{x}}_1^{n+1}, t^{n+\frac{1}{2}}] \\
\dot{\underline{x}}_3^{n+1} &\triangleq \underline{f}[\underline{x}^n + \frac{T}{2}\dot{\underline{x}}_2^{n+1}, t^{n+\frac{1}{2}}] \\
\dot{\underline{x}}_4^{n+1} &\triangleq \underline{f}[\underline{x}^n + T\dot{\underline{x}}_3^{n+1}, t^{n+1}] \\
\underline{x}^{n+1} &= \frac{T}{6}[\dot{\underline{x}}_1^{n+1} + 2\dot{\underline{x}}_2^{n+1} + 2\dot{\underline{x}}_3^{n+1} + \dot{\underline{x}}_4^{n+1}] + \underline{x}^n
\end{aligned}
\tag{9.20}
$$

The strategy underlying the above is to obtain an exceptionally good estimate of $\dot{\underline{x}}^{n+1/2}$ and then use a weighted average of $\underline{x}$ over the interval $[\underline{x}^n, \underline{x}^{n+1}]$ as in (9.20). The error in this Runge-Kutta procedure is related to the fifth derivative of x_i, a reasonable precision for most analysis problems.

The stability of the Runge-Kutta procedure can be determined by substituting the linear equation $\dot{\underline{x}} = \underset{\sim}{A}\,\underline{x}$ into (9.20). It can be shown that the entire updating process can be represented in the form of

$$\underline{x}^{n+1} = \underset{\sim}{H}(\underset{\sim}{A}T)\underline{x}^n$$

where $\underset{\sim}{H}(\underset{\sim}{A}T)$ is the series expansion of (9.16). Therefore, the stability regions associated with the Runge-Kutta procedure are identical to those shown in Figure 9.1. For this reason, Runge-Kutta is rarely used for circuit analysis.

9.5 CHOICE OF INTEGRATION VARIABLES

9.5.1 Effects on Solution Error [9.11]

The terms "stability" and "error growth" are usually synonymous when dealing with a set of linear differential equations. However, when equations are nonlinear, viz,

$$\dot{\underline{x}} = \underline{f}(\underline{x}, t) \tag{9.21}$$

we can usefully distinguish between the global concept of stability and the local growth of errors. Clearly, the latter concept is of more concern for numerical integration, which is performed over a finite time interval.

To estimate the growth of errors in solution of (9.21) consider the differential equation resulting from a perturbation $\underline{\delta}_0$ applied to the $\underline{x}$ vector at $t = 0$. We have

$$\frac{d}{dt}(\underline{x}(t) + \delta\underline{x}(t)) = \underline{f}(\underline{x}(t) + \delta\underline{x}(t), t) \qquad \delta\underline{x}(0) = \underline{\delta}_0$$

$$\approx \underline{f}(\underline{x}(t), t) + \frac{\partial \underline{f}}{\partial \underline{x}} \delta \underline{x}(t) \tag{9.22}$$

Here, we have linearized the perturbed set of equations around the trajectory of the original equations. We can subtract from (9.22) the equation $\dot{\underline{x}} = \underline{f}(\underline{x}, t)$ to yield simply

$$\delta \dot{\underline{x}} = \left(\frac{\partial \underline{f}}{\partial \underline{x}}\right) \delta \underline{x} \qquad \delta \underline{x}(0) = \underline{\delta}_0 \tag{9.23}$$

$$= \underset{\sim}{J}_{\underline{x}}(t)\, \delta \underline{x}$$

This linear, time-varying equation describes the error growth due to an initial error $\underline{\delta}_0$. Unfortunately, because $\underset{\sim}{J}_{\underline{x}}$ is time-varying, no simple solution of (9.23) exists [9.25]. For a single equation, however,

$$\delta x(t) = \delta_0 e^{\int_0^t J_x(\tau)\, d\tau} \tag{9.24}$$

In practice, we want to determine not the effect of a single perturbation δ_0, but the aggregate effects of errors made at all time steps. Let us assume that the error made per unit time (ϵ_t) is constant. The error made in an interval of time dt_1 is then $\epsilon_t dt_1$, which grows according to (9.24). The growth of the error for $t < t_1$ is given by (9.24) as

$$\epsilon^{t_1, t} = (\epsilon_t dt_1)\, e^{\int_{t_1}^{t} J_x(\tau)\, d\tau}$$

The aggregate error E^{0, t_f} at time t_f is then the sum of errors originating for all times $t_1 < t_f$, so that [9.7]

$$E^{0, t_f} = (\epsilon_t \int_0^{t_f} e^{\int_{t_1}^{t_f} J_x(\tau)\, d\tau} dt_1 \tag{9.25}$$

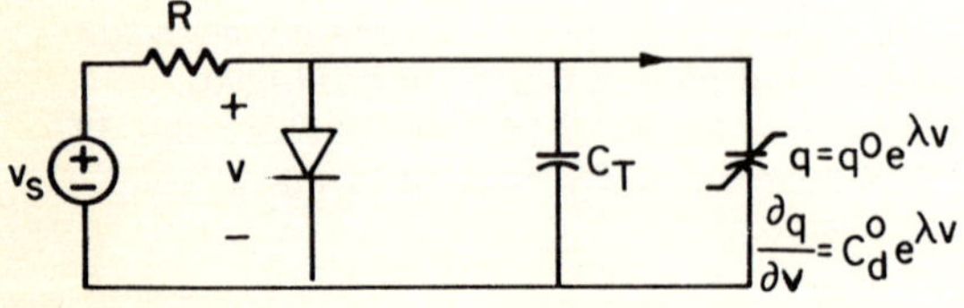

FIG. 9.2 Circuit to Demonstrate Error Growth

To study the significance of this formula, consider the analysis of a series voltage source-resistor-diode circuit (Figure 9.2) described by

$$(C_T + C_d^o e^{\lambda v}) \frac{dv}{dt} = -\frac{v}{R} + \frac{v_s}{R} - I_s(e^{\lambda v} - 1), \quad \lambda = 40$$

with a Jacobian given as

$$J_v = \frac{\lambda C_d^o e^{\lambda v} \frac{v - v_s}{R} + I_s (e^{\lambda v} - 1)}{(C_T + C_d^o e^{\lambda v})^2} - \frac{(\frac{1}{R} + \lambda I_s e^{\lambda v})}{(C_T + C_d^o e^{\lambda v})} \tag{9.26}$$

If $v_S < 0$ and $v > 0$ (i.e., the diode is turned on but being driven toward turnoff), the first term shifts the eigenvalue of J_V toward the right half plane. This eigenvalue can be shifted arbitrarily far into the right half plane by choosing v_S sufficiently negative.

Inspection of (9.26) shows that after switching has been completed ($dv/dt = 0$), the eigenvalue of J_V returns to the left half plane. As long as $J_V > 0$ during switching, however, truncation errors introduced by the integration formula are amplified according to (9.25).

To illustrate the effect this growth can have on the numerical integration, a number of numerical experiments have been performed (corresponding to a diode f_c of 4.5 ghtz) with the following parameters:

C_T = 0.01, 0.1, 1 pf $\qquad$ $I_s = 10^{-9}$ amperes

$C_d^o = 2 \times 10^{-6}$ pf $\qquad$ v_s = -15, 15 volts

R = 50 Ω

The results are given in Table 9.2. The integration is well behaved in the turnon analysis but is far from the intended in the turnoff solution. A plot of the transient response would show that the error is largely a phase error; i.e., although the error in v is large for a fixed t, the error in t to achieve a fixed level of v is small. This discrepancy is possible because $|dv/dt|$ is large in the region of concern.

The origin of this error growth is the selection of voltage as the integration variable. To show this, represent the circuit equations in the form

$$\underset{\sim}{S}\dot{\underline{x}} = \underline{f}(\underline{x}, t) \tag{9.27}$$

where the voltage dependence of capacitances is now explicitly accounted for in $\underset{\sim}{S}$. Then, the Jacobian $\partial \dot{\underline{x}}/\partial \underline{x}$ is given by

$$\underset{\sim}{J}_{\underline{x}} = \frac{\partial \dot{\underline{x}}}{\partial \underline{x}} = -\underset{\sim}{S}^{-1} \frac{\partial \underset{\sim}{S}}{\partial \underline{x}} \underset{\sim}{S}^{-1} \underline{f} + \underset{\sim}{S}^{-1} \frac{\partial \underline{f}}{\partial \underline{x}} \tag{9.28}$$

TABLE 9.2 Integration Errors (C_T = .01 pf; final time = t_f)

	Total Error		
	Intended ($\epsilon_t \times t_f$)	Experimental	From (9.25)
Turnoff (0.4 to 0 volt)	$\pm 10^{-2}$	-2.62	-5.85
	$\pm 10^{-3}$	-0.319	-0.585
	$\pm 10^{-4}$	-0.0312	-0.0585
Turnon (-0.5 to 0.4 volt)	$\pm 10^{-2}$	0.8×10^{-3}	0
	$\pm 10^{-3}$	1.7×10^{-4}	0
	$\pm 10^{-4}$	2×10^{-5}	0

The first term of (9.26) accounts for the linearization of capacitances and the second the linearization of resistances. The former evidently is responsible for the error growth, since the network of (positive) capacitances and linearized resistances represented by the second term assuredly has left half plane natural frequencies (indeed, this is the common small signal network model).

This unnatural error growth can be avoided by redefining the integration variables. If we let

$$\dot{\underline{q}} = \underset{\sim}{S}(\underline{x})\dot{\underline{x}} \tag{9.29}$$

then

$$\underset{\sim}{J}_{\underline{q}} = \frac{\partial \dot{\underline{q}}}{\partial \underline{q}} = \frac{\partial \underline{f}}{\partial \underline{x}} \underset{\sim}{S}^{-1}$$

$$= \underset{\sim}{S}\left(\underset{\sim}{S}^{-1} \frac{\partial \underline{f}}{\partial \underline{x}}\right) \underset{\sim}{S}^{-1}$$

Since $\underset{\sim}{J}_{\underline{q}}$ is related to $\underset{\sim}{S}^{-1} \frac{\partial \underline{f}}{\partial \underline{x}}$ by a similarity transformation, the eigenvalues of $\underset{\sim}{J}_{\underline{q}}$ and $\underset{\sim}{S}^{-1}\partial \underline{f}/\partial \underline{x}$ are identical and are the familiar small signal frequencies [9.21].

The $\underline{q}$ vector is easily recognized as a vector of capacitor charges (or inductor flux linkages), since, if $\underline{x} = \underline{v}$,

$$\frac{d\,\underline{q}(\underline{v})}{dt} = \frac{\partial \underline{q}}{\partial \underline{v}} \frac{d\underline{v}}{dt}$$

which is identical to (9.29) if $\underset{\sim}{S}(\underline{v}) = \partial\underline{q}/d\underline{v}$. We conclude that solution in the charge variable with an A-stable integration method will not result in error growth.

The significance of the error growth in transient analysis of common circuits is not known. It has been conjectured* that this phenomena occurs when a nonlinear capacitance is forced to changed value rapidly, e.g., by a step voltage source. If this is correct, then this problem will occur principally at the input stage of a complicated circuit; for circuit capacitances can be expected to degrade an input signal sufficiently so that its rise time in later stages will not far exceed the ability of the individual nonlinear capacitances to respond.

9.5.2 Effects on Solution Efficiency

Integration in charge variables is usually far more efficient than in voltage variables because most electronic components are current rather than voltage driven. To demonstrate this claim, consider the transient solution of the circuit of Figure 9.2, from an initial condition of v = -2.6 volts. The normalized derivatives in the voltage (x) and charge (q) variables are displayed in Table 9.3. The higher derivatives of x are consistently larger than the derivatives of q. Therefore, an integration method with a truncation error proportional to one of these derivatives will make a far larger error if integration is performed in the x variables, or alternatively, will require a smaller step size to maintain the same truncation error.

We conclude that, given an equation of the form

$$\underset{\sim}{S}\dot{x} = \underline{f}(\underline{x}, t)$$

it is preferable to solve this as two equations, viz,

$$\dot{\underline{q}} = \underline{f}(\underline{x}(\underline{q}), t) \tag{9.30}$$

and

$$\begin{aligned} \underline{q} &= \int \underset{\sim}{S}\, d\underline{x} + \underline{q}_0 \\ &= g(\underline{x}) \end{aligned} \tag{9.31}$$

Discretization is performed on the charge variable in (9.30) with small error (from Table 9.3), whereas (9.31) is solved precisely by iteration.

The problem with integration in charge is that the error in voltage - which has more physical significance - is not easily related to charge error. The reader may want to consider this question again after discussion of error control in Section 9.6.4.

9.6 MULTISTEP IMPLICIT INTEGRATION

9.6.1 Introduction

The Gear integration algorithm [9.2] [9.3] [9.5] is a very powerful method for solving stiff systems of differential and (as we will later show)

---

*G. D. Hachtel, private communication.

TABLE 9.3 Comparison of $\underline{x}$- and $\underline{q}$- variable Integration

Diode Voltage v	Normalized Derivatives* (Absolute Values) $(\dot{x}, \dot{q})$	$(\ddot{x}, \ddot{q})$	$(\dddot{x}, \dddot{q})$	$(\ddddot{x}, \ddddot{q})$
-2.6	(3.3, 3.3)	(.55, .55)	(.1, .1)	
.203	(3.8, 3.9)	(2.0, .64)	(8.0, .16)	
.258	(2.2, 2.9)	(14., .40)	(120., .93)	
.304	(.78, 2.1)	(4.9, .14)	(76., 1.1)	
.337	(.20, 1.4)	(2.8, .073)	(42., .128)	(480, 3.8)
.345	(.18, 1.2)	(1.0, .045)	(13., .055)	(180, .28)
.351	(.124, 1.0)	(.21, .036)	(.59, .018)	(5.6, .016)
.362	(.086, .8)	(.11, .028)	(.15, .005)	(2.1, .03)
.386	(.03, .38)	(.02, .012)	(.016, .0003)	(.1, .0015)
.414	(.008, .13)	(.0015, .0037)	(.0002, .0001)	(.0005, .00001)
.420	(.0037, .061)	$(4 \times 10^{-4}, 2 \times 10^{-3})$	$(3 \times 10^{-5}, 5 \times 10^{-5})$	$(3 \times 10^{-5}, 2 \times 10^{-6})$
.448	$(7 \times 10^{-4}, .013)$	$(4 \times 10^{-5}, 4 \times 10^{-4})$	$(4 \times 10^{-7}, 10^{-5})$	$(4 \times 10^{-7}, 4 \times 10^{-7})$
.454	$(3 \times 10^{-4}, .005)$	$(10^{-5}, 10^{-4})$	$(3 \times 10^{-7}, 4 \times 10^{-6})$	$(4 \times 10^{-8}, 2 \times 10^{-7})$

* The true derivatives are divided by the value of the variable to yield the values shown.

algebraic equations. Its utility lies not only in its relative stability when widely separated time constants are present. Gear has also implemented the basic integration rule with outstanding error control features: (1) variable step size, to preserve a prescribed error per unit time, and (2) variable order, to maximize the step size with prescribed error.

For purposes of discussion, we will extract from Gear's method those topics which are applicable to any (implicit) multistep method. This will leave the determination of certain coefficients for later consideration in Section 9.7.

We will consider equations of the general form

$$\underline{F}(\underline{x}, \dot{\underline{x}}, t) = \underline{0} \tag{9.32}$$

This includes not only differential equations of the form $\underline{f}(\underline{x}, t) - \dot{\underline{x}} = \underline{0}$ but also algebraic equations, coupled sets of algebraic and differential equations, and finally sets of equations involving mixtures of differential and algebraic variables (as obtained, for example, from node analysis). With the assurance that any complete set of network equations can be similarly solved, we will then be free to choose a formulation procedure to accommodate other numerical or programming considerations.

Before discussing the details of a general algorithm, we must define new terminology and review a number of related theoretical results that motivated the development of implicit integration rules. We will depart from our narrative style in order to survey this background material efficiently [9.31].

(1) Multistep integration. Given equations of the form of (9.32), consider an integration rule of the form

$$\alpha_0 \underline{x}^{n+1} = -T\beta_0 \dot{\underline{x}}^{n+1} - \sum_{j=1}^{k} (\alpha_j \underline{x}^{n-j+1} + T\beta_j \dot{\underline{x}}^{n-j+1}) \tag{9.33}$$

where $\alpha_0 = 1$. In contrast to single step methods such as Runge-Kutta that achieve their accuracy by estimating higher derivatives in the time interval $[t^n, t^{n+1}]$, multistep methods accumulate derivative information from the values of $\underline{x}$ and $\dot{\underline{x}}$ occurring before t^n. Accordingly, these methods are most useful when $\underline{x}$ and its derivatives are at least continuous (see Section 3.6). The backward Euler integration algorithm can be viewed as a multistep method with $\beta_0 = \alpha_1 = -1$ and $\beta_i = \alpha_{i+1} = 0$ for $i \geq 1$.

(2) Solution of the equation. If $\beta_0 \neq 0$ in (9.33), $\underline{x}^{n+1}$ cannot be found explicitly since $\dot{\underline{x}}^{n+1}$ appears on the right hand side. Therefore, the equation must be solved by iteration or must be "corrected." Such a formula is accordingly called a corrector formula, in contrast to a predictor formula where $\beta_0 = 0$. The iteration can be explicit.* If we try to solve (9.33) by assigning known values (at the m-th iteration) to the right hand side, then we could write the explicit formula

---

*These terms are not vigorously defined in the mathematical literature. An implicit iteration is generally recognized by the necessity of solving a set of simultaneous equations.

$$\alpha_0 \underline{x}^{n+1,m+1} = -T\beta_0 \dot{\underline{x}}^{n+1,m} - \sum_{j=1}^{k} (\alpha_j \underline{x}^{n-j+1} + T\beta_j \dot{\underline{x}}^{n-j+1})$$

For a set of linear differential equations $\dot{\underline{x}} = \underset{\sim}{A}\underline{x}$, it is easily shown using the methods of Section 9.3 that divergence will occur when

$$|Ts_i| > 1$$

where s_i is an eigenvalue of $\underset{\sim}{A}$. Thus, this common type of explicit iteration is wholly unacceptable for stiff systems of equations. In contrast, the implicit iteration methods which we will shortly study have no such restriction.

(3) Stability. The method will be A-stable for a linear equation of the form $\dot{\underline{x}} = \underset{\sim}{A}x$ if and only if all roots of the polynomial equation [9.2]

$$\eta(\xi) = \sum_{j=0}^{k} (\alpha_j + Ts_i\beta_j)\, \xi^{k-j} = 0 \tag{9.34}$$

lie within the unit circle in the complex ξ-plane, where the s_i are the eigenvalues of $\underset{\sim}{A}$. This stability is concerned with the behavior at successive time steps as discussed in Section 9.3 for single step methods; it should not be confused with the question of divergence of the corrector iteration considered in (2).

(4) Order. The order p of a multistep method is defined as the highest degree of polynomial $x_i(t)$ for which the method yields an exact step by step solution. If we define

$$\sigma(\xi) = \sum_{j=0}^{k} \beta_j \xi^{k-j} \quad \text{and} \quad \zeta(\xi) = \sum_{j=0}^{k} \alpha_j \xi^{k-j} \tag{9.35}$$

then Henrici has shown [9.7] that a multistep method of the order p satisfies an equation of the form

$$\zeta(\xi) - \log \xi \; \sigma(\xi) = O(\xi - 1)^{p+1} \quad \text{as} \quad \xi \to 1 \tag{9.36}$$

(5) Stability versus order. Dahlquist has shown [9.8] that if a multistep method is to be A-stable, then $p \leq 2$. The method of order 2 with smallest truncation error is the trapezoidal rule.

The later results would seem to rule out highly accurate stable multistep methods. However, let us examine again the necessity for A-stability. It happens that this apparent requirement can be compromised without seriously imparing the integration of simultaneous equations.

Consider the contribution of the poles shown in Figure 9.3a to a transient response and the subsequent requirements imposed on a numerical

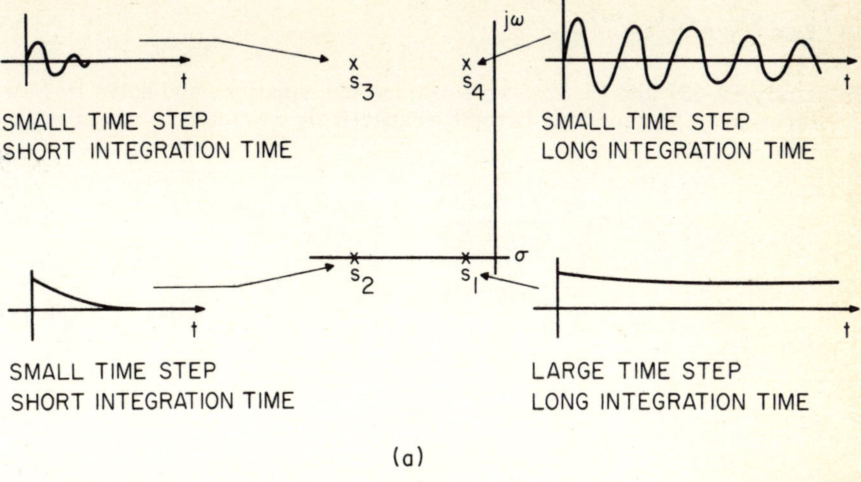

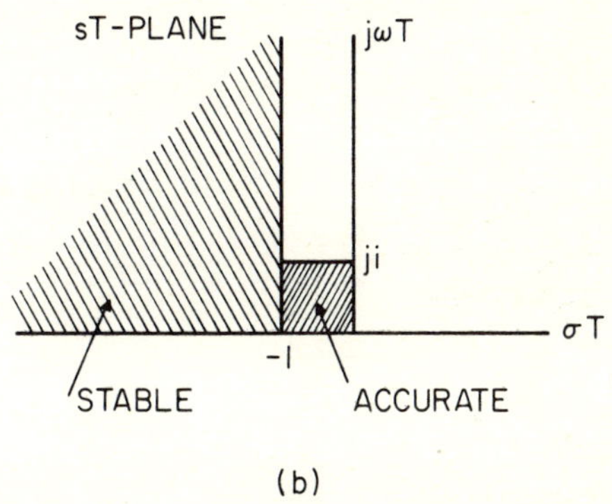

FIG. 9.3 Illustration to Accompany Instability Discussion

integration procedure. The responses associated with each eigenvalue are displayed, and the step size and integration time necessary to maintain accuracy are given. Clearly, the most demanding pole location is s_4, requiring many integration steps for accurate integration. The integration method need not be stable in this region, since the time step must always be small to insure accuracy. On the other hand, if only s_1 and s_3 (or s_1 and s_2) are present, we could take a small time step initially to capture the response for s_3 (or s_2) and then a large step to accommodate s_1. Thus the stability region should include s_2 and s_3 or, more generally, regions far removed from the imaginary axis. In terms of the normalized Ts plane, a method with the stability regions shown in Figure 9.3b would be acceptable.

9.6.2 Implicit Corrector Iteration

An integration step will be judged completed if the integration rule of (9.33) is satisfied and in addition

$$\underline{F}(\underline{x}^{n+1}, \dot{\underline{x}}^{n+1}, t^{n+1}) = \underline{0} \tag{9.37}$$

To satisfy (9. 33) and (9. 37), we linearize the equations and solve by Newton iteration. The equations become respectively

$$\frac{\partial \underline{F}^m}{\partial \underline{x}}(\Delta \underline{x}^m) + \frac{\partial \underline{F}^m}{\partial \dot{\underline{x}}}(\Delta \dot{\underline{x}}^m) = -\underline{F}^m \tag{9.38}$$

$$-\alpha_0 \Delta \underline{x}^m - \beta_0 T \Delta \dot{\underline{x}}^m = \alpha_0 \underline{x}^m + \beta_0 T \dot{\underline{x}}^m + \sum_{j=1}^{k} (\alpha_j \underline{x}^{n-j+1} + T\beta_j \dot{\underline{x}}^{n-j+1}) \tag{9.39}$$

where the n+1 superscripts have been dropped for variables to be iterated and where

$$\begin{aligned} \Delta \underline{x}^m &= \underline{x}^{m+1} - \underline{x}^m \\ \Delta \dot{\underline{x}}^m &= \dot{\underline{x}}^{m+1} - \dot{\underline{x}}^m \\ \underline{F}^m &= \underline{F}(\underline{x}^m, \dot{\underline{x}}^m, t^{n+1}) \\ \frac{\partial \underline{F}^m}{\partial \underline{x}} &= \left.\frac{\partial \underline{F}}{\partial \underline{x}}\right|_{\underline{x} = \underline{x}^m,\ \dot{\underline{x}} = \dot{\underline{x}}^m,\ t = t^{n+1}} \end{aligned} \tag{9.40}$$

We can remove the summation from (9. 39) by subtracting from it a similar expression for the (m-1)st iteration. This operation yields the simple result that

$$\Delta \dot{\underline{x}}^m = \frac{\alpha_0}{\beta_0 T} \Delta \underline{x}^m \tag{9.41}$$

Substituting (9. 41) into (9. 39) and solving for $\Delta \underline{x}^m$ gives

$$\left(\frac{\partial \underline{F}^m}{\partial \underline{x}} - \frac{\alpha_0}{\beta_0 T} \frac{\partial \underline{F}^m}{\partial \dot{\underline{x}}}\right) \Delta \underline{x}^m = -\underline{F}^m \tag{9.42}$$

Equations (9. 41) and (9. 42) provide us the updating formulae for both $\underline{x}$ and $\dot{\underline{x}}$, which are necessary to evaluate $\underline{F}(\underline{x}, \dot{\underline{x}}, t)$. To maintain the consistency of this equation with the corrector formula (9. 33) we must begin the iteration with an $\underline{x}$ and an $\dot{\underline{x}}$ that satisfy (9.33) for m = 0. It is clear that, given $\underline{x}^0$, we would have

$$\dot{\underline{x}}^0 = \frac{\alpha_o}{T\beta_o}\underline{x}^0 - \frac{1}{T\beta_o}\sum_{j=1}^{k}(\alpha_j\underline{x}^{n-j+1} + T\beta_j\dot{\underline{x}}^{n-j+1}) \tag{9.43}$$

Equations (9.41), (9.42), and (9.43) comprise the basic computation algorithm. Surveying the development, several points are worthy of note.

(1) The procedure is valid for <u>any</u> set of α's and β's.

(2) The Jacobians $\underset{\sim}{J}_x = \partial\underline{F}/\partial\underline{x}$ and $J_{\dot{x}} = \partial\underline{F}/\partial\dot{\underline{x}}$ are introduced soley for iteration of the corrector equation. We can in fact choose any approximation to these Jacobians and, <u>so long as the iteration converges</u> ($\underline{x}^{m+1} = \underline{x}^m$), the corrector formula is satisfied.

This permits the initial use of crude approximate techniques for computing the Jacobian, or the use of the same Jacobian for several corrector iterations or several time steps. If slow corrector convergence is detected, the Jacobian is re-evaluated.

<u>Example 9.1.</u> We have commented that the backward Euler formula is a special case of (9.33) with $\alpha_o = 1$, $\beta_o = -1$. Let us apply the above formulae to the solution of the circuit of Figure 9.4.

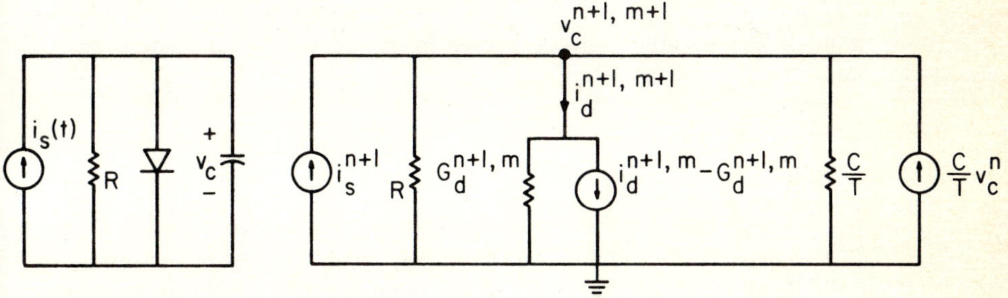

FIG. 9.4 Diode Circuit and Companion Model

The circuit can be described by the single nonlinear differential equation

$$F(v_c, \dot{v}_c, t) = C\dot{v}_c + \frac{v_c}{R} + I_s(e^{\lambda v_c} - 1) + i_s(t) = 0$$

Equations (9.41), (9.42), and (9.43) become

$$\left(\frac{\partial F(v_c, \dot{v}_c, t)^{n+1,m}}{\partial v_c} + \frac{1}{T}\frac{\partial F(v_c, \dot{v}_c, t)^{n+1,m}}{\partial \dot{v}_c}\right)\Delta v_c^{n+1,m} = -F(v_c, \dot{v}_c, t)^{n+1,m} \tag{9.44}$$

$$\Delta \dot{v}_c^{n+1,m} = \frac{1}{T} \Delta v_c^{n+1,m} \tag{9.45}$$

$$\dot{v}_c^{n+1,0} = \frac{1}{T} v_c^{n+1,0} - \frac{1}{T} v_c^n \tag{9.46}$$

$$= \frac{v_c^{n+1,0} - v_c^n}{T} \tag{9.47}$$

If as in Chapter IV, we choose $v_c^{n+1,0} = v_c^n$, (9.47) yields $\dot{v}_c^{n+1,0} = 0$. Equation (9.44) then becomes

$$\left(\frac{1}{R} + \lambda I_s e^{\lambda v_c^{n+1,m}} + \frac{C}{T}\right) \Delta v_c^{n+1,m}$$

$$= -C\dot{v}_c^{n+1,m} - \frac{v_c^{n+1,m}}{R} - I_s\left(e^{\lambda v_c^{n+1,m}} - 1\right) - i_s^{n+1}$$

where, on the right hand side, $\dot{v}_c^{n+1,m}$ is updated from (9.45). Inspection of the left hand side reveals a summation of "conductance-like" terms as would result from a companion network model studies in Chapter IV. Indeed, Equations (9.44-9.47) together can be shown to be identical to the equation obtained from a companion model.

9.6.3 Prediction

Without prediction of $\underline{x}$ and its derivatives, typically 3-5 corrector iterations are required with (9.42) to satisfy the integration formula (9.33). With prediction, convergence is likely in one iteration, although an additional iteration may be necessary to check convergence. This halving of computational effort should be adequate motivation for incorporating a predictor in an integration algorithm.

The prediction process must of course be based on values of $\underline{x}$ and its derivatives at previous time steps. There are two common approaches to storing this information.

(1) Store $\underline{x}$ and $\dot{\underline{x}}$ at each time step. The predictor formula then appears similar to the corrector in form, viz,

$$\alpha_o' \underline{x}_p^{n+1,0} = \sum_{j=1}^{k'} \alpha_j' \underline{x}^{n-j+1} + \beta_j' T \dot{\underline{x}}^{n-j+1} \tag{9.48}$$

where β_o is missing to avoid iteration.

(2) Store estimated values of its derivatives at each time step. The predictor for $\underline{x}$ then has the form of a Taylor's series expansion about $\underline{x}^n$, viz,

$$\underline{x}^{n+1,0} = \underline{x}^n + \sum_{j=1}^{k'} \frac{T^j(\underline{x}^{(j)})^n}{j!} \tag{9.49}$$

where $\underline{x}^{(j)}$ is the estimated j-th time derivative. The resulting matrix $\underset{\sim}{X} = [\underline{x}\ T\underline{x}^{(1)}\ \frac{T^2}{2}\underline{x}^{(2)} \ldots \frac{T^{k'}}{k'!}\underline{x}^{(k')}]$ is known as the Nordseick matrix [9.12]. It is not difficult to relate (9.48) and (9.49);* however, the estimation of the higher derivatives required at each time step is an involved matter. Nonetheless, Gear has shown the rather remarkable fact that the same vector that updates $\Delta\underline{x}^{n+1,m}$ and $\Delta\dot{\underline{x}}^{n+1,m}$ in (9.41) and (9.42) can be used to update estimates of higher derivatives. That is, the j-th component of the Nordseick vector can be updated according to the formula

$$\frac{T^j}{j!}(\Delta\underline{x}^{(j)})^{n+1,m} = -\frac{a_j}{a_0}\left[\frac{\partial \underline{F}}{\partial \underline{x}} + \frac{1}{a_0 T}\frac{\partial \underline{F}}{\partial \dot{\underline{x}}}\right]^{-1}\underline{F}^{n+1,m} \qquad j=0,1,\ldots k \tag{9.50}$$

where a_j depends on the order k. Here, $a_0 = -\beta_0/\alpha_0$ and $a_1 = 1$ to be consistent with (9.41) and (9.42).

Equation (9.50) is the central formula in the integration algorithm we will adopt. We can state the entire prediction-correction process as follows.

(1) From estimates of $(\underline{x}^{(j)})^n$, $j = 0, 1, 2, \ldots k$, predict $(\underline{x}^{(j)})^{n+1,0}$ for $j = 0$ using (9.49) and the method of Problem 9.11 for $j = 1, 2, .. k$.

(2) At each corrector iteration, update $(x^{(j)})^{n+1,m}$ $j = 0, 1$ using (9.50). Use $(x^{(0)})^{n+1,m+1}$ and $(x^{(1)})^{n+1,m+1}$ to update the partial derivatives required to evaluate $F^{n+1,m+1}$ (9.51)

(3) After convergence of the corrector iteration, form

$$\underline{S}^{n+1} = \sum_m (\Delta\underline{x}^{(1)})^{n+1,m} \tag{9.52}$$

and update higher derivatives with

$$\frac{T^j}{j!}(\underline{x}^{(j)})^{n+1} = (\underline{x}^{(j)})^n + \frac{a_j}{a_0}\underline{S}^{n+1} \qquad j = 2, 3, \ldots \tag{9.53}$$

One omission has been necessary in the above development: the procedure for obtaining the a_j's from the yet-to-be-derived α_j's and β_j's has not been described. The interested reader should see [9.26].

---

*The Nordseick form of (9.49) may or may not be equivalent to (9.48) if T changes. The latter method can be shown to be more stable if T changes but the former is in a natural form for interpolating between solution points for display purposes. See [9.18] for details.

9.6.4 Error Control [9.16]

We initially limit our error control analysis to a single differential equation

$$\dot{x} = a\,x$$

Let us assume that over a time interval $0 = t = L$, we can tolerate a total error in x of E_{max}, a positive number. The allowable error per unit time is then E_{max}/L; if a step size T is being used, then the allowable error per step is $E_{max}\,T/L$.

For a p-th order method, the approximate error per time step is of the form

$$T^{p+1} C_{p+1} x^{(p+1)} \tag{9.54}$$

where C_{p+1} is a constant dependent on the integration method. It would appear that to calculate the proper step size we need only equate

$$T^{p+1} C_{p+1} x^{(p+1)} = E_{max} T/L \tag{9.55}$$

and calculate T.

The solution of (9.55) is displayed graphically in Figure 9.5 for several values of p. We note that although the error associated with a higher order method is necessarily smaller for sufficiently small T, the error increase in a higher order method is more rapid. Thus, a larger T may very well be allowed by a lower order method, if a relatively large stepwise error (E_1 of Figure 9.5) is permitted. If C_{p+1} is relatively large (as it is for stable implicit methods [9.28]), the error at which a lower order method is to advantage (e.g., E_2 and E_3) occurs at levels of common "engineering accuracy." Our conclusion: that a good integration algorithm should have a capability to continuously adjust both order and step size in order to maximize the step size.

The development of our integration error control scheme should therefore be aimed at finding the order that maximizes T and then using this

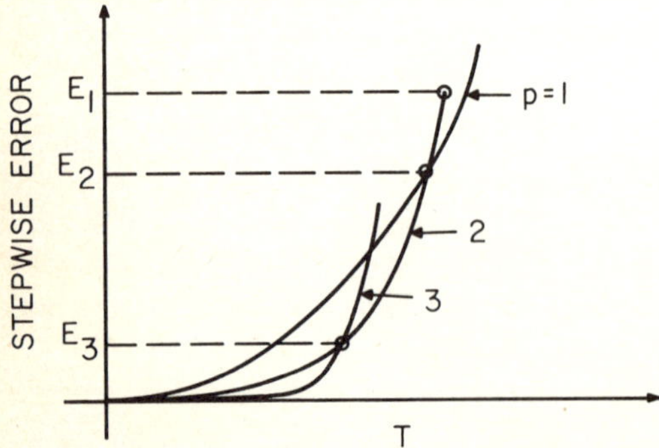

FIG. 9.5 Relation Between Stepwise Error and Order (p)

value T_{max}. We must first, however, be able to estimate the derivative $x^{(p+1)}$ for each order p of potential value. In particular, if the order p at the n-th time step is p_2^n with step size T_2^n satisfying (9.55), then it is common to examine orders $p_1^n = p_2^n - 1$ and $p_3^n = p_2^n + 1$ with step sizes T_1^n and T_3^n respectively, satisfying (9.55). Then

$$T_{max}^n = \max_{1 \leq i \leq 3} (T_i^n) \tag{9.56}$$

from which the optimum order is elected. Thus, orders are allowed to change by at most one during each time step.

Now assume that we have an estimate of the derivatives

$$\left(x^{(p+1)}\right)^n = \frac{\left(x^{(p)}\right)^n - \left(x^{(p)}\right)^{n-1}}{T_2^n}$$
$$\left(x^{(p+2)}\right)^n = \frac{\left(x^{(p+1)}\right)^n - \left(x^{(p+1)}\right)^{n-1}}{T_2^n} \tag{9.57}$$

Using (9.55), we can write for the three orders of interest

$$(1)\quad \left(T_1^n\right)^p C_p \left|\left(x^{(p)}\right)^n\right| = E_{max} T_1^n / L \tag{9.58}$$

or

$$T_1^n = \left(E_{max}/LC_p \left|\left(x^{(p)}\right)^n\right|\right)^{1/(p-1)} \tag{9.59}$$

$$(2)\quad T_2^n = \left(E_{max}/\,LC_{p+1} \left|\left(x^{(p+1)}\right)^n\right|\right)^{1/p} \tag{9.60}$$

$$(3)\quad T_3^n = \left(\dot{E}_{max}/\,LC_{p+2} \left|\left(x^{(p+2)}\right)^n\right|\right)^{1/(p+1)} \tag{9.61}$$

The choice of order from (9.56) and (9.60-9.61) defines a new order p_2^{n+1} and a new step size T_2^{n+1}. This value of T is often multiplied by a constant $\gamma > 1$ to assure the quality of the above approximations. Further, after the step is taken then $(x^{(p+1)})^{n+1}$ is estimated again and (9.60) checked by requiring

$$\left| x^{(p+1)} \right|^{n+1} \leq E_{max} / LC_{p+1} \left(T_2^{n+1}\right)^p \tag{9.62}$$

If (9. 62) is not satisfied, the step is considered unsuccessful, the stepwise error criterion having been violated during the integration step.

9. 6. 5 Algebraic and Differential Variables

It might be noted that no distinction has been made between algebraic and differential variables in the entire discussion of derivative storage, prediction, and error control. In a formal sense, we are seemingly free to choose any set of variables to appear in the $\underline{x}$ vector as solution variables. Practically, however, we are usually forced to choose at least the state variables (capacitor voltages and inductor currents) as solution variables for error control purposes. The simple example of Figure 9. 6 illustrates the pitfall of neglecting error on these variables. A pulse applied to a lumped approximation to an RC line will not produce a discernible output until some time t_1 later. If error control were exerted only on the output variable v_o, then a large time step could apparently be taken initially until v_o began to change. This would of course be a poor strategy, since the capacitances on the input side of the line would be affected by the pulse far sooner and would require an appropriately small integration step for $t < t_1$.

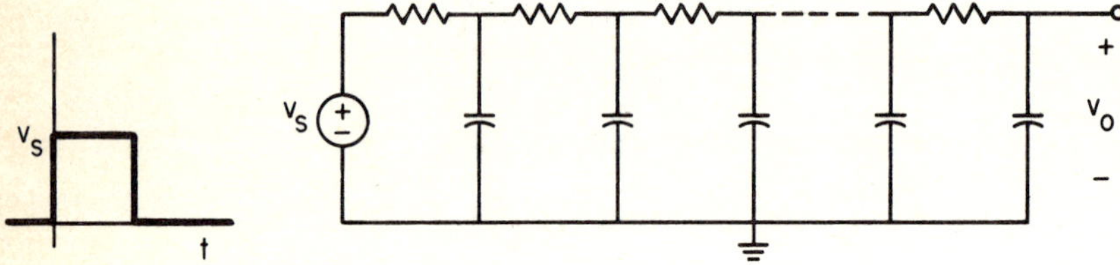

FIG. 9. 6 RC Line and Pulse Response

9. 6. 6 Explicit Integration Methods

Although of limited usefulness for the solutions of large circuits, which inevitably have widely separated time constants, an explicit corrector iteration using (9. 33) can be implemented with the same error control, the same Nordseick derivative updating and storage scheme, and the same order selection procedures as implicit methods. Since a matrix inversion (factorization) is unnecessary, explicit integration is faster per time step. By making both types of algorithms available in the same subroutine [9. 16], the user can first attempt explicit integration and, if instability results, the slower implicit integration formula can be used. However, algebraic relationships are usually solved implicitly, thus detracting from the appeal of an explicit corrector formula as a general solution algorithm.

9. 7 MULTISTEP COMPANION MODELS

Although the implicit multistep algorithm can be applied to any complete set of differential-algebraic circuit equations, the assembly of these equations can be complicated. As we realize from Chapter III and IV, the linearized, discretized network equations can also be assembled on a component basis by developing companion models for each branch. These models would then be inserted in place of the original network component; the solution of the resultant network - using any method of analysis - would then be equivalent to solving (9. 50) in the solution variables of the analysis procedure.

In this section, we apply the corrector formula of (9. 50) to several common circuit elements. For notational convenience, we will examine only the updating of $\underline{x}$ (i. e., j=0), which we represent as

$$\left(\frac{1}{a_o T} \frac{\partial F^m}{\partial \dot{\underline{x}}} + \frac{\partial \underline{F}^m}{\partial \underline{x}} \right) \Delta \underline{x}^m = -F^m \tag{9.63}$$

(1) Exponential Diode Model. We recognize two possible solution variables i_d and v_d related by the algebraic equation

$$F(v_d, i_d) = I_s (e^{\lambda v_d} - 1) - i_d = 0$$

Application of (9. 63) then routinely yields

$$\frac{\partial F}{\partial v_d} (v_d^{n+1, m+1} - v_d^{n+1, m}) + \frac{\partial F}{\partial i_d} (i_d^{n+1, m+1} - i_d^{n+1, m}) = -F^{n+1, m} \tag{9.64}$$

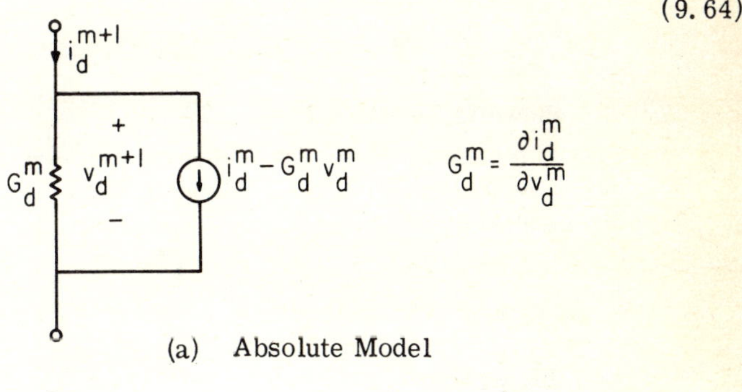

(a) Absolute Model

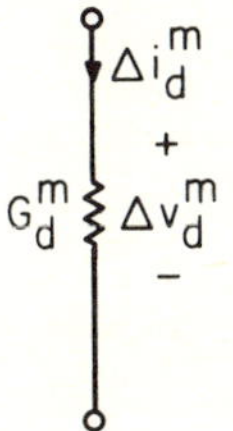

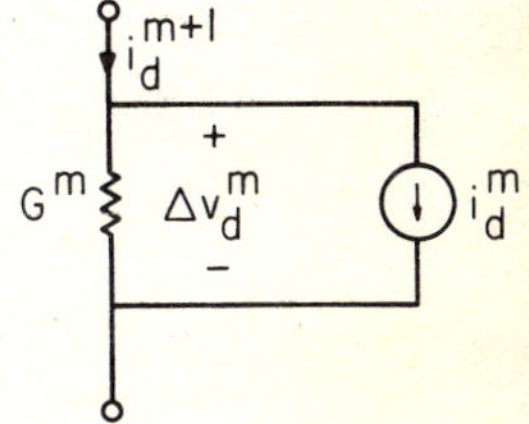

(b) Incremental Model (c) Mixed Model

FIG. 9. 7 Diode Companion Models

We can express this relationship in a number of forms:

(a) in absolute form by solving for i_d^{m+1}, as

$$i_d^{m+1} = i_d^m + \frac{\partial i_d}{\partial v_d}^m (v_d^{m+1} - v_d^m)$$

$$= \frac{\partial i_d}{\partial v_d}^m v_d^{m+1} + i_d^m - \frac{\partial i_d}{\partial v_d}^m v_d^m \ ;$$

this form is of course familiar and is represented by the model of Figure 9.7a;

(b) in incremental form, as

$$\Delta i_d^m = \frac{\partial i_d}{\partial v_d}^m \Delta v_d^m;$$

the network representation of this form contains simply the conductance of Figure 9.7b;

(c) in mixed form, involving the absolute value of i_d and the incremental value of v_d, as

$$i_d^{m+1} = \frac{\partial i_d}{\partial v_d}^m \Delta v_d^m + i_d^m$$

shown in Figure 9.7c.

(2) Linear Capacitor Model. We identify two variables - v_c, a differential variable, and i_c, an algebraic variable - related by the single differential equation

$$F(v_c, i_c, \dot{v}_c) = C\dot{v}_c - i_c = 0 \tag{9.65}$$

For this equation, the left hand side of (9.63) becomes

$$\left(\frac{C}{a_0 T} + \frac{\partial F}{\partial v_c}\right)(v_c^{n+1,m+1} - v_c^{n+1,m}) + \left(\frac{\partial F}{\partial i_c}\right)(i_c^{n+1,m+1} - i_c^{n+1,m})$$

$$= \frac{C}{a_0 T}(v_c^{n+1,m+1} - v_c^{n+1,m}) - (i_c^{n+1,m+1} - i_c^{n+1,m})$$

Dropping the n+1 superscript and equating this expression to $-\underline{F}^m = i_c^m - C\dot{v}_c^m$, we again have an

(a) absolute form,

$$i_c^{m+1} = \frac{C}{a_0 T} v_c^{m+1} + (C\dot{v}_c^m - \frac{C}{a_0 T} v_c^m)$$

(b) incremental form,

$$\frac{C}{a_0 T} \Delta v_c^{\ m} - \Delta i_c^{\ m} = i_c^{\ m} - C\dot{v}_c^{\ m}$$

(c) mixed form,

$$i_c^{\ m+1} = \frac{C}{a_0 T} \Delta v_c^{\ m} + C\dot{v}_c^{\ m}$$

Before proceeding with more models, it is worth commenting on the application of each form. The incremental form is of use when all the network current and voltage variables are solution variables, since i_c^m and $\dot{v}_c^m$ are both required to be known. This occurs in the network tableau approach of Hachtel (see Section 8. 3. 2). The mixed form is of special value in node analysis if all independent network sources are current sources* - since these can be represented in absolute form - while at the same time the solution variables - the node voltages - appear in incremental form as required in (9. 50). This avoids potentially large terms such as Cv_c^m/a_0T on the right hand side of the equations and the corresponding cancellations in the forward and back substitutions noted in Example 2. 5b [9. 30]. The absolute form (which we have used previously) does not require that resistors and other linear resistive elements be addressed in updating either the network matrix or the right hand side excitation vector in nonlinear d. c. or transient analysis. In contrast, the incremental or mixed form both require updating of the right hand side vector for such elements. This can result in considerable savings both in repetitive formation of the network equations and also in the sparsity of the resultant right hand side vector (see Section 10. 33). The absolute form is therefore preferred except in those cases when the accuracy advantage of the other form is required (commonly one or two significant figures).

(3) Nonlinear Capacitor Model. We recognize three variables q_c, v_c, and i_c related by the two equations

$$f_2(q_c, v_c) = q_c - f_c(v_c) = 0 \tag{9.66}$$

$$f_1(i_c, \dot{q}_c) = \dot{q}_c - i_c = 0 \tag{9.67}$$

It is left as an exercise to show that a mixed form model for this element is described by the network model of Figure 9. 8, where a gyrator-like element connects charge and voltage variables.

For components as complicated as the above, the value of a network model is questionable. When we examine components with more internal variables - such as the representation of a Laplace Transform in Problem 9.15 - the linearized, discretized component equations should probably be obtained directly using (9. 50).

---

*Gyrators can be used as indicated in Problem 3. 12.

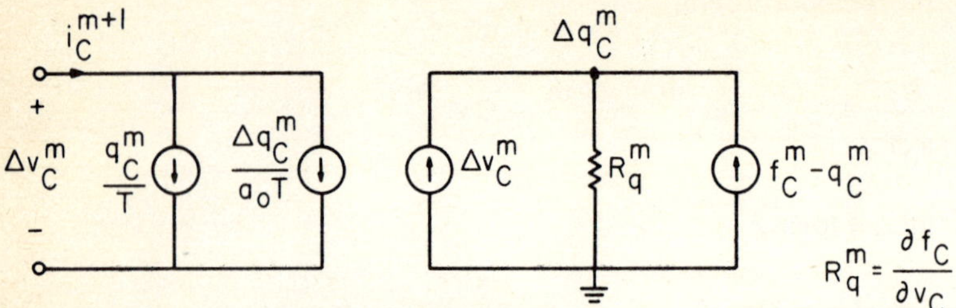

FIG. 9. 8 Nonlinear Capacitor Mixed Model

9. 8 THE GEAR COEFFICIENTS [9. 31]

Besides enunciating most of the preceding algorithms for prediction and error control, Gear developed a set of coefficients (α_i, β_i) for the corrector formula of (9. 33) which yield excellent stability properties. The argument by which he developed this unique set proceeds as follows (see Eq. (9. 34)).

(1) If the roots of the polynomial

$$\eta(\xi) = \sum_{j=0}^{k} (\alpha_j + Ts_i \beta_j)\xi^{k-j} \tag{9.68}$$

$$= \sigma(\xi) + Ts_i \zeta(\xi) \tag{9.69}$$

are to be within the unit circle for all $T > 0$, then the roots of $\sigma(\xi)$ and $\zeta(\xi)$ must all lie within the unit circle (why?)

(2) If the roots of $\zeta(\xi)$ are near the origin in the ξ-plane, (Figure 9. 9) then the roots of $\eta(\zeta)$ will be attracted toward the origin (and away from the stability boundary) as quickly as possible as T increases from zero. Accordingly, the general form of the Gear algorithm is

$$\alpha_0 \underline{x}^{n+1} = -\beta_0 \dot{\underline{x}}^n - \sum_{j=1}^{k} \alpha_j \underline{x}^{n-j+1} \tag{9.70}$$

where $\beta_i = 0, \; i > 0$.

This choice has a marked effect on the accuracy possible with a multi-step method. With a k-th degree method, it is ordinarily possible to achieve an order of 2k; however, by eliminating all β_i except β_0, the order is immediately limited to k. This can readily be checked for k=1, by applying the accuracy test of (9. 36). Let = - 1 and $\alpha_0 = 1$; then for k = 1

$$\xi + \alpha_1 - (\log \xi)(\beta_0 \xi)$$

$$= 1 + \theta + \alpha_1 - (\theta - \frac{1}{2}\theta^2 + \frac{1}{3}\theta^3 + \ldots)(\beta_0 + \beta_0 \theta)$$

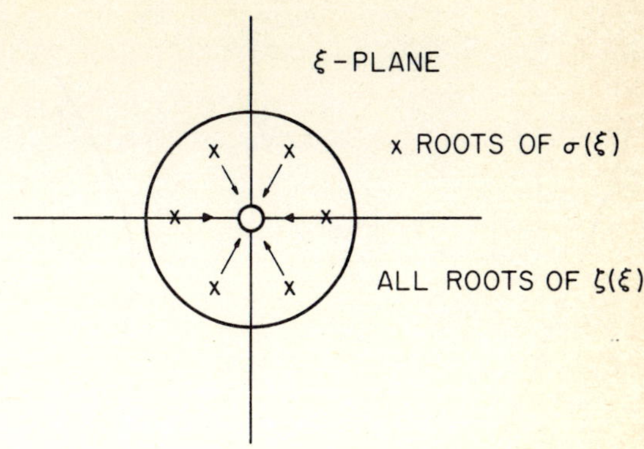

FIG. 9.9 Root Locus of $\eta(\xi)$ as a Function of T

$$= 1 + \alpha_1 + \theta\,(1 - \beta_0) + \theta^2\,(\frac{1}{2}\,\beta_0 - \beta_0) + \dots$$

Attempting to set as many θ^i coefficients to zero as possible yields $\alpha_1 = -1$ and $\beta_0 = 1$, and leaves a residual which grows as p+1 where p = 1. The coefficients for other k's are given in Table 9.4a.

TABLE 9.4 Gear Coefficients

k	β_0	α_1	α_2	α_3	α_4	α_5	α_6
2	2/3	4/3	-1/3	0	0	0	0
3	6/11	18/11	-9/11	2/11	0	0	0
4	12/25	48/25	-36/25	16/25	-3/25	0	0
5	60/137	300/137	-300/137	200/137	-75/137	12/137	0
6	60/137	360/147	-450/147	400/147	-225/147	72/147	-10/147

(a)

k	a_0	a_1	a_2	a_3	a_4	a_5	a_6
1	1	1					
2	2/3	1	1/3				
3	6/11	1	6/11	1/11			
4	24/50	1	35/50	10/50	1/50		
5	120/274	1	225/274	85/274	15/274	1/274	
6	720/1764	1	1624/1764	735/1764	175/1764	21/1764	1/1764

(b)

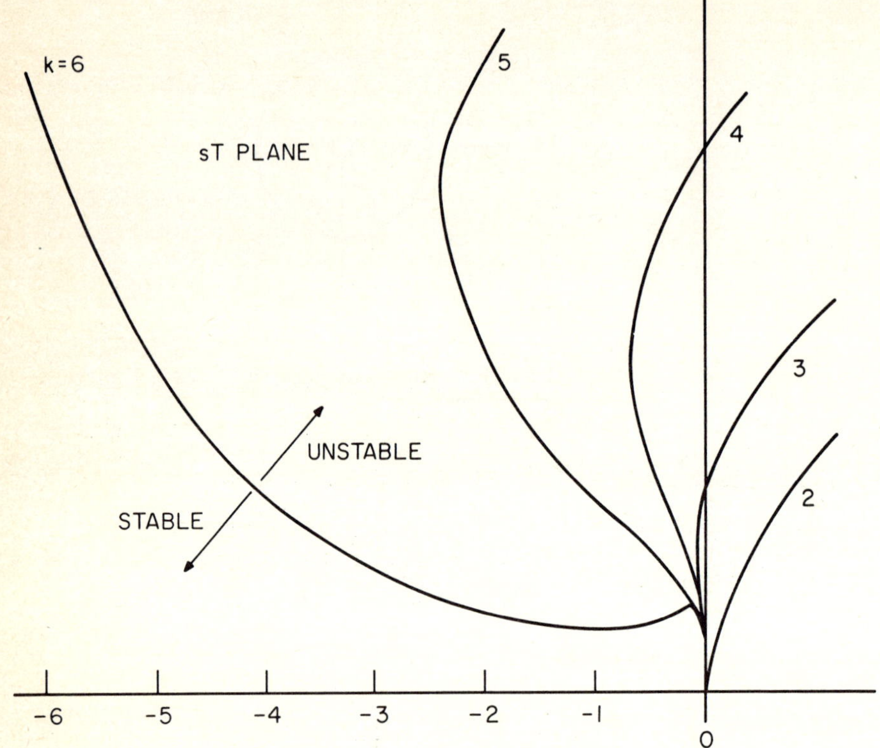

FIG. 9.10 Stability Regions Using Gear Coefficients

From Section 9.6.3 and Equation (9.50), we recall that the a_j used in updating the Nordseick vector can be obtained from the α's and β's [9.26]. These are given in Table 9.4b for the coefficients of Table 9.4a.

Having argued only heuristically in establishing the stability requirement that $\beta_i = 0$ $(i > 0)$, we must plot the resultant stability region as a check. These are shown in Figure 9.10 for various k. They compare favorably with the ideal A-stability characteristic, especially in the important feature of negative real axis stability.

9.9 OTHER SELECTED TOPICS

9.9.1 Analysis of Lossless Lines [9.14] [9.15]

In performing transient analysis of distributed networks our first inclination might be to construct the familiar lumped RC, LC, or similar approximate network model.

Such models, however, yield surprisingly poor time-domain approximation to the response of the distributed line. For illustration, Figure 9.11 shows the transient response to a sinusoidal pulse of a 50 section approximation to a lossless line. Although the ragged appearance of the sinusoids is due to the method of display, it should be observed that both the leading and trailing edges of the pulse are ill-defined, the latter displaying a ringin

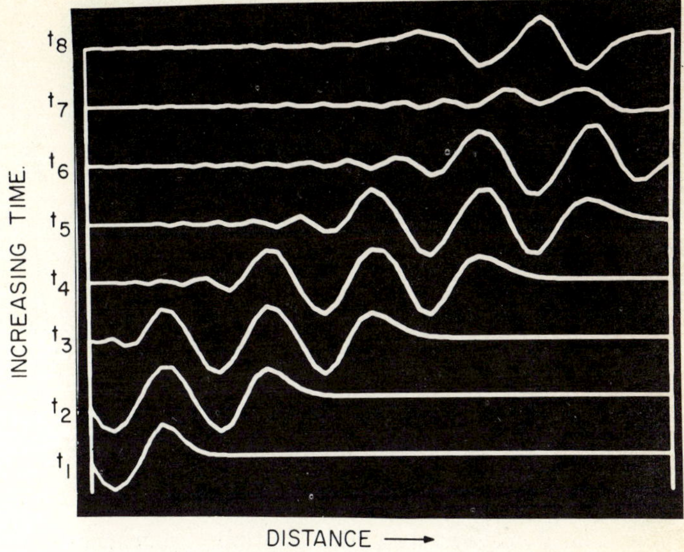

FIG. 9.11 Response of Lumped Approximation to Lossless Line

long after the pulse has passed. In the ideal distributed lossless line, the pulse would propagate undistorted.

In this section, we will devise methods of analyzing the lossless line precisely. Unfortunately, these methods do not extend to general lossy lines [9.27].

In developing the solution procedure we could appeal to physical arguments based on the so-called "characteristics" of the solution of the associated partial differential equation. This is the approach taken in [9.15]. Instead, we will begin with the Laplace transformed solution of the equations as represented by the two-port admittance matrix of Eq. (2.42). The rationale for this approach is that it emphasizes manipulation of and numerical solution with the delay operator, which has utility beyond this single application.

As mentioned in Chapter II, the two-port admittance description of a lossless line is

$$Y_o \begin{bmatrix} \coth \tau s & -\operatorname{csch} \tau s \\ -\operatorname{csch} \tau s & \coth \tau s \end{bmatrix} \begin{bmatrix} V_1 \\ V_2 \end{bmatrix} = \begin{bmatrix} I_1 \\ I_2 \end{bmatrix} \tag{9.71}$$

where $Y_o = \sqrt{c/\ell}$, the characteristic admittance

$\tau = d\sqrt{\ell c}$, the time delay

d = line length

ℓ and c are inductance and capacitance per unit length

To expose the delay operator, we re-write (9.71) in the form

$$Y_o \begin{bmatrix} \frac{1+e^{-2\tau s}}{1-e^{-2\tau s}} & \frac{-2e^{-\tau s}}{1-e^{-2\tau s}} \\ \frac{-2e^{-\tau s}}{1-e^{-2\tau s}} & \frac{1+e^{-2\tau s}}{1-e^{-2\tau s}} \end{bmatrix} \begin{bmatrix} V_1 \\ V_2 \end{bmatrix} = \begin{bmatrix} I_1 \\ I_2 \end{bmatrix} \tag{9.72}$$

or

$$\frac{1}{1-e^{-2\tau s}} \begin{bmatrix} 1 & -e^{-\tau s} \\ -e^{-\tau s} & 1 \end{bmatrix} \begin{bmatrix} Y_o & -Y_o e^{-\tau s} \\ -Y_o e^{-\tau s} & Y_o \end{bmatrix} \begin{bmatrix} V_1 \\ V_2 \end{bmatrix} = \begin{bmatrix} I_1 \\ I_2 \end{bmatrix} \tag{9.73}$$

The left hand side and its scalar multiplier represent the inverse of a matrix; pre-multiplying both sides by this matrix we obtain

$$\begin{bmatrix} Y_o & -Y_o e^{-\tau s} \\ -Y_o e^{-\tau s} & Y_o \end{bmatrix} \begin{bmatrix} V_1 \\ V_2 \end{bmatrix} = \begin{bmatrix} 1 & e^{-\tau s} \\ e^{-\tau s} & 1 \end{bmatrix} \begin{bmatrix} I_1 \\ I_2 \end{bmatrix} \tag{9.74}$$

In the time domain, this becomes

$$\begin{aligned} Y_o v_1(t) - Y_o v_2(t-\tau) &= i_1(t) + i_2(t-\tau) \\ -Y_o v_1(t-\tau) + Y_o v_2(t) &= i_1(t-\tau) + i_2(t) \end{aligned} \tag{9.75}$$

or alternatively as

$$\begin{aligned} i_1(t) &= Y_o v_1(t) - i_2^{\tau} \\ i_2(t) &= Y_o v_2(t) - i_1^{\tau} \end{aligned} \tag{9.76}$$

where

$$\begin{aligned} i_1^{\tau} &= Y_o v_1(t-\tau) + i_1(t-\tau) \\ i_2^{\tau} &= Y_o v_2(t-\tau) + i_2(t-\tau) \end{aligned} \tag{9.77}$$

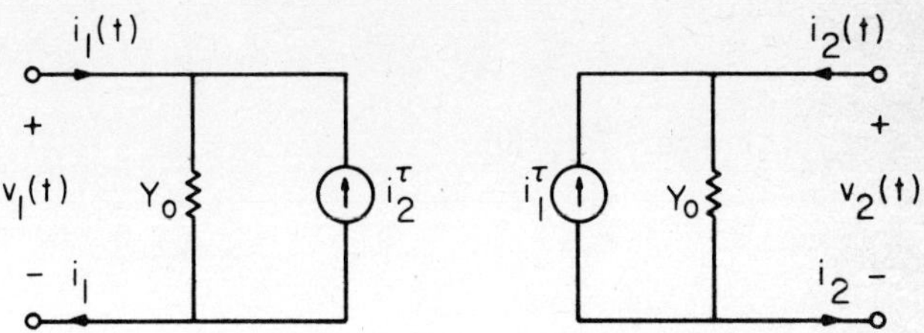

FIG. 9.12 Network Model of Lossless Line

These equations are purely algebraic, and are satisfied by the resistive model of Figure 9.12. This model, unlike previous companion models, represents an exact solution of the lossless line.

The difficulty in implementing this formula is the necessity of evaluating i_1^τ and i_2^τ. Since τ is determined by the line parameters, we must either (1) integrate with a fixed step size T that is a submultiple of τ and save $i_1(t)$ and $i_2(t)$ for the last τ/T time steps, or, (2) if T varies, be prepared to evaluate i_1^τ and i_2^τ between time steps if necessary.

Fortunately, the Gear integration algorithm provides us with each solution variable and its estimated derivatives at every time step, offering a simple means of interpolating i_1^τ and i_2^τ. Storing the four variables $i_1(t)$, $i_2(t)$, $v_1(t)$, and $v_2(t)$ plus their derivatives at each time step suffices to construct the model of Figure 9.12 and therefore the contributions of the lossless line to the node equations at each corrector iteration.

9.9.2 Steady State Periodic Response

If a circuit input to a linear time-invariant stable circuit is periodic, i.e.,

$$\underline{u}(t^n + \Delta) = \underline{u}(t^n) \qquad n = 1, 2, \ldots . \quad \Delta \geq 0 \tag{9.78}$$

then the response $\underline{x}(t)$ may be represented as the sum of a steady state term $\underline{x}_{ss}(t)$ and a transient term $\underline{x}_{tr}(t)$, viz,

$$\underline{x}(t) = \underline{x}_{tr}(t) + \underline{x}_{ss}(t) \tag{9.79}$$

where

$$\underline{x}_{ss}(t^n + \Delta) = \underline{x}_{ss}(t^n) \tag{9.80}$$

Without loss of generality, we can select $t^n = 0$ since the network is time-invariant. By manipulation of (9.9) using (9.80), it can be shown that

$$\underline{x}_{ss}(\Delta) = e^{\underset{\sim}{A}\Delta}\underline{x}_{ss}(0) + e^{\underset{\sim}{A}\Delta}\int_0^\Delta e^{-\underset{\sim}{A}\lambda}\underset{\sim}{B}\,\underline{u}(\lambda)\,d\lambda \tag{9.81}$$

and

$$\underline{x}_{tr}(\Delta) = e^{\underset{\sim}{A}\Delta}\underline{x}_{tr}(0)$$

where

$$\underline{x}_{ss}(0) = [e^{-\underset{\sim}{A}\Delta} - \underset{\sim}{I}]^{-1} \int_0^{\Delta} e^{-\underset{\sim}{A}\lambda}\underset{\sim}{B}\underline{u}(\lambda)\,d\lambda \tag{9.82}$$

$$\underline{x}_{tr}(0) = \underline{x}(0) - \underline{x}_{ss}(0)$$

This derivation is left as a problem.

This result is useful only for linear circuits with nonsinusoidal inputs, since the sinusoidal steady state response could be calculated far more simply using operational methods. However, most pulse circuits, which are often excited by such periodic inputs, are commonly modeled with nonlinear component models. The requirement of linearity is thus violated and (9.81) has marginal practical value.

Aprille and Trick [9.29] have recently proposed an iterative procedure for finding the steady state response for nonlinear equations of the form

$$\dot{\underline{x}} = \underline{f}(\underline{x}, t) \tag{9.83}$$

Following the procedure of Section 9.51 for studying error growth around the solution vector $\underline{x}(t)$, we linearize (9.83) to obtain

$$\begin{aligned}\frac{d}{dt}(\underline{x}(t) + \delta\underline{x}(t)) &= \underline{f}(\underline{x}(t) + \delta\underline{x}(t), t) \\ &\approx f(\underline{x}(t), t) + \frac{\partial \underline{f}}{\partial \underline{x}}\,\delta\underline{x}(t)\end{aligned} \tag{9.84}$$

Subtracting (9.83) from (9.84) gives the linearized description

$$\delta\dot{\underline{x}}(t) = \underset{\sim}{J}_{\underline{x}}(\underline{x}(t))\,\delta\underline{x}(t) \tag{9.85}$$

This equation represents the original equation in the vicinity of the original trajectory $\underline{x}(t)$. As in Newton iteration, then, let the perturbed vector $\underline{x}(t) + \delta\underline{x}(t)$ be the (m+1)st guess of the desired steady state periodic response $\underline{x}_{ss}(t)$. We accordingly adopt the notation

$$\delta\dot{\underline{x}}\overset{m}{(t)}_{ss} = \underline{J}_{\underline{x}}(\underline{x}\overset{m}{(t)}_{ss})\,\delta\underline{x}\overset{m}{(t)}_{ss} \tag{9.86}$$

where

$$\underline{x}\overset{m+1}{(t)}_{ss} = \underline{x}\overset{m}{(t)}_{ss} + \delta\underline{x}\overset{m}{(t)}_{ss} \tag{9.87}$$

The solution of (9.86) may be represented as

$$\delta \underline{x}_{ss}^{m}(\Delta) = \underset{\sim}{\phi}^{m}\,\delta \underline{x}_{ss}^{m}(0) \tag{9.88}$$

where $\underset{\sim}{\phi}^{m}$ is termed the state transition matrix [9.25]. In expanded form, this equation becomes

$$(\underline{x}_{ss}^{m+1}(\Delta) - \underline{x}_{ss}^{m}(\Delta)) = \underset{\sim}{\phi}^{m}(\underline{x}_{ss}^{m+1}(0) - \underline{x}_{ss}^{m}(0)) \tag{9.89}$$

We are now in a position to impose the periodicity requirement $\underline{x}_{ss}^{m+1}(\Delta) = \underline{x}_{ss}^{m+1}(0)$. Equation (9.89) then may be solved to yield

$$\underline{x}_{ss}^{m+1}(0) = [\underset{\sim}{I} - \underset{\sim}{\phi}^{m}]^{-1}[\underline{x}_{ss}^{m}(\Delta) - \underset{\sim}{\phi}^{m}\,\underline{x}_{ss}^{m}(0)] \tag{9.90}$$

The iteration process to converge to the steady state response may be summarized as follows.

(1) Guess an initial state $\underline{x}_{ss}^{0}(0)$; This could be obtained from insight or by integrating the original network equations (9.83) until $\underline{x}$ evidenced a near-periodic behavior.

(2) Integrate the original equations (9.83) to obtain $\underline{x}_{ss}^{m}(\Delta)$

(3) Determine $\underset{\sim}{\phi}$, and find $\underline{x}_{ss}^{m+1}(0)$ from -9.90)

(4) Increment m and return to (2).

We must now address the problem of computing $\underset{\sim}{\phi}^{m}$. We could approximate $\underset{\sim}{\phi}$ directly in the initial equation (9.90); however, the expansion of $\underset{\sim}{\phi}$ over the interval $0 \le t \le \Delta$, although in theory absolutely and uniformly convergent, is not feasible in practice because of the large number of expansion terms necessary [9.25]. Rather, we consider stepwise solution in time that will be consistent with other dc transient solution methods.

We return to (9.85) to perform our approximation. We first divide the period Δ into N sub-intervals. Assuming a backward Euler derivative approximation at the (n+1)st sub-interval, we have the approximate stepwise integration formula

$$\delta \underline{x}_{ss}^{m,n+1} = [\underset{\sim}{I} - T\underset{\sim}{J}_{\underline{x}}(\underline{x}_{ss}^{m,n+1})]^{-1}\,\delta \underline{x}_{ss}^{m,n}$$

Here it is assumed that the Jacobian is known at time t^{n+1} from the solution of (9.83). Integration of (9.86) over the interval $[0, \Delta]$ then has the form

$$\delta \underline{x}_{ss}^{m,N} = \prod_{i=0}^{N} [\underset{\sim}{I} - T\underset{\sim}{J}_{\underline{x}}(\underline{x}_{ss}^{m,i+1})]^{-1}\,\delta \underline{x}_{ss}^{m,0} \tag{9.91}$$

Comparing (9. 88) and (9. 91), we can write

$$\underset{\sim}{\phi}^m \approx \prod_{i=0}^{N} [\underset{\sim}{I} - T\underset{\sim}{J}_{\underline{x}}(\underline{x}_{ss}^{m,i+1})]^{-1}$$

This approximation can now be used in (9. 90) to determine $\underline{x}_{ss}^{m+1}(0)$.

The above procedure is of course not an accurate one since the backward Euler formula has been used to approximate $\underset{\sim}{\phi}$. At the time of this writing a higher order approximation has not been found. Ideally, such a procedure should be consistent with a multistep formula and should exploit the sparsity of the Jacobian matrix. Some of the difficulties encountered in developing such a formula can be appreciated by studying Problem 9.14. An alternative method for iteration to the periodic steady state response has been proposed in [9. 32]; this method uses the methods of Chapter XII to compute the sensitivity of the final state vector $\underline{x}_{ss}^m(\Delta)$ to the initial state vector $\underline{x}_{ss}^m(0)$ and then utilizes optimization methods similar to those of Chapter XI to iterate to the steady state. Standard solution methods may then be used to evolve the requisite gradient information.

Problems

9. 1 If the input $\underline{u}(t)$ is regarded as constant between nT and (n+1)T, show that we can write

$$\underline{x}^{n+1} = e^{\underset{\sim}{A}T}\underline{x}^n + (e^{\underset{\sim}{A}T} - I)\underset{\sim}{A}^{-1}\underset{\sim}{B}\underline{u}^n$$

9. 2 If $\underline{u}(t)$ can be written

$$\underline{u}(t^n + \lambda) = \underline{u}(t^n) + \lambda\frac{d\underline{u}(t^n)}{dt}$$

show that the integral term in Eq. (9. 9) can be written

$$e^{\underset{\sim}{A}T}\int_0^T e^{-\underset{\sim}{A}\lambda}\underset{\sim}{B}\left[\underline{u}(t^n) + \lambda\frac{d\underline{u}(t^n)}{dt}\right]d\lambda$$

$$- (e^{\underset{\sim}{A}T} - \underset{\sim}{I})\underset{\sim}{A}^{-1}\underset{\sim}{B}\underline{u}(t^n) + [(e^{\underset{\sim}{A}T} - \underset{\sim}{I})\underset{\sim}{A}^{-2} - T\underset{\sim}{A}^{-1}]\underset{\sim}{B}\frac{d\underline{u}(t^n)}{dt}$$

9.3 Show how the fourth-order Runge-Kutta procedure is related to the $e^{\underset{\sim}{A}T}$ expansion in solution of linear time-invariant circuits. Consider only the homogeneous case.

9.4 Assume that the input vector $\underline{u}(t)$ is constant and that the Pade approximants are to be used for numerical integration. Show that the contribution of the input to $\underline{x}((n+1)T)$ is given by the formulae of Table P1.1. Now give the complete expression for $\underline{x}((n+1)T)$.

TABLE P1.1

$n = 2$	$T\left[\underset{\sim}{I} - \frac{\underset{\sim}{A}T}{2}\right]^{-1} \underset{\sim}{B}\underline{u}$
$n = 3$	$T\left[\underset{\sim}{I} - \frac{\underset{\sim}{A}T}{2} + \frac{(\underset{\sim}{A}T)^2}{12}\right]^{-1} \underset{\sim}{B}\underline{u}$
$n = 4$	$T\left[\underset{\sim}{I} - \frac{\underset{\sim}{A}T}{2} + \frac{(\underset{\sim}{A}T)^2}{10} - \frac{(\underset{\sim}{A}T)^3}{120}\right]^{-1} \left[\underset{\sim}{I} + \frac{(\underset{\sim}{A}T)^2}{60}\right] \underset{\sim}{B}\underline{u}$

9.5 Find the largest integration step size resulting in stability in the numerical solution of the differential equation

$$\frac{d^2x}{dt^2} + 4\frac{dx}{dt} + 4x = 7 \qquad x(0) = \dot{x}(0) = 0$$

using the backward Euler formula.

9.6 Repeat Problem 9.5 for the fourth-order Runge-Kutta formula.

9.7 A certain well-known network analysis program uses the approximation

$$e^{\underset{\sim}{A}T} \approx \underset{\sim}{I} + \underset{\sim}{A}T + \frac{(\underset{\sim}{A}T)^2}{6}$$

Compare the associated region of stability (a) on the negative real axis and (b) in the left hand plane with first and second order Runge-Kutta methods.

9.8 Using the criterion of (9.33), show that BEI is a first-order method and trapezoidal integration is second order.

9.9 Gear requires that

$$\frac{T^{p+1}}{(p+1)!} \left|(x^{(p+1)})^{n+1}\right| \leq E_{max}\, T/(LC_{p+1}(p+1)!)$$

be satisfied for a successful integration step. Show that this is the same requirement as (9.66).

9.10 Complete the Pascal matrix, which is used for predicting $(x^{(i)})^{n+1,0}$ from the $(x^{(i)})^n$, i=0, 1,... k.

$$\begin{bmatrix} (x^{(0)})^{n+1,0} \\ (x^{(1)})^{n+1,0} \\ \frac{1}{2}(x^{(2)})^{n+1,0} \\ \cdot \\ \cdot \\ \frac{1}{k!}(x^{(k)})^{n+1,0} \end{bmatrix} = \begin{bmatrix} 1 & 1 & 1 & \cdot & \cdot & \cdot & \cdot & \cdot & \cdot & 1 \\ 0 & 1 & 2 & 3 & \cdot & \cdot & \cdot & \cdot & \cdot & \\ 0 & 0 & 1 & \cdot & 3 & \cdot & \cdot & \cdot & \cdot & \\ \cdot & \cdot & & & \ddots & & & & & \\ \cdot & \cdot & & & & & \ddots & & & \\ 0 & 0 & & & & & & & & 1 \end{bmatrix} \begin{bmatrix} (x^{(0)})^{n} \\ (x^{(1)})^{n} \\ \frac{1}{2}(x^{(2)})^{n} \\ \cdot \\ \cdot \\ \frac{1}{k!}(x^{(k)})^{n} \end{bmatrix}$$

9.11 Develop a corrector formula from Equation (9.42) for the coupled set of differential and algebraic equations

$$\underline{F}(\underline{y}, \dot{\underline{y}}, t) = \begin{Bmatrix} \dot{\underline{x}} - \underline{f}_1(\underline{x}, \underline{u}, t) \\ -\underline{f}_2(\underline{x}, \underline{u}, t) \end{Bmatrix} = \underline{0}$$

where $\underline{x}$ in (9.42) has been replaced by

$$\underline{y} = \begin{bmatrix} \underline{x} \\ \underline{u} \end{bmatrix}$$

9.12 Consider a set of simultaneous ODE's of the form

$$S(\underline{x})\dot{\underline{x}} = \underline{f}(\underline{x}, t)$$

where $\underset{\sim}{S}(\underline{x})$ is nonsingular for all $\underline{x}$. Develop an updating formula of the form of (9.42). Also comment on the possibility of an alternate updating formula by defining auxiliary algebraic variables in the manner of (9.29).

9.13 Consider a linear, time-varying set of equations of the form

$$\delta\dot{\underline{x}} = \underset{\sim}{J}(\underline{x}(t))\,\delta\underline{x} \qquad \delta\underline{x}(0) = \underline{\delta}_0$$

where $\underline{x}(t)$ is given. Let this set of equations be integrated for four steps using the multistep formula of (9.33) with $\beta_i = 0$ $i \geq 1$. If a first order procedure is used for the first two integration steps and a second order for the last two steps, we can represent the entire process in matrix form as

$$\begin{bmatrix} \underset{\sim}{I} + \beta_0 \underset{\sim}{J}^1 & \underset{\sim}{0} & \underset{\sim}{0} & \underset{\sim}{0} \\ \alpha_{21} \underset{\sim}{I} & \underset{\sim}{I} + \beta_0 \underset{\sim}{J}^2 & \underset{\sim}{0} & \underset{\sim}{0} \\ \alpha_{31} \underset{\sim}{I} & \alpha_{32} \underset{\sim}{I} & \underset{\sim}{I} + \beta_0 \underset{\sim}{J}^3 & \underset{\sim}{0} \\ 0 & \alpha_{42} \underset{\sim}{I} & \alpha_{43} \underset{\sim}{I} & \underset{\sim}{I} + \beta_0 \underset{\sim}{J}^4 \end{bmatrix} \begin{bmatrix} \delta \underline{x}^1 \\ \delta \underline{x}^2 \\ \delta \underline{x}^3 \\ \delta \underline{x}^4 \end{bmatrix} = \begin{bmatrix} \alpha_1 \underline{\delta}^o \\ \underline{0} \\ \underline{0} \\ \underline{0} \end{bmatrix}$$

where $\delta \underline{x}^r = \delta \underline{x}(t^r)$ and $\underset{\sim}{J}^r = \underset{\sim}{J}(\underline{x}^r, t^r)$.

(a) Find the state transition matrix $\underset{\sim}{\phi}$ that relates $\delta \underline{x}^4$ and δ^0, i.e.,

$$\delta \underline{x}^4 = \underset{\sim}{\phi}\, \delta^0$$

(b) Find the solution of the iteration formula of (9.90) in terms of the parameters of the above formula.

9.14 It is often not feasible to represent a complicated circuit in component form. Rather, its input - output nonlinear dc characteristic is approximated by a spline-fitting algorithm (see Problem 3.10) and its dynamic characteristic by a Laplace Transform.

Consider, then, the differential equation

$$a_0 v_i + a_1 \dot{v}_i + a_2 \ddot{v}_i + \ldots\ldots + a_n v_i = b_0 v_o + b_1 \dot{v}_o + \ldots\ldots + b_n v_o$$

(a) Show that this equation can be represented in the form of first order equations as follows

$$b_n \dot{v}_o - a_n \dot{v}_i - v_1 = 0$$

$$b_{n-1} \dot{v}_o + \dot{v}_1 - a_{n-1} \dot{v}_i - v_2 = 0$$

$$\vdots$$

$$b_1 \dot{v}_o + v_{n-1} - a_1 \dot{v}_i - v_n = 0$$

$$b_0 v_o + v_n - a_0 v_i = 0$$

(b) Show that corrector formula of (9.42) involves a matrix equation with the left hand side

$$\begin{bmatrix} Rb_n & -1 & 0 & 0 & . & . & -Ra_n \\ Rb_{n-1} & R & -1 & 0 & . & . & -Ra_{n-1} \\ Rb_{n-2} & 0 & R & -1 & . & . & -Ra_{n-2} \\ . & & & & & & . \\ . & & & & & & . \\ Rb_1 & . & . & . & R & -1 & -Ra_1 \\ b_0 & . & . & . & 0 & 1 & -a_0 \end{bmatrix} \begin{bmatrix} v_o^{n+1} \\ v_1^{n+1} \\ v_2^{n+1} \\ . \\ . \\ v_n^{n+1} \\ v_i^{n+1} \end{bmatrix}$$

where $R = 1/a_0T$.

(1) Represent the Laplace Transform of the above n-th order differential equation in the form of poles and zeros rather than the coefficients (recall from Chapter II that this is a numerically better - conditioned problem). Develop a corrector formula for this formulation, where each equation (or group of equations) represents a pole or a zero. There are several solutions to this problem.

9.15 Show that the formula for obtaining $\underline{x}_{ss}(0)$ for the linear case (Eq. 9.82) can be regarded as a special case of the nonlinear updating formula of (9.90).

REFERENCES

9.1 Curtiss, C.F., and J.O. Hirschfelder, "Integration of Stiff Equations," Proc. Nat. Acad. Sci., U.S.A., vol. 38, pp. 62-78, 138-150; 1952.

9.2 Gear, C.W., "Numerical Integration of Stiff Ordinary Differential Equations," Report No. 221, Department of Computer Science, University of Illinois; January, 1967.

9.3 Gear, C.W., "The Control of Parameters in the Automatic Integration of Ordinary Differential Equations," Internal Report, Department of Computer Science, University of Illinois; May, 1968.

9.4 Jain, M.K., and V.K. Srivastava, "High Order Stiffly Stable Methods for Ordinary Differential Equations," Report No. 394, Department of Computer Science, University of Illinois; April, 1970.

9.5 Gear, C.W., "The Automatic Integration of Stiff Ordinary Differential Equations," Proc. IFIP Congress, Edinburgh; 1968.

9.6 Liniger, W., and R.A. Willoughby, "Efficient Integration Methods for Stiff Systems of Ordinary Differential Equations," SIAM J. Numer. Anal., vol. 17, no. 1; 1970.

9.7 Henrici, P., Discrete Variable Methods for Ordinary Differential Equations, Wiley; 1963.

9.8 Dahlquist, G.G., "A Special Stability Problem for Linear Multistep Methods," BIT, vol. 3, pp. 27-43; 1963.

9.9 Calahan, D.A., "Efficient Numerical Analysis of Nonlinear Circuits," Proc. Sixth Annual Allerton Conference on Circuit and System Theory, pp. 321-331; October, 1968.

9.10 Allen, R., and C. Pottle, "Stable Integration Methods for Electronic Circuit Analysis with Widely Separated Time Constants," Proc. Sixth Annual Allerton Conference on Circuit and System Theory, pp. 311-320; October, 1968.

9.11 Calahan, D. A., "Numerical Consideration for Implementation of a Nonlinear Transient Circuit Analysis Program," Trans. IEEE, vol. CT-18, no. 1, pp. 66-72; January, 1971.

9.12 Nordseick, A., "On Numerical Integration of Ordinary Equations," Math. Comp., vol. 16, pp. 22-49; 1962.

9.13 Gear, C. W., "Simultaneous Numerical Solution of Differential-Algebraic Equations," Trans. IEEE, vol. CT-18, no. 1, pp. 89-94; January, 1971.

9.14 Dommel, H. W., "Digital Computer Solution of Electromagnetic Transient in Single-and Multiphase Networks," Trans. IEEE, vol. PAS-88, no. 4, pp. 388 ≤ 398; April, 1969.

9.15 Branin, F. H., Jr., "Transient Analysis of Lossless Transmission Lines," Proc. IEEE, vol. 55, pp. 2012-2013; November, 1967.

9.16 Gear, C. W., "The Automatic Integration of Ordinary Differential Equations," CACM, vol. 14, no. 3, pp. 176-179; March, 1971.

9.17 Odeh, F., and W. Liniger, "A Note on Unconditional Fixed Stability of Linear, Multistep Formulae," to appear in Computing.

9.18 Brayton, R. K., F. G. Gustavson, and G. D. Hachtel, "A New Efficient Algorithm for Solving Differential - Algebraic Systems Using Implicit Backward Differential Formulae," to appear in Proc. IEEE, vol. 72, no. 1; January, 1972.

9.19 Pope, D. A., "An Exponential Method of Numerical Integration of Ordinary Differential Equations," Comm. ACM; August, 1963.

9.20 Liou, M. L., "Response of Linear, Time-Invariant Systems Due to Periodic Inputs," Proc. IEEE, vol. 55, pp. 241-243; February, 1967.

9.21 Hohn, F., Elementary Matrix Algebra, Macmillan; 1964.

9.22 Van Valkenburg, M. F., Introduction to Modern Network Synthesis (Chapter 4), Wiley; 1960.

9.23 Varga, R. S., Matrix Iterative Analysis, Prentice-Hall, Inc.; 1962.

9.24 Calahan, D. A., "Numerical Solution of Linear Systems with Widely Separated Time Constants," Proc. IEEE, vol. 55, no. 1, pp. 2016-2017; November, 1967.

9.25 DeRusso, P. M., R. J. Roy, and C. M. Chase, State Variables for Engineers, Wiley; 1965.

9.26 Gear, C. W., "Numerical Solution of Ordinary Differential Equations of Various Orders," Argonne National Laboratory Report ANL 7126, Argonne, Illinois; 1966.

9.27 Haack, G. R., "Comment on Transient Analysis of Lossy Transmission Lines," Proc. IEEE, vol. 59, no. 6, pp. 1022-1023; June, 1971.

9.28 Widlund, O. B., "A Note on Unconditionally Stable Linear Multistep Methods," BIT, vol. 7, pp. 65-70; 1967.

9.29 Aprille, T. J., and T. N. Trick, "Computer-Aided Steady State Analysis of Nonlinear Circuits," Proc. Fourteenth Midwest Symposium on Circuit Theory; May, 1971; also see Proc. IEEE, Jan., 1972.

9.30 Wilkinson, J. H., Rounding Errors in Algebraic Processes, Prentice-Hall; 1963.

9.31 Gear, C. W., Numerical Initial Value Problems in Ordinary Differential Equations, Prentice-Hall; 1971.

9.32 Director, S. W., A. J. Broderson, and D. A. Wayne, "A Method for Quick Determination of the Periodic Steady-State in Nonlinear Networks," Proc. Ninth Annual Allerton Conf. on Circuit and System Theory, Univ. of Illinois (Urbana); October, 1971.

10

Sparse Matrix Procedures and Related Topics

10.1 INTRODUCTION

Consideration of matrix sparsity is quite important to the implementation of most network analysis techniques: dc analysis using Newton"s method, frequency domain analysis, and, especially, transient analysis using implicit integration. Indeed, it often happens that the ability to ignore the number of equations completely changes the equation formulation procedure itself [10.13].

Because this topic involves considerable "technique" rather than concept, we will give only an overview of the field as applied to circuit analysis. The variables of implementation, such as machine type and storage facilities suggest that a detailed study can involve many disciplines.

The growth in use of sparse matrix techniques has accounted for a rethinking of several related computational procedures. Two of these are also included in this chapter: (1) the solution of large problems by partitioning, and (2) the computation of eigenvalues of sparse matrices as required in pole-zero analysis.

10.2 SPARSE MATRIX TECHNIQUES

10.2.1 Introduction

In Section 2.4.3, we first became aware that attention to matrix sparsity and ordering of equations can yield significant computational rewards. However, even the rather impressive savings in fill displayed in Table 2.5 is of little value if the matrix is processed as a full matrix. Each of the zero valued positions would still have to be addressed in the factorization process to discover that it is indeed zero valued.

The solution of sparse systems of equations involves two kinds of techniques:

(1) efficient storage and retrieval of the nonzero matrix entries;
(2) efficient implementation of the logic of the factorization algorithm to exploit storage efficiencies.

10.2.2 Storing and Referencing Sparse Matrix Data

When less than 30% of the matrix entries are nonzero, it is considered inefficient to store the matrix entries as a two dimensional array. Instead, the structural information (the positions of the nonzero entries) is stored in compacted form.

The most obvious method of compacting of data is simply to store the row and column numbers of each nonzero valued position in two vectors and the value in a third vector. This is termed "i - j" ordering and is a natural method of initially storing node matrix data from a user-supplied list of network elements, since the entries may be stored in any order.

In the threaded list (or chained list) method the columns (or rows) of the original matrix are viewed as arranged in a single array in a left to right column (row) order. Zero valued entries are again eliminated. Structural information in the form of either row or column positions are stored for each member of the compacted array. Further, to avoid ambiguities such as completely vacant columns (rows), a vector is carried which points to the first column (row) position in the row (column) and value vectors. This description of position information - often called simply column (row) ordering - is the most natural for processing by factorization algorithms, since the latter usually proceeds a row or column at a time.

Another procedure consists in forming a bit-map of the matrix. The matrix can again be viewed in column or row order, but only the value vector is compacted. Each (i, j) position of the original matrix is represented by a one bit if the entry is nonzero and by a zero bit otherwise. This method is less frequently used since the size of the bit-map increases with the square of the matrix size, whereas the size of the threaded list increases as the number of nonzero entries.

In summary, the i - j ordering is usually most convenient and can be converted to other orderings efficiently by sorting procedures (see Problem 10.1).

10.2.3 Sparse Matrix Factorization Techniques

It is rare indeed that only one set of simultaneous equations is to be solved in the computer analysis of a circuit. DC, transient, and frequency domain analysis all require repeated solution of network equations. Moreover, these equations retain the same structure during the network solution, while the values of entries change from solution to solution (e.g., ω is stepped during frequency domain analysis). In this context, we examine the factorization, forward, and backward substitution processes for sparse matrices to determine which operations need be performed only once and which must be performed at each solution.

In most network problems, the ordering of equations need be performed only once if attention is given to the possibility of certain positions becoming

TABLE 10.1 Comparison of Threaded List and Bit-map Methods of Representing a Sparse Matrix for Ladder Network Example

Row Vector	Column Vector	Value Vector
1	1	$y_1 + y_2$
1	2	$-y_2$
3	3	$y_4 + y_5 + y_6$
3	4	$-y_6$
2	1	$-y_2$
3	2	$-y_4$
2	2	$y_2 + y_3 + y_4$
4	3	$-y_6$
2	3	$-y_4$
4	4	$y_6 + y_7 + y_8$

(a) (i, j) list (unordered)

Column Break Vector (position of initial entry in each column	Row Position Vector	Value Vector (common to (b) and (c))
1	1	$y_1 + y_2$
3	2	$-y_2$
6	1	$-y_2$
9	2	$y_2 + y_3 + y_4$
	3	$-y_4$
	2	$-y_4$
	3	$y_4 + y_5 + y_6$
	4	$-y_6$
	3	$-y_6$
	4	$y_6 + y_7 + y_8$

(b) Threaded list (column ordering)

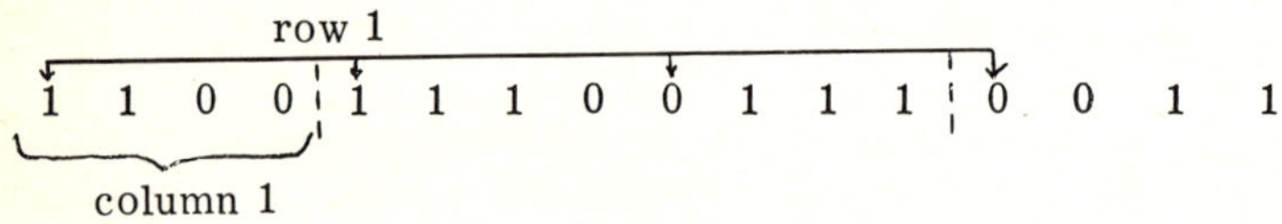

1 represents a non-zero position

32 bits (one or zero) per word

(c) Bit-map (column ordering)

zero valued as the network solution proceeds. For example, during the iteration of a dc network solution, a diode conductance G_d could become small or zero valued. This could result in excessive roundoff error (Section 2.5.2) or a division by zero if G_d appears in a pivot position. This problem can be accommodated by considering numerical data as well as fill information in selecting pivot positions. Assuming, for example, a worst-case zero value for all diode conductances in selecting pivots, we could avoid all risk of later pivoting on diode-related matrix locations, provided a dc path existed through the network with diodes removed.

During the factorization of compacted matrix data, one of the most time-consuming operations results from the need to address packed entries in the partially formed $\underset{\sim}{L}$ and $\underset{\sim}{U}$. For example, consider the factorization of the matrix

$$\underset{\sim}{A} = \begin{bmatrix} a_{11} & a_{12} & a_{13} & 0 \\ a_{21} & a_{22} & a_{23} & a_{24} \\ 0 & 0 & a_{33} & a_{34} \\ a_{41} & 0 & a_{43} & a_{44} \end{bmatrix} \tag{10.1}$$

using diagonal pivoting. Assuming column ordering, each column of $\underset{\sim}{U}$ and $\underset{\sim}{L}$ can be formed in turn and the entire column packed in ascending row number ordering. However, the general formulae for the j-th column, viz,

$$u_{ij} = (a_{ij} - \sum_{k=1}^{i-1} \ell_{ik} u_{kj})/\ell_{ij} \qquad i < j \tag{10.2}$$

$$\ell_{ij} = a_{ij} - \sum_{k=1}^{j-1} \ell_{ij} u_{kj} \qquad i \geq j \tag{10.3}$$

require retrieval of an entire row from the partially formed column-ordered $\underset{\sim}{L}$, creating an addressing problem.

There are several ways of avoiding this problem, but each requires effort beyond the formation of the j-th column of $\underset{\sim}{U}$ and $\underset{\sim}{L}$. One method involves fixing k in (10.2) and (10.3) and partially forming the entire j-th column of $\underset{\sim}{U}$ and $\underset{\sim}{L}$. The algorithm proceeds as follows. Assume that the first j - 1 columns of $\underset{\sim}{L}$ and $\underset{\sim}{U}$ have already been formed and packed, denoted by $\underline{\ell}_1, \underline{\ell}_2, \ldots, \underline{\ell}_{j-1}$. Now define

$$u_{ij}^{(0)} = a_{ij} \quad i < j \tag{10.4}$$

$$\ell_{ij}^{(0)} = a_{ij} \quad i \geq j \tag{10.5}$$

where this j-th column $\underline{\ell}_j$ is temporarily regarded as a full (unpacked) vector. We now form, for $k = 1, 2, \ldots, j-1$,

$$u_{ij}^{(k)} = u_{ij}^{(k-1)} / \ell_{kk} \qquad \text{for } i = k \tag{10.6}$$

$$u_{ij}^{(k)} = u_{ij}^{(k-1)} - \ell_{ik} u_{kj}^{(k)} \qquad k < i < j \tag{10.7}$$

$$\ell_{ij}^{(k)} = \ell_{ij}^{(k-1)} - \ell_{ik} u_{kj}^{(k)} \qquad i \geq j \tag{10.8}$$

where i ranges over those row positions of $\underset{\sim}{L}$ represented in the packed array $\underline{\ell}_k$. This procedure allows us to process an entire column $\underline{\ell}_k$ at one time. The complete annotated process is shown in Table 10.2 for the matrix of (10.1).

Note that the vector $\underline{\ell}_j$ must apparently be treated as a full vector during its formation to allow for fill and then must be packed after computation. This can be avoided by a process known as "garbage collection" [10.8] [10.9]. Here, the u_{ij} (or ℓ_{ij}) as well as the subscript i are saved in two large arrays in the order calculated, ignoring the subscript i. After u_{ij} and ℓ_{ij} are completely determined for a given j, a sort of these unordered "garbage" arrays is made on the subscript i, adding terms with the same subscript to simulate the general formulae of (10.2) and (10.3).

TABLE 10.2 Factorization of $\underset{\sim}{A}$ of (10.1)

Load first column of $\underset{\sim}{A}$

$$\begin{cases} \ell_{11}^{(0)} = a_{11} \\ \ell_{21}^{(0)} = a_{21} \\ \ell_{31}^{(0)} = 0 \\ \ell_{41}^{(0)} = a_{41} \end{cases}$$

Pack first column of $\underset{\sim}{L}\underset{\sim}{U}$ $\Rightarrow$

$$\underline{\ell}_1 = \begin{bmatrix} \ell_{11} \\ \ell_{21} \\ \ell_{41} \end{bmatrix} = \begin{bmatrix} \ell_{11}^{(0)} \\ \ell_{21}^{(0)} \\ \ell_{41}^{(0)} \end{bmatrix}$$

Load second column of $\underset{\sim}{A}$

$$\begin{cases} u_{12}^{(0)} = a_{12} \\ \ell_{22}^{(0)} = a_{22} \\ \ell_{32}^{(0)} = 0 \\ \ell_{42}^{(0)} = 0 \end{cases}$$

Retrieve and process $\underline{\ell}_1$

$$\begin{cases} u_{12}^{(1)} = u_{12}^{(0)} / \ell_{11} \\ \ell_{22}^{(1)} = \ell_{22}^{(0)} - \ell_{21} u_{12}^{(1)} \\ \ell_{42}^{(1)} = \ell_{42}^{(0)} - \ell_{41} i_{12}^{(1)} \end{cases}$$

TABLE 10.2 (Continued)

Pack second column of $\underset{\sim}{L}\underset{\sim}{U}$ $\Longrightarrow$ $\underline{\ell}_2 = \begin{bmatrix} u_{12} \\ \ell_{22} \\ \ell_{42} \end{bmatrix} = \begin{bmatrix} u_{12}^{(1)} \\ \ell_{22}^{(1)} \\ \ell_{42}^{(1)} \end{bmatrix}$

Load third column of $\underset{\sim}{A}$
$$\begin{cases} u_{13}^{(0)} = a_{13} \\ u_{23}^{(0)} = a_{23} \\ \ell_{33}^{(0)} = a_{33} \\ \ell_{43}^{(0)} = a_{43} \end{cases}$$

Load and process $\underline{\ell}_1$
$$\begin{cases} u_{13}^{(1)} = u_{13}^{(0)}/\ell_{11} \\ u_{23}^{(1)} = u_{23}^{(0)} - \ell_{21}\, u_{13}^{(0)} \\ \ell_{43}^{(1)} = \ell_{43}^{(0)} - \ell_{41}\, u_{13}^{(1)} \end{cases}$$

Load and process $\underline{\ell}_2$
$$\begin{cases} u_{23}^{(2)} = u_{23}^{(1)}/\ell_{22} \\ \ell_{43}^{(2)} = \ell_{43}^{(1)} - \ell_{42}\, u_{23}^{(2)} \end{cases}$$

Pack third column of $\underset{\sim}{L}\underset{\sim}{U}$ $\Longrightarrow$ $\underline{\ell}_3 = \begin{bmatrix} u_{13} \\ u_{23} \\ \ell_{33} \\ \ell_{43} \end{bmatrix} = \begin{bmatrix} u_{13}^{(1)} \\ u_{23}^{(2)} \\ \ell_{33}^{(0)} \\ \ell_{43}^{(2)} \end{bmatrix}$

Load fourth column of $\underset{\sim}{A}$
$$\begin{cases} u_{14}^{(0)} = 0 \\ u_{24}^{(0)} = a_{24} \\ u_{34}^{(0)} = a_{34} \\ \ell_{44}^{(0)} = a_{44} \end{cases}$$

Skip $\underline{\ell}_1$, since $u_{14}^{(0)} = 0$

Load and process $\underline{\ell}_2$
$$\begin{cases} u_{24}^{(2)} = u_{24}^{(1)}/\ell_{22} \\ \ell_{44}^{(2)} = \ell_{44}^{(0)} - \ell_{42}\, u_{24}^{(2)} \end{cases}$$

Load and process $\underline{\ell}_3$
$$\begin{cases} u_{34}^{(3)} = u_{34}^{(0)}/\ell_{33} \\ \ell_{44}^{(3)} = \ell_{44}^{(2)} - \ell_{43}\, u_{34}^{(2)} \end{cases}$$

Pack fourth column of $\underset{\sim}{L}\underset{\sim}{U}$ $\Longrightarrow$ $\underline{\ell}_4 = \begin{bmatrix} u_{24} \\ u_{34} \\ \ell_{44} \end{bmatrix} = \begin{bmatrix} u_{24}^{(2)} \\ u_{34}^{(3)} \\ \ell_{44}^{(3)} \end{bmatrix}$

10.2.4 Compiled Code

The most time-consuming part of the above process for a large sparse matrix is the recall of column pointers to $\underline{\ell}_k$'s which do not contribute to the formation of $\underline{\ell}_j$ because they contain no ℓ_{ij} for $i \geq j$. Although one can adopt more sophisticated procedures to avoid even this inefficiency, an alternative procedure has been proposed by Gustavson et al., [10.4] which makes the organization of the factorization algorithm a secondary consideration.

Gustavson et al., suggested that knowledge of the matrix structure permitted the writing of an explicit Fortran (or machine language) program which would factor $\underset{\sim}{A}$ and solve $\underset{\sim}{A}\underline{x} = \underline{b}$, taking into account matrix sparsity. For example, factorization of the matrix

$$\underset{\sim}{A} = \begin{bmatrix} a_{11} & a_{12} & a_{13} \\ 0 & a_{22} & 0 \\ a_{31} & 0 & a_{33} \end{bmatrix} \tag{10.9}$$

is represented by column (a) of Table 10.3. An associated Fortran program is shown in Column (b). The latter will factor any matrix $\underset{\sim}{A}$ having the structure of (10.9) and arranged in column order in the packed array A(I), I = 1, 2, . . . , 6. Performed in this manner, the factorization is executed as rapidly as possible on most computers.

The program to generate such Fortran code is not significantly more difficult to write than the factorization algorithm of Section 10.2.3. As each nonzero component of the j-th column is finally packed into $\underline{\ell}_j$, a counter r is incremented and stored. This component would then be denoted C(r) in the code of Table 10.3b. In determining which components of C appear in any expression of Table 10.3, the r's corresponding to each $u^{(k)}$ of (10.6), (10.7), and (10, 8) are saved for each k. These r's are collected after $\underline{\ell}_j$ is formed and the compacted Fortran statements of Table 10.3b generated.

TABLE 10.3 Explicit Factorization Program for Column-Ordered Matrix of Equation (10.9)

(a) Compacted Factorization	(b) Fortran Code
$\ell_{11} = a_{11}$	C(1) = A(1)
$\ell_{31} = a_{31}$	C(2) = A(2)
$u_{12} = a_{12}/\ell_{11}$	C(3) = A(3)/C(1)
$\ell_{22} = a_{22}$	C(4) = A(4)
$\ell_{32} = -\ell_{31}u_{12}$	C(5) = - C(2)* C(3)
$u_{13} = a_{13}/\ell_{22}$	C(6) = A(5)/C(4)
$\ell_{33} = a_{33} - \ell_{31}u_{13}$	C(7) = A(6) - C(2) * C(6)

10.2.5 Interpretable Code [10.28] [10.15]

Although the Fortran program shown above can be generated directly in machine code (thus avoiding the necessity for a lengthy compilation), the storage requirements of the object program are often so large as to necessitate exchange between high-speed core and slow-speed peripheral storage devices. This of course slows the program execution.

A number of compromises have been suggested to reduce this storage requirement at some cost in computation speed [10.14]. The "interpretive" approach exploits the fact that operations are more efficiently stored in groups than individually. For example, the need to multiply and add occurs often in matrix factorization. Rather than represent these machine operations in the stored explicit program as two sets of machine instructions, a single higher-level instruction may be stored. During execution, this code must be "interpreted" or decomposed into two sets of machine instructions.

The decision between compiled and interpretable code is often made on the basis of the type of and precision of arithmetic involved. When single precision real arithmetic is to be performed, the time to interpret code is approximately equal to the time to perform the arithmetic operations; interpretable code then requires roughly twice the execution time of compiled code and consumes approximately half the storage. The latter feature could be quite important in a time-shared environment. For complex arithmetic, the cost of interpreting a higher level language would be a fraction of the time to execute the arithmetic operations while requiring a fraction of the core necessary to store the compiled code. Obviously, then, the appeal of interpretable code increases with the complexity of the arithmetic operations.

Other features of interpretable code are:

(1) machines with a parallel processing capability (such as the CDC 6600 with dual arithmetic units) can be made to interpret and execute complex arithmetic faster than compiled code;
(2) certain machines (such as the PDP-10) that do not have double precision hardware require considerable execution time to perform this higher precision arithmetic with software; again, the time to interpret code is a fraction of the total time during execution.

The notion of "interpreting" being a broad concept, there are a number of sparse matrix implementations that qualify as "interpreters." One common form is illustrated in Table 10.4. Although all reduction steps can be viewed as a form of inner product, it is necessary to identify the variables ($\underset{\sim}{L}$, $\underset{\sim}{U}$, $\underline{Y}$, or $\underline{X}$) involved in the product; also it may be convenient to identify whether an inner product

(1) produces a fill in the factorization and so does not involve $\underset{\sim}{A}$,
(2) must be divided by ℓ_{ii}, as necessitated in the computation of u_{ij}, and in the forward substitution.

This information can be collected in the "type" code of Table 10.5. It is common to identify 6-12 types of inner product operations, depending on a variety of hardware and algorithm considerations. The remainder of the list is devoted to indexes associated with appropriate arrays.

The result of using an interpretable list and a symmetric factorization algorithm to solve a large complex-valued symmetric matrix are shown in Table 10.5 [10.15]. Considering that more recent scientific computers are 10-20 times faster than the machine quoted, it is clear that solution of 5,000-10,000 node circuits is entirely feasible.

TABLE 10.4 Comparison of Compiled Code, Interpretable List

C(205) = (-C(12) * C(55) - C(61) * C(331) + A(197)) /C(150)

```
      ┌─ no. of products
  1 | 2 | 12  55  61  331  197  150  205
type┘      └────────────┬──────────────┘
                        └──────────── indexes
```

X(13) = (- C(57) * X(10) + X(13)) /C(41)

```
  2 | 1 | 57  10  13  41  13
```

TABLE 10.5 Results of Using Interpretable List (IBM 360/67 running times; double-precision, complex arithmetic).

Order	2,020	3,304
Nonzero Values*	28,190	60,854
Circuit Elements:		
Resistors	1,337	2,196
Inductors	849	1,331
Capacitors	2,183	4,001
Mutual Inductors	2,297	4,622
Preprocessing Time (Sec)	263	626
Nonzero with Fill*	43,150	105,470
Unique Values in Matrix	5,298	9,930
Storage Locations for LU	4,867	11,514
Storage Req. for List (Megabytes)	1.01	3.60
Core Storage - Execution (Bytes)	350K	550K
Soln Time/Freq. (Sec)	17.4	65.7

*Including both upper and lower half

10.2.6 Ordering of Equations [10.1] [10.29]

In Section 2.4.3, we indicated that ordering of equations could drastically affect the number of nonzero positions in $\underset{\sim}{L}$ and $\underset{\sim}{U}$. If explicit solution code is generated as shown in Table 10.3, then not only the execution time but also the code generation time and memory size for the code storage are affected by the number of fills.

The question of minimizing is an interesting one. Keeping track of the number of fills can be a time-consuming task, requiring significantly more time than a single solution and approximately as much effort as the generation of the code of Table 10.3. The reason for this is clear. The detection of a fill requires the same order of effort as filling the position numerically, since we must inspect unreduced rows and columns to determine whether a nonzero $u_{ik}l_{kj}$ product fills an empty position or merely adds to one already filled. Moreover, this check must be made on all remaining nonzero positions, increasing the total effort well beyond a single numerical solution.

In contrast to a strict fill minimization policy, it is common instead to minimize the number of multiplications involved in Eqs. (10.2/10.3). If we are examining the (i, j) position as a potential pivot, then only the i-th row and j-th column need be retrieved; the number of multiplications is then simply determined.

10.3 PARTITIONING

10.3.1 Introduction

The concept of partitioning a large problem into a set of tractable subproblems is a natural and necessary procedure. Usually, this division process is based on either geographical (topological) or functional properties. The optimal ordering procedures discussed above are based primarily on the former; in contrast, we observe that the algorithm requiring the matrix inversion may result in parts of the matrix having properties inviting further exploitation.

We observed that the need for recursive factorization of a sequence of matrices with the same structure may justify the writing of explicit solution code. Now we suggest that it may not be necessary to execute the entire factorization code if only part of the matrix has changed from one solution to the next. This would occur if the sequence of matrices represented

(1) networks with varying design parameters (Figure 10.1a), or
(2) network models of an iteration process with elemental models requiring updating (Figure 10.1b).

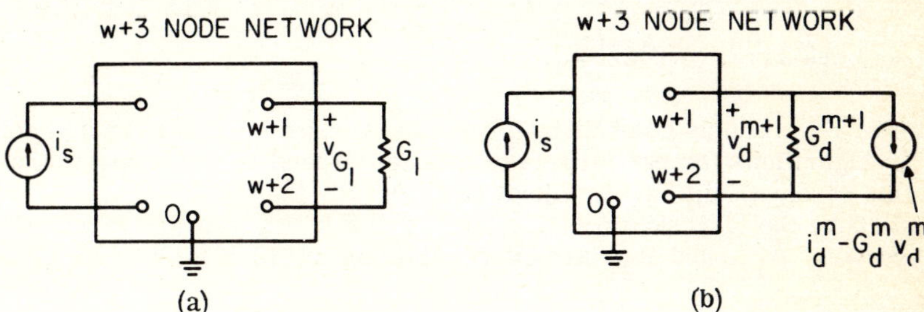

FIG. 10.1 Partitioned Networks

For example, let $\underset{\sim}{G}_N$ be the node conductance matrix of the circuit of Figure 10.1a. The node equations are of the form

$$\underset{\sim}{G}_N \underline{v} = \underline{i}$$

where $\underline{v}$ is a node voltage vector and $\underline{i}$ a source vector. Without knowing the contents of network N, we can still write

$$\left[\begin{array}{cccc|cc} g_{11} & g_{12} & \cdot & \cdot & g_{1,w+1} & g_{1,w+2} \\ g_{21} & g_{22} & \cdot & \cdot & g_{2,w+1} & g_{2,w+2} \\ \cdot & & & g_{w,w} & \cdot & \cdot \\ \hline g_{w+1,1} & g_{w+1,2} & \cdot & & g_{w+1,w+1}+G_1 & g_{w+1,w+2}-G_1 \\ g_{w+2,1} & g_{w+2,2} & \cdot & & g_{w+2,w+1}-G_1 & g_{w+2,w+2}+G_1 \end{array}\right] \left[\begin{array}{c} v_1 \\ v_2 \\ \cdot \\ \hline v_{w+1} \\ v_{w+2} \end{array}\right] = \left[\begin{array}{c} \text{current} \\ \text{sources} \end{array}\right] \tag{10.10}$$

Here the partition is shown to distinguish that part of $\underset{\sim}{G}_N$ which contains G_1. We can write this equation more succinctly as

$$\begin{bmatrix} \underset{\sim}{G}_{11} & \underset{\sim}{G}_{12} \\ \underset{\sim}{G}_{21} & \underset{\sim}{G}_{22}(G_1) \end{bmatrix} \begin{bmatrix} \underline{v}_1 \\ \underline{v}_2 \end{bmatrix} = \begin{bmatrix} \underline{i}_1 \\ \underline{i}_2 \end{bmatrix}$$

where the dependence of $\underset{\sim}{G}_{22}$ on G_1 is explicitly shown. In performing an $\underset{\sim}{L}\underset{\sim}{U}$ factorization of $\underset{\sim}{G}$ when G_1 changes, we sense that a complete recalculation of $\underset{\sim}{L}$ and $\underset{\sim}{U}$ is unnecessary and, if w is large, quite costly.

10.3.2 Partitioning Algorithms

We assume a prior knowledge of the location of all possible parameters $\underline{p} = [p_1, p_2, \ldots, p_r]$ so that at least the upper left hand partition of the matrix

$$\underset{\sim}{A} = \begin{bmatrix} \underset{\sim}{A}_{11} & \underset{\sim}{A}_{12}(\underline{p}) \\ \underset{\sim}{A}_{21}(\underline{p}) & \underset{\sim}{A}_{22}(\underline{p}) \end{bmatrix} \qquad (10.11)$$

can be regarded as constant by proper re-arrangement of rows and columns.

Algorithms for partial refactorization depend on the property that calculation of ℓ_{ij} or u_{ij} involves knowledge of only the ℓ_{ik} or u_{kj} to the left of and above the (i, j) position (i. e., $k < j$ and $k < i$, respectively). Now suppose $\underset{\sim}{A}_{11}$, $\underset{\sim}{A}_{12}$, and $\underset{\sim}{A}_{21}$ are all constant as in (10.10). We can write

$$\begin{bmatrix} \underset{\sim}{A}_{11} & \underset{\sim}{A}_{12} \\ \underset{\sim}{A}_{21} & \underset{\sim}{A}_{22}(\underline{p}) \end{bmatrix} = \begin{bmatrix} \underset{\sim}{L}_{11} & \underset{\sim}{0} \\ \underset{\sim}{L}_{21} & \underset{\sim}{L}_{22}(\underline{p}) \end{bmatrix} \begin{bmatrix} \underset{\sim}{U}_{11} & \underset{\sim}{U}_{12} \\ \underset{\sim}{0} & \underset{\sim}{U}_{22}(\underline{p}) \end{bmatrix} \qquad (10.12)$$

where the $\underset{\sim}{L}$ and $\underset{\sim}{U}$ are partitioned at the same row and column as $\underset{\sim}{A}$ (e.g., $\underset{\sim}{A}_{11} = \underset{\sim}{L}_{11}\underset{\sim}{U}_{11}$). The entries in $\underset{\sim}{L}_{11}$, $\underset{\sim}{L}_{21}$, $\underset{\sim}{U}_{11}$, and $\underset{\sim}{U}_{12}$ are clearly independent of $\underline{p}$ and may be calculated once and for all. This suggests that the ℓ_{ij} and u_{ij} be calculated with the ordering of Figure 10.2. Then the factorization existing after 2w steps will remain a valid factorization of the first w rows and columns when $\underline{p}$ changes. Other useful factorization orderings are given in Problem 10.5.

We can be even more efficient in updating $\underset{\sim}{L}$ and $\underset{\sim}{U}$. Consider the form of the factorization of $\underset{\sim}{G}_{22}$ of (10.10).

$$\ell_{w+1,w+1} = g_{w+1,w+1} + G_1 - \sum_{k=1}^{w} \ell_{w+1,k} u_{k,w+1}$$

$$\ell_{w+2,w+1} = g_{w+2,w+1} - G_1 - \sum_{k=1}^{w} \ell_{w+2,k} u_{k,w+1}$$

$$u_{w+1,\,w+2} = (g_{w+1,\,w+2} - G_1 - \sum_{k=1}^{w} \ell_{w+1,\,k} u_{k,\,w+2}) / \ell_{w+1,\,w+1}$$

$$\ell_{w+2,\,w+2} = g_{w+2,\,w+2} + G_1 - \sum_{k=1}^{w} \ell_{w+2,\,k} u_{k,\,w+2} - \ell_{w+2,\,w+1} u_{w+1,\,w+2}$$

All of the summations $\sum_{k=1}^{w} \ldots$ are independent of G_1 and need to be formed only once. Indeed, only two multiplication - division operations must be performed in this case to completely update the node matrix factorization, regardless of the size of the matrix.

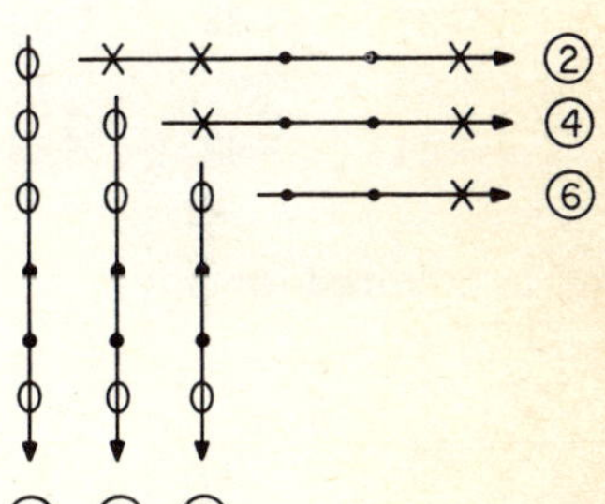

FIG. 10. 2 Column-Row (Crout) Order Factorization ($\underset{\sim}{A}_{11}$, $\underset{\sim}{A}_{12}$, $\underset{\sim}{A}_{21}$ constant)

10. 3. 3 Partitioning the Forward and Back Substitution

In most iteration schemes, it is necessary to change only certain source values from iteration to iteration. At the same time, not all solution variables need to be re-calculated at each iteration; for example, only diode voltages must be determined in the examples of Chapter III. Both of these observations suggest that partitioning of the source and solution vectors would produce savings in the forward and back substitution steps, apart from those which may be realized in factorization.

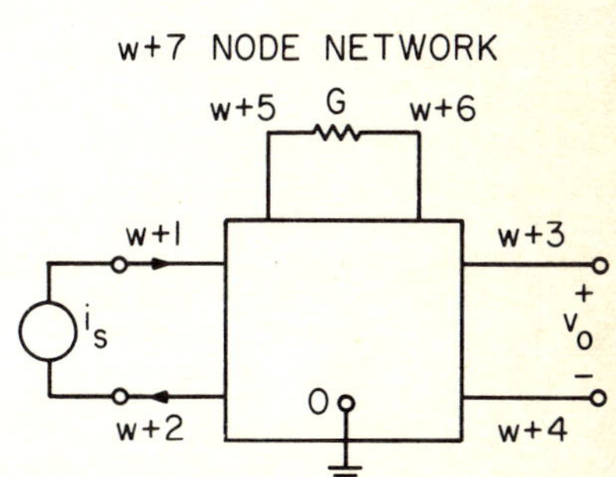

FIG. 10. 3 A Simple Partitioning Model

These potential efficiencies are realized as follows. With the model of Figure 10. 3 as an example, we define

$\underline{i}_s$ - independent current sources
$\underline{v}_2$* - node voltages associated with current sources contained in $\underline{i}_s$
$\underline{v}_3$ - output node voltages to be determined
$\underline{v}_4$ - node voltages associated with parameters $\underline{p}$
$\underline{v}_1$ - node voltages not included in $\underline{v}_{2-4}$.

We can write the node equations as

$$\begin{bmatrix} \tilde{G}_{11} & \tilde{G}_{12} & \tilde{G}_{13} & \tilde{G}_{14} \\ \tilde{G}_{21} & \tilde{G}_{22} & \tilde{G}_{23} & \tilde{G}_{24} \\ \tilde{G}_{31} & \tilde{G}_{32} & \tilde{G}_{33} & \tilde{G}_{34} \\ \tilde{G}_{41} & \tilde{G}_{42} & \tilde{G}_{43} & \tilde{G}_{44}(\underline{p}) \end{bmatrix} \begin{bmatrix} \underline{v}_1 \\ \underline{v}_2 \\ \underline{v}_3 \\ \underline{v}_4 \end{bmatrix} = \begin{bmatrix} \underline{0} \\ \underline{i}_s \\ \underline{0} \\ \underline{0} \end{bmatrix} \quad (10.13)$$

or in factored form

$$\begin{bmatrix} \tilde{L}_{11} & \tilde{0} & \tilde{0} & \tilde{0} \\ \tilde{L}_{21} & \tilde{L}_{22} & \tilde{0} & \tilde{0} \\ \tilde{L}_{31} & \tilde{L}_{32} & \tilde{L}_{33} & \tilde{0} \\ \tilde{L}_{41} & \tilde{L}_{42} & \tilde{L}_{43} & \tilde{L}_{44}(\underline{p}) \end{bmatrix} \begin{bmatrix} \tilde{I} & \tilde{U}_{12} & \tilde{U}_{13} & \tilde{U}_{14} \\ \tilde{0} & \tilde{I} & \tilde{U}_{23} & \tilde{U}_{24} \\ \tilde{0} & \tilde{0} & \tilde{I} & \tilde{U}_{34} \\ \tilde{0} & \tilde{0} & \tilde{0} & \tilde{I} \end{bmatrix} \begin{bmatrix} \underline{v}_1 \\ \underline{v}_2 \\ \underline{v}_3 \\ \underline{v}_4 \end{bmatrix} = \begin{bmatrix} \underline{0} \\ \underline{i}_s \\ \underline{0} \\ \underline{0} \end{bmatrix} \quad (10.14)$$

The forward substitution can be shortened to simply

$$\underline{y}_1 = 0$$

and

$$\begin{bmatrix} \tilde{L}_{22} & \tilde{0} & \tilde{0} \\ \tilde{L}_{32} & \tilde{L}_{33} & \tilde{0} \\ L_{42} & L_{43} & L_{44}(\underline{p}) \end{bmatrix} \begin{bmatrix} \underline{y}_2 \\ \underline{y}_3 \\ \underline{y}_4 \end{bmatrix} = \begin{bmatrix} \underline{i}_s \\ \underline{0} \\ \underline{0} \end{bmatrix} \quad (10.15)$$

Similarly, since only $\underline{v}_3$ is of interest, the back substitution can be written as

*Strictly speaking, this vector is unnecessary. Why?

$$\begin{bmatrix} \tilde{I} & \tilde{U}_{34} \\ \tilde{0} & \tilde{I} \end{bmatrix} \begin{bmatrix} \underline{v}_3 \\ \underline{v}_4 \end{bmatrix} = \begin{bmatrix} \underline{y}_3 \\ \underline{y}_4 \end{bmatrix} \quad (10.16)$$

This formulation preserves the position of the updated part of $\tilde{G}_N$ in the lower right hand corner of the matrix so that the partial re-factorization algorithm of Section 10.3.2 can still be applied. Moreover, the node voltages of $\underline{v}_4$ are available for updating of the $\underline{v}$ vector as required in nonlinear dc and transient analysis.

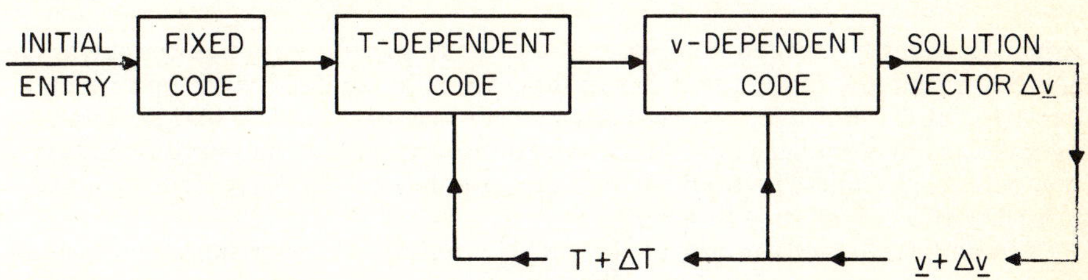

FIG. 10.4 Use of Partitioned Code for Transient Analysis

10.3.4 Elemental Partitioning [10.11] [10.12] [10.13]

One of the most recent and intriguing suggestions relative to matrix partitioning is the notion that, if the factorization is to be separated as described above, then there is no reason to restrict the separation to row and column partitions [10.11]. Rather, it is possible to keep track of those ℓ_{ij} and u_{mn} which depend on $\underline{p}$. When $\underline{p}$ changes, only the operations involving these ℓ's and u's are performed.

This procedure has been shown to be especially effective in performing transient analysis of nonlinear circuits. Certain elements (such as capacitors) depend on the time step T while other elements (such as diodes) depend on the solution vector $\underline{v}$. Therefore, if compiled or interpretable code is generated, only that partition of the code which depends on T or $\underline{v}$ is entered. Since $\underline{v}$ is changed at every iteration, this partition is always used, as Figure 10.4 shows.

10.3.5 Choice of Solution Variables

A node equation formulation usually does not yield minimal-sized partitions of either the matrix, the source vector, or the solution vector. In (10.10), for example, the 2 x 2 partition to separate G_1 was necessary because the voltage across G_1 involved two node voltages. A smaller (1 x 1) partition would result if v_{G_1} were a solution variable. Similar remarks apply to the choice of outputs as solution variables to compact the forward and back substitution steps.

The Hachtel tableau of Chapter VIII, being the most flexible in choice of solution variables, offers the possibility of minimal partition sizes. Here, every branch current and voltage is a potential solution variable. It is possible to form partitions of explicitly those network variables that are involved in the updating process [10.11]. For example, if diode conductances $G_{d_i}^{m+1,m}$, i=1,2,...r, require updating, the v_{d_i} can in general be chosen as the final r solution variables.

10.3.6 Partitioning and Diakoptics*

The above approach to partitioning emphasizes the separation of $\tilde{L}$ and $\tilde{U}$ into parts. This in fact is a rather recent viewpoint [10.6] [10.8]. For many years, it had been customary to partition or "tear" $\tilde{G}$ itself; such procedures known as "diakoptics" were based on the original work of Kron [10.10].

To exemplify the distinction between these methods, it is easy to show that the solution of (10.11) for $\underline{v}_2$ can be represented in "torn" form as

$$[\tilde{G}_{22}(p) - \tilde{G}_{21}\tilde{G}_{11}^{-1}\tilde{G}_{12}]\,\underline{v}_2 = \underline{i}_2 - \tilde{G}_{21}\tilde{G}_{11}^{-1}\underline{i}_1$$

Here the identity of each component of $\tilde{G}$ is retained. This has the advantage of preserving any insight which may have been associated with the original partition of $\tilde{G}$. However, at best it does not suggest a numerical solution procedure; at worst it could be construed to imply that the explicit matrix inverse $\tilde{A}_{11}^{-1}$ must be formed to obtain a solution. We know from Figure 2.4 that this can be disasterous.

In contrast, partitioning of $\tilde{L}$ and $\tilde{U}$ implies a proper numerical solution process; i.e., matrix factorization. In addition, it suggests partitioning strategies that involve not only separation along row and column boundaries but also groupings of individual elements of the $\tilde{L}$ and $\tilde{U}$ matrices to affect more efficient orderings and updating, as discussed in Section 10.3.4.

10.4 ASSEMBLY OF NETWORK EQUATIONS [10.8] [10.9]

Compacting the network matrix into row or column order would appear to represent an ultimate storage efficiency. However, as we are about to discover, it is possible to solve the network equations without even storing the complete network matrix. Instead, we assemble and reduce only a row of the matrix at a time and then discard it.

Given a set of equations of the form

$$\tilde{A}\underline{x} = \tilde{L}(\tilde{U}\underline{x}) = \tilde{L}\underline{y} = \underline{b}$$

where A is n x n, we view the matrix $\tilde{A}$ as arranged in row order. Next, assume that we are at the i-th step of a process for solving these equations, and that we have the following information at our disposal.

(1) u_{jk} and y_j, $j = 1, 2, \ldots i-1$
$\qquad k = j+1, \ldots n$

(2) a_{ik}, $k = 1, 2, \ldots n$

(3) b_i

These assumptions are illustrated in Figure 10.5. Note that only the i-th row of $\tilde{A}$ is assumed available. The elements of the i-th row of $\tilde{L}$ (l_{ik}, $k = 1, 2, \ldots i-1$) can now be determined without access to any other components of $\tilde{L}$, since

*It should be admitted that there are many views of the meaning of Kron's work.

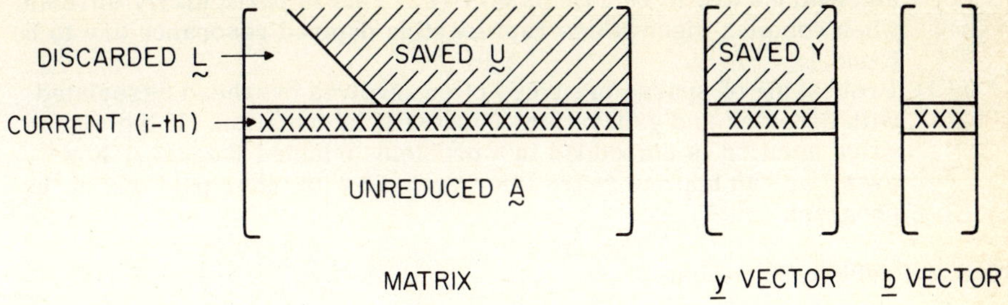

FIG. 10.5 Matrix and Vector Partitions

$$l_{ik} = a_{ik} - \sum_{j=1}^{k-1} l_{ij}\, u_{jk} \qquad k \leq i$$

These l_{ik} may now be used to find y_i, using

$$y_i = (b_i - \sum_{k=1}^{i-1} l_{ik}\, y_k)/\, l_{ii}$$

This completes the i-th step. If repeated for $i = 1, 2, \ldots n$, we are left with the complete $\underline{y}$ vector; we can then proceed with the back substitution phase of solving $\underset{\sim}{U}\underline{x} = \underline{y}$.

This procedure requires storage of only $\underset{\sim}{U}$ and $\underline{y}$ rather than $\underset{\sim}{A}$, $\underset{\sim}{L}$, $\underset{\sim}{U}$, $\underline{y}$, and $\underline{b}$. It has the disadvantage of requiring complete refactorization and re-substititions when any element in $\underset{\sim}{A}$ or $\underline{b}$ changes; in this sense, it is similar to Gauss elimination [10.8].

The requirement that a row of $\underset{\sim}{A}$ be dynamically created and discarded demands efficient ways be developed to scan the branch list of elements to locate those branch quantities that affect the i-th row of $\underset{\sim}{A}$. The reader is invited to consider this "backward pointer" problem as well as the problem of adapting the code generation method of Table 10.3 to handle a row-at-a-time solution procedure.

Even more storage may be saved when symmetric matrices are involved. See Problem 10.9 and Reference [10.15], or write to author for notes.

10.5 COMPUTATION OF EIGENVALUES

10.5.1 Introduction

Sparsity introduces special considerations in selection of eigenvalue solution methods.

(1) Sparsity requirements typically preclude any form of matrix-matrix operations such as are involved in the plane rotations used in the QR algorithm [10.20]. These operations, performed repetitively, are costly with sparse matrices and usually produce intolerable fill of zero valued positions.

(2) Beyond 50-100th order systems, it is unusual to require calculation of all eigenvalues. This poses the problem of specifying which

eigenvalues are to be calculated first. This is particularly difficult when complex eigenvalues representing damped resonance are to be found.

(3) Processing of sparse matrices often involves overhead associated with ordering and generation of indexing information. If the eigenvalue solution is embedded in a problem-oriented language, however, we can hope to share this overhead with other portions of the program.

10.5.2 Problem Statement

Given the n x n matrices $\underset{\sim}{A}$ and $\underset{\sim}{B}$, find the non trivial solution of

$$(\underset{\sim}{A} - s_i \underset{\sim}{B})\underline{X}_i = 0 \tag{10.17}$$

where s_i is the ith of n eigenvalues and $\underline{X}_i$ is a corresponding eigenvector. This is the generalized eigenproblem; if $\underset{\sim}{B} = \underset{\sim}{I}$, it is termed simply the eigenproblem.

10.5.3 Methods Not Requiring an Inverse

Several sparsity-oriented eigenvalue calculation procedures exist which do not necessitate determination of a matrix inverse, or, more properly, a triangular decomposition. Even the system equations need not be assembled, an appealing characteristic when thousands of equations are involved.

The simplest method that could conceivably meet the requirements of sparsity is the power method, given by

$$\underline{Y}^{k+1} = (\underset{\sim}{A} - \rho^k \underset{\sim}{I})\underline{Y}^k$$

where ρ^k representa an origin shift and $\underline{Y}^0$ is an arbitrary vector. Unfortunately, the convergence of this method is slow even with the origin shift and it has the undesirable property of extracting the largest (real) eigenvalue first.

When the matrices $\underset{\sim}{A}$ and $\underset{\sim}{B}$ in (10.17) are symmetric, then several authors have shown that constrained function minimization methods can be used to obtain the (real) s_i, in ascending order of magnitude. The procedure involves minimization of the Rayleigh quotient [10.23] [10.24].

$$r = \frac{\underline{X}_i^T \underset{\sim}{A} \underline{X}_i}{\underline{X}_i^T \underset{\sim}{B} \underline{X}_i} \tag{10.18}$$

The gradient of r is given by

$$\nabla r = \frac{2(\underset{\sim}{A}\underline{X} - r\underset{\sim}{B}\underline{X})}{\underline{X}^T \underset{\sim}{B} \underline{X}}$$

so that $r = s_i$ when $\nabla r = 0$. Deflation is accomplished by constraining $\underline{X}_i$ to be orthogonal to the already determined eigenvectors (a consequence of symmetry). The difficulty with this method is that the number of iterations required per eigenvalue appears to rise sharply with the problem size, thus negating the appeal of the method to solve large problems. However, it is apparently the only globally convergent sparsity-oriented method not requiring an inverse.

10.5.4 Methods Requiring an Inverse

Introduction. Eigenvalue methods that require either (1) solution of a set of simultaneous equations or (2) evaluation of a determinant $\Delta(s)$ and its derivatives $d^r\Delta(s)/ds^r$ are amendable to sparse matrix implementation. We will presume that eigher compiled or interpretable code can be generated for any structure-dependent sequence of operations to be performed repetitively. As previously noted, this is usually worthwhile when 5-10 repetitions are to be carried out.

Inverse Iteration. When $\underset{\sim}{B} = \underset{\sim}{I}$ and $\underset{\sim}{A}$ is symmetric, then a process known as inverse iteration with origin shift can be quite effective. The basic formula is

$$(\underset{\sim}{A} - \rho^k \underset{\sim}{I})\underline{Y}^{k+1} = \ell^k \underline{Y}^k$$

where ℓ^k is a normalizing scalar. If the origin shift is given by

$$\rho^k = \frac{(\underline{X}^k)^T \underline{\underline{A}}\underline{X}^k}{(\underline{X}^k)^T \underline{X}^k}$$

then the process is known as Rayleigh quotient iteration and, under certain conditions [10.25] converges cubically to an eigenvalue. The generalization of the above to unsymmetric matrices [10.26] requires matrix-matrix operations and so is not sparsity-oriented.

Muller's Method. The problem of solving (10.17) can be replaced by the problem of finding the zeros of the determinant

$$\Delta(s) = ||\underset{\sim}{A} - s\,\underset{\sim}{B}||$$

Since the coefficients of Δ are not obtainable, only those root-finding methods which employ function iteration can be used. This unfortunately eliminates several powerful globally convergent methods [10.22]; however, two methods with excellent but unproven convergence properties remain: Muller's method [10.21] and Laguerre's method [10.19].

Of the two, only Muller's method is applicable to the general eigenproblem. It requires only the evaluation of the determinant for a given value of s; from these values, the first and second derivatives $d\Delta/ds$ and $d^2\Delta/ds^2$ are estimated and an interpolation to $\Delta=0$ is made. This leaves such questions as fill minimization, partitioning, etc., to be decided on other grounds. The method is somewhat less than quadratically convergent (since derivatives are not explicitly computed).

Muller's method does not require prior knowledge of the number of eigenvalues. Eigenvalues are extracted until the value of Δ remains within a prescribed tolerance on successive iterations. This value is then used to find the leading coefficient of Δ. After the poles are found the same procedure can be used to find the roots of any cofactor of $[\underset{\sim}{A} - \lambda\underset{\sim}{B}]$; these "zeros" are necessary in applications where the complete transient response is sought.

Muller's method has been used extensively for modal analysis of electronic circuits [10.27] using modal analysis.

The number of iterations per eigenvalue has been found relatively independent of problem sizes. The statistics associated with calculation of all the natural modes of an amplifier circuit is shown in Table 10.6. The sparse matrix factorization program used achieved approximately 1/2 the efficiency of an interpretive program.

TABLE 10.6 Amplifier Example (Supplied by W. J. McCalla)

No. of node equations	50
No. of modes (poles)	
real	26
complex pairs	7
Error (relative)	$<5 \times 10^{-7}$
$\lvert s_i \rvert_{max}$, $\lvert s_i \rvert_{min}$	$10^{10}, 10^{3}$
Iterations/eigenvalue	8
Total time (CDC 6400, single precision)	
Experimental, partially exploiting sparsity	40 sec
Estimated, using interpreter	20 sec

Conclusions. Considering the results of Table 10.6 as state-of-the-art, we can see that there is considerable work to be done in development of new algorithms for eigenvalue calculation. Although there is current effort being applied to the solution of the general eigenproblem for symmetric matrices [10.17] there is relatively little interest in other branches of engineering in the unsymmetric case.

10.6 SPARSE MATRIX METHODS IN GENERAL NETWORK DESIGN PROGRAMS

There is a current trend in network design programs to include all types of analysis - dc, transient, and ac - and appropriate optimization algorithms within a single program. This would appear to place extraordinary demands on sparse matrix algorithms.

For example, it might appear that separate pivot selections have to be made for ac and dc analysis. However, if the network has a dc solution, then the dc pivot order can be used throughout.

The question of the type of code generator - interpreter or compiler - to be used is not quite so transparent. If only the dc and transient analyses are to be performed, then the same compiled code can service both analyses by assuming the sparsity structure of the transient analysis Jacobian and selecting pivots on the basis of the dc analysis matrix (i.e., "holes" are left in positions dependent on 1/T). When ac analysis is to be performed, then a single interpretable code is preferred; only the interpreter changes, using either real or complex arithmetic. This has the additional feature of permitting the user to switch from single to double precision without regenerating code, as a compiler would require.

Also, one can easily permit options which use single precision for a Jacobian calculation and double precision for the right hand side vector, as might be preferred for transient analysis (see Section 9.6.2).

Problems

10.1 It is sometimes necessary to rearrange unordered position data (as in Table 10.1a) into row or column ordering. This is achieved by assigning "weights" to each (i, j) position such that an arrangement or sort of the weights in ascending order accomplishes the desired row or column ordering. Devise weighting schemes for an $n \times n$ matrix which results in these reorderings.

10.2 Repeat (10.1) for Crout ordering.

10.3 Complete Table 10.3 to perform the complete solution of

$$\underset{\sim}{A}\underline{x} = \underline{b}$$

where $\underline{b}$ is a full vector. Use only the arrays A(I), C(I), and B(I), where B(I) contains the entries of $\underline{b}$ and $\underline{x}$.

10.4 Suppose that the matrix

$$\underset{\sim}{A} = \begin{bmatrix} a_{11} & a_{12} & 0 & a_{14} \\ a_{21} & a_{22} & a_{23} & a_{24} \\ 0 & a_{32} & a_{33} & a_{34} \\ a_{41} & a_{42} & a_{43} & a_{44} \end{bmatrix}$$

is the zero-sum indefinite node matrix of a network of resistors. Write an explicit factorization algorithm using diagonal (1-2-3) ordering which avoids cancellation.

10.5 Match the following factorization orderings with assumptions on the variability of $\underset{\sim}{A}_{11}$, $\underset{\sim}{A}_{12}$, $\underset{\sim}{A}_{21}$, and $\underset{\sim}{A}_{22}$ as in Figure 10.2.

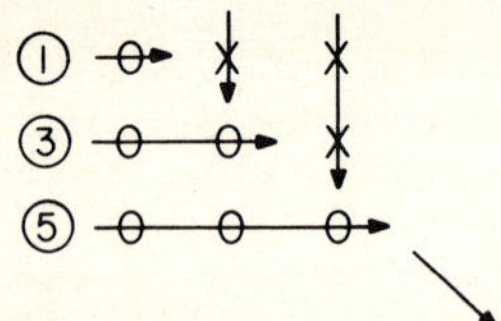

(a) BLOCK ORDER FACTORIZATION

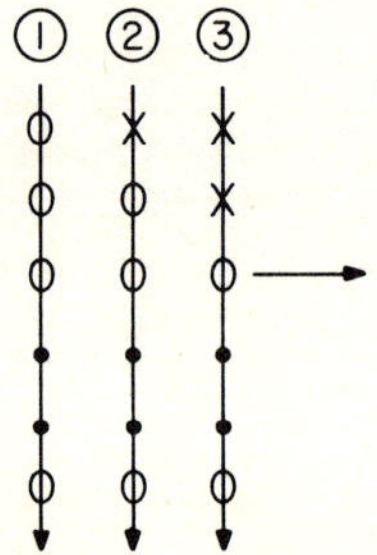

(b) COLUMN ORDER FACTORIZATION

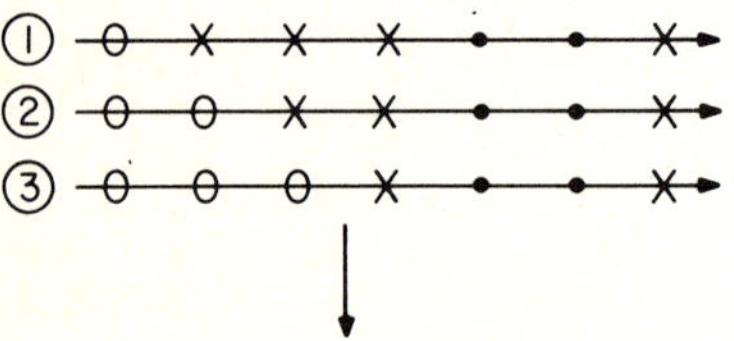

(c) ROW ORDER FACTORIZATION

10.6 Assume in the network model of Figure 10.3 that the unchanging part of the factorization of $\underset{\sim}{G}$ has been determined. If both i_S and R change, what is the greatest number of multiplication-division operations required to recalculate v_o.

10.7 It may be shown that

$$\Delta = \prod_{j=1}^{n} \ell_{jj}$$

where Δ is the determinant of a matrix and the ℓ_{ij} are diagonal positions of $\underset{\sim}{L}$ when diagonal pivoting is used. A similar formula applies when off-diagonal pivoting is used, except that the sign of the determinant must be independently computed from a knowledge of the pivot positions. Develop an algorithm to compute this sign.

10.8 As a rule of thumb, it is found that code generation for factorization of a matrix requires five times the computational effort of a single factorization. Therefore, it is sometimes to an advantage to factor parts of a sparse matrix by techniques other than code generation. Develop a complete solution algorithm in terms of the partitions of Eq. (10.12).

10.9 A symmetric matrix $\underset{\sim}{A} = \underset{\sim}{A}^T$ can always be factored in the form

$$\underset{\sim}{A} = \underset{\sim}{L}\,\underset{\sim}{D}\,\underset{\sim}{L}^T$$

where $\underset{\sim}{L}$ is a lower triangular matrix

$$D = \text{diag}\,[d_{11}\; d_{22} \ldots d_{nn}]$$

Develop a factorization algorithm using methods similar to (a) $\underset{\sim}{L}\underset{\sim}{U}$ factorization and (b) Gauss elimination, assuming that you are not permitted duplicate storage for symmetric parts of $\underset{\sim}{A}$ and that only $\underset{\sim}{L}^T$ and $\underset{\sim}{D}$ (plus a modest work vector) may be stored. Show that Gauss elimination can be made more efficient in the number of multiplications required.

10.10 In Figure 10.5, consider the case when

(1) only the last n_1 b_i's are nonzero, and
(2) only the last n_2 x_i's are to be found

Which parts of the $\underset{\sim}{U}$ matrix can be discarded during processing if only one solution is desired?

10.11 According to Section 9.5 the eigenvalues of the Jacobian depends on whether $\underline{q}$ or $\underline{v}$ is selected as an integration variable. Suppose we were interested in performing an ac or small-signal analysis about a dc operating point. Would the ac response depend on the same choice? Illustrate your answer with a resistor-diode network that includes the nonlinear diode capacitance.

10.12 A cofactor is required by Muller's method in computing the zeros of a transfer function. If compiled or interpreted code is being used to find $\Delta(s_i)$, then a row and column cannot be formally deleted without re-generation of the code. Suggest one or more ways of avoiding this problem.

REFERENCES

10.1 Tinney, W. F., and J. W. Walker, "Direct Solutions of Sparse Network Equations by Optimally Ordered Triangular Factorization," Proc. IEEE, vol. 55, pp. 1801-1809; November, 1967.

10.2 Tewarson, R. P., "Solution of a System of Simultaneous Linear Equations with a Sparse Coefficient Matrix by Elimination Methods," BIT, vol. 7, pp. 226-239; 1967.

10.3 Spillers, W. R., and N. Hickerson, "Optimal Elimination for Sparse Symmetric Systems as a Graph Problem," Quart. Appl. Math, vol. 26, pp. 425-432; 1968.

10.4 Gustavson, F., W. Liniger, and R. Willoughby, "Symbolic Generation of an Optimal Crout Algorithm for Sparse Systems of Linear

Equations," IBM Report RC 1852, IBM Thomas Watson Research Center, Yorktown Heights, New York; June, 1967.

10.5 Van Bree, K., "Frequency Analysis of Linear Circuits Using Sparse Matrix Techniques," MS Dissertation, MIT; September, 1969.

10.6 Dommel, H. W., "Digital Computer Solution of Electromagnetic Transients in Single-Multiple Networks," Trans. IEEE, vol. PAS-88, no. 4, pp. 388-398; April, 1969.

10.7 McNamee, J. M., "Algorithm 408: a Sparse Matrix Package," CACM, vol. 14, no. 4, pp. 265; April, 1971.

10.8 Ogbuobiri, E. C., W. F. Tinney, and J. W. Walker, "Sparsity-Directed Decomposition for Gaussian Elimination On Matrices," Trans. IEEE, vol. PAS-89, no. 1, pp. 141-150; January, 1970.

10.9 Ogbuobiri, E. C., "Dynamic Storage and Retrieval in Sparsity Programming," Trans. IEEE, vol. PAS-89, no. 1, pp. 150-155; January, 1970.

10.10 Kron, G., "A Set of Principles to Interconnect the Solutions of Physical Systems," J. Appl. Physics, vol. 24, pp. 965-980; 1953.

10.11 Hachtel, G. D., R. Brayton, and F. Gustavson, A Sparse Matrix Approach to Network Analysis," Proc. Second Cornell Conf. on Computerized Electronics, Ithaca, N. Y., pp. 68-82; August, 1969.

10.12 Norin, R. S., and C. Pottle, "Effective Ordering of Sparse Matrices Arising from Nonlinear Electrical Networks," Trans. IEEE, vol. CT-18, no. 1, pp. 139-145; January, 1971.

10.13 Hachtel, G. D., R. K. Brayton, and F. G. Gustavson, "The Sparse Tableau Approach to Network Analysis and Design," Trans. IEEE, vol. CT-18, no. 1, pp. 101-113; January, 1971.

10.14 Also see Proc. Sparse Matrix Symposium held at IBM Watson Research Center, Yorktown, New York, September 9-10, 1968; second conference held at same location, September, 1971; for extended bibliography see "Some Results on Sparse Matrices," IBM Report 2332, same address; February, 1969.

10.15 Erisman, A. M., "Sparse Matrix Approach to the Frequency Domain Analysis of Linear Passive Electrical Networks," IBM Symposium on Sparse Matrices and Their Application, Yorktown Hgts., N. Y.; 1971. (Also published in book form by Plenum Publishing Co.)

10.16 Wilkinson, J. H., The Algebraic Eigenvalue Problem, Clarendon Press; 1965.

10.17 Tewarson, R. P., "On the Reduction of a Sparse Matrix to Hessenberg Form," International J. of Comp. Math., vol. 2, pp. 283-295; 1970.

10.18 Peters, G., and J. H. Wilkinson, "Eigenvalues of $Ax = \lambda Bx$ with Band Symmetric A and B," Comp. Jour., vol. 14, pp. 398-404; 1969.

10.19 Parlett, B., "Laguerre's Method Applied to the Matrix Eigenvalue Problem," Math. Comp., vol. 18, pp. 464-485; 1964.

10.20 Frances, J. G. F., "The QR Transformation - A Unitary Analogue to the LR Transformation," Comp. Jour., vol. 4, pp. 265-271, pp. 332-345; 1961.

10.21 Muller, D. E., "A Method for Solving Algebraic Equations Using an Automatic Computer," MTAC, vol. 10, pp. 208-215; 1956.

10.22 Traub, J. F., "A Class of Globally Convergent Iteration Functions for the Solution of Polynomial Equations," Math. Comp., vol. 20, pp. 113-138; 1966.

10.23 Bradbury, W. W., and R. Fletcher, "New Iterative Methods for Solution of the Eigenproblem," Numerische Mathematick, vol. 9, pp. 259-267; 1966.

10.24 Fry, R. L., and M. P. Kapoor, "A Minimization Method for the Selection of the Eigenproblem Arising in Structural Dynamics," Proc. Second Conf. on Matrix Methods in Structural Mechanics, Wright-Patterson Air Force Base, Dayton, Ohio, pp. 271-304; 1969.

10.25 Ostrowski, A. M., "On the Convergence of the Rayleigh Quotient Iteration for the Computation of the Characteristic Roots and Vectors," Parts I-VI, Arch. Rat. Math. and Math. Physics, vols. 1-4; 1958-1959.

10.26 Parlett, B. N., and W. Kahan, "On the Convergence of a Practical QR Algorithm," Proc. IFIP Congress, pp. A25-A30; 1968.

10.27 McCalla, W. J., and D. O. Pederson, "Elements of Computer-Aided Circuit Analysis," Trans. IEEE, vol. CT-15, no. 1, pp. 14-26; January, 1971.

10.28 Lee, H. B., "An Interpretive Implementation of Gaussian Elimination," Digest of the Sparse Matrix Symposium, pp. 75-81, IBM Thomas Watson Research Center, Yorktown Heights, N. Y. September, 1968.

10.29 Markowitz, H. M., "The Elimination Form of the Inverse and its Application to Linear Programming," Management Science, vol. 3, pp. 255-269; 1957.

11

Network Optimization Methods

11.1 INTRODUCTION

In Chapter VI, we introduced the notion that the computer could be programmed to accomplish certain design objectives that can both replace and extend even a knowledgeable designer. There were, however, at least three major limitations to the previously proposed design algorithms.

(1) The design problem had to be formulated as a problem of matching a calculated response function to a discrete-valued specification function; as we shall see in Chapter XII, there is often a need to achieve a continuous-valued response, especially in time-domain design.

(2) The iteration methods offered the strong possibility of divergence especially if the iteration was initiated far removed from the design goal.

(3) Equality constraints could be imposed in element values by simple parameter mapping techniques; however, responses could not be similarly constrained.

In this chapter, we will develop concepts and simple iteration schemes related to the minimization of the function $P(\underline{p})$, where $\underline{p}$ is a set of design parameters. This problem is a generalization of the least squares problem of Chapter VI to include any function P that is a well-behaved function of $\underline{p}$ and has certain other properties we will shortly discuss.

11.2 FUNDAMENTALS OF OPTIMIZATION

11.2.1 Definition of a Minimum

The requirements for a minimum of a function P of a single variable p_j are well known, namely

$$\frac{\partial P}{\partial p_j} = 0 \tag{11.1}$$

$$\frac{\partial^2 P}{\partial p_j^2} > 0 \tag{11.2}$$

Now let P be a scalar function of two variables written as a vector so that we can write $P(\underline{p})$ where $p = [p_1\ p_2]^T$. If P is expanded in a Taylor's series about the minimum $\underline{p} = \hat{\underline{p}}$, the first two terms are

$$P(\underline{p}) = P(\hat{\underline{p}}) + (p_1-\hat{p}_1)\frac{\partial P(\hat{\underline{p}})}{\partial p_1} + (p_2-\hat{p}_2)\frac{\partial P(\hat{\underline{p}})}{\partial p_2}$$

$$+ \frac{1}{2}\left[(p_1-\hat{p}_1)^2\frac{\partial^2 P(\hat{\underline{p}})}{\partial p_1^2} + 2(p_2-\hat{p}_2)(p_1-\hat{p}_1)\frac{\partial^2 P(\hat{\underline{p}})}{\partial p_1 \partial p_2} + (p_2-\hat{p}_2)^2\frac{\partial^2 P(\hat{\underline{p}})}{\partial p_2^2}\right] \tag{11.3}$$

The key to generalizing (11.1) and (11.2) is to note that the second term of (11.3) can be written more succinctly, viz,

$$P(\underline{p}) = P(\hat{\underline{p}}) + \begin{bmatrix}\frac{\partial P}{\partial p_1} & \frac{\partial P}{\partial p_2}\end{bmatrix}\underline{\Delta p} + \frac{1}{2}\,\underline{\Delta p}^T\begin{bmatrix}\frac{\partial^2 P}{\partial p_1^2} & \frac{\partial^2 P}{\partial p_1 \partial p_2} \\ \frac{\partial^2 P}{\partial p_2 \partial p_1} & \frac{\partial^2 P}{\partial p_2^2}\end{bmatrix}\underline{\Delta p} \tag{11.4}$$

where $\underline{\Delta p}^T = [p_1-\hat{p}_1\ p_2-\hat{p}_2]$. This notation will further be abbreviated

$$P(\underline{p}) = P(\hat{\underline{p}}) + (\underline{\nabla}P(\underline{p}))^T\,\underline{\Delta p} + \frac{1}{2}\underline{\Delta p}^T\left[\frac{\partial^2 P}{\partial p_i \partial p_j}\right]\underline{\Delta p} \tag{11.5}$$

where $\underline{\nabla}P(\underline{p})$ denotes the gradient of $P(\underline{p})$ and is a column vector. Since $\hat{\underline{p}}$ is a minimum of $P(\underline{p})$, then any small change $\underline{\Delta p}$ away from $\hat{\underline{p}}$ must increase

$P(\underline{p})$. This is equivalent to requiring the term depending linearly on $\Delta\underline{p}$ be zero and the second term depending quadratically on $\Delta\underline{p}$ be positive. The former requires the gradient to be zero, viz,

$$\nabla P(\underline{p}) \Big|_{\underline{p} = \hat{\underline{p}}} = \underline{0} \tag{11.6}$$

The latter requires that the matrix of second partial derivatives (known as the Hessian of P) be positive definite, i.e.,

$$\Delta\underline{p}^T \left[\frac{\partial^2 P}{\partial p_i \partial p_j} \right] \Delta\underline{p} > 0 \tag{11.7}$$

for all $\Delta\underline{p}$. This may appear stronger than required, since $\Delta\underline{p}$ was originally assumed small. However, if the term is negative for any $\Delta\underline{p}$, then it will be negative for $\gamma \Delta p$, where $|\gamma|$ is arbitrarily small. The above conditions degenerate to (11.1) and (11.2) for the single variable case.

Note that:

(1) we have ignored the case when the second term in the Taylor series expansion is zero at the minimum; higher order terms must then be examined to determine the nature of this stationary point;

(2) $P(\underline{p})$ behaves quadratically near the minimum, since $\nabla P(\hat{\underline{p}}) = \underline{0}$.

11.2.2 Local Minima

The polynomial

$$P(p_1) = \frac{p_1^4}{4} - \frac{7}{3} p_1^3 + 7p_1^2 - 8p_1 + 5 \tag{11.8}$$

is sketched versus p_1 in Figure 11.1. Clearly, two minima exist. The value of the global minimum, $P(\hat{p}'_1)$, satisfies the condition $\min_1 P(p_1)$. All other minima are termed local minima.

The possibility (or better, likelihood) of local minima introduces an uncertainty in practical optimization problems. There being no local test to distinguish local from global minima in general, the assumption usually made or implied in minimization procedures and convergence proofs is that

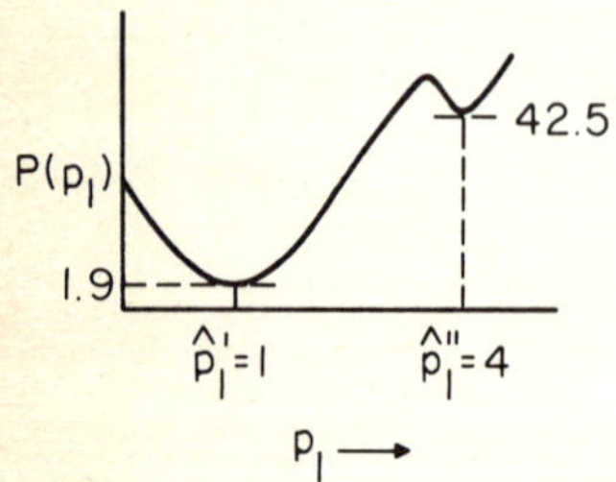

FIG. 11.1 Demonstration of Local Minima

any local minimum is the global minimum. This is justifiable in a practical sense if the minimization is initiated near the global minimum (a fact established from other considerations) or near a satisfactory local minimum. Without this assumption, one is forced to sample $P(p)$ over the entire range of interest—usually in a statistical sense—a prohibitive process when $\underline{p}$ is a vector.

11.2.3 Convexity

In certain cases, the Hessian matrix can be shown to be positive definite over an entire region R in the parameter space of $\underline{p}$. P is then termed <u>convex</u> in R.

From Figure 11.1, it is clear for the one-dimensional case that minima must be separated by regions where the second derivative is negative. This may be formally shown by expanding $P(p_1)$ in a Taylor's series about two adjacent minima $\hat{p}'_1$ and $\hat{p}''_1$ and truncating after a single term; the Taylor's formula with remainder then yields

$$P(p_1) = P(\hat{p}'_1) + (p_1 - \hat{p}'_1)\frac{\partial P(\hat{p}'_1)}{\partial p_1} + \frac{(p_1 - \hat{p}'_1)^2}{2}\frac{\partial^2 P(\xi')}{\partial p_1^2} \tag{11.9}$$

$$P(p_1) = P(\hat{p}''_1) + (p_1 - \hat{p}''_1)\frac{\partial P(\hat{p}''_1)}{\partial p_1} + \frac{(p_1 - \hat{p}''_1)^2}{2}\frac{\partial^2 P(\xi'')}{\partial p_1^2} \tag{11.10}$$

where ξ' and ξ'' lie in the intervals $[p'_1, p_1]$ and $[\hat{p}''_1, p_1]$ respectively. Evaluating (11.9) at $p_1 = \hat{p}''_1$ and (11.10) at $p_1 = p'_1$ and noting that the first derivatives are then identically zero, we have

$$P(\hat{p}''_1) = P(p_1) + \frac{(p''_1 - \hat{p}'_1)^2}{2}\frac{\partial^2 P(\xi')}{\partial p_1^2} \tag{11.11}$$

$$P(\hat{p}'_1) = P(\hat{p}''_1) + \frac{(\hat{p}'_1 - \hat{p}''_1)^2}{2}\frac{\partial^2 P(\xi'')}{\partial p_1^2} \tag{11.12}$$

Equations (11.11) and (11.12) together require one of the second derivatives to be negative in the interval $[\hat{p}'_1, \hat{p}''_1]$.

The n-dimensional version of the above requires that the matrix of second partial derivatives cannot be positive definite along the straight line connecting minima.

<u>Example 11.1.</u> Consider the quadratic function

$$P(\underline{p}) = \underline{C}^T\underline{p} + \frac{1}{2}\,\underline{p}^T\underset{\sim}{A}\,\underline{p} \tag{11.13}$$

where $\underset{\sim}{A}$ is a real, symmetric matrix and $\underline{C}$ is a column vector. Then setting

$$\underline{\nabla}P(\underline{p}) = \underline{0} \tag{11.14}$$

requires

$$\underline{C} + \underset{\sim}{A}\underline{p} = \underline{0} \tag{11.15}$$

so that

$$\underline{p} = -\underset{\sim}{A}^{-1}\underline{C} \tag{11.16}$$

The matrix of second partial derivatives of P is simply $\underset{\sim}{A}$. Therefore, the solution of (11.14) is a minimum if $\underset{\sim}{A}$ is positive definite. If $\underset{\sim}{A}$ is not positive definite, then the solution is either a maximum or a saddle point (i.e., $\underline{\nabla} P = \underline{0}$, and P either increases or decreases depending on the direction we move away from the $\hat{\underline{p}}$ of (11.3)). Further, since the Hessian of P is independent of $\underline{p}$, positive definiteness of $\underset{\sim}{A}$ implies the solution is a unique minimum. If, for example

$$P(\underline{p}) = p_1^2 + 2p_1 p_2 + 4p_2 p_2 + p_1 \tag{11.17}$$

$$= [1 \quad 0] \begin{bmatrix} p_1 \\ p_2 \end{bmatrix} + [p_1 \; p_2] \begin{bmatrix} 1 & 1 \\ 1 & 4 \end{bmatrix} \begin{bmatrix} p_1 \\ p_2 \end{bmatrix} \tag{11.18}$$

then

$$\begin{bmatrix} \hat{p}_1 \\ \hat{p}_2 \end{bmatrix} = \frac{-1}{6} \begin{bmatrix} 4 & -1 \\ -1 & 1 \end{bmatrix} \begin{bmatrix} 1 \\ 0 \end{bmatrix} = \begin{bmatrix} \frac{-2}{3} \\ \frac{1}{6} \end{bmatrix} \tag{11.19}$$

11.2.4 Rate of Convergence

The usual method of comparing iterative methods is to demonstrate their behavior in the vicinity of the minimum. In preparation for comparing n-dimensional procedures, we consider the one-dimensional Newton-Raphson procedure at the i-th iteration*

$$p^{i+1} = p^i - \frac{P(p^i)}{P'(p^i)} \tag{11.20}$$

---

*Recall the superscript is equal to the number of iterations which have been performed.

where $P' = dP/dp$. We expand the continuous function $P(p)$ in a Taylor series about p. In remainder form, we may write

$$P(p) = P(p^i) + (p - p^i)\, P'(p^i) + \frac{1}{2}(p - p^i)^2\, P''(\xi) \tag{11.21}$$

where ξ is contained in $[p^i, \hat{p}]$. Hence, from (11.20) and (11.21).

$$\hat{p} - p^{i+1} = \hat{p} - p^i + \frac{P(p^i)}{P'(p^i)} \tag{11.22}$$

$$= \frac{(\hat{p} - p^i)\, P'(p^i) + P(p^i)}{P'(p^i)} \tag{11.23}$$

$$= \frac{-1}{2}(\hat{p} - p^i)^2 \frac{P''(\xi)}{P'(p^i)} \tag{11.24}$$

where ξ lies in $[p^i, \hat{p}]$. Thus, the error in p^{i+1} tends to be proportional to the square of the error in p^i. Such a method is termed a second order process (or has quadratic convergence) and characteristically converges rapidly near the minimum.

11.3 MINIMIZATION PROCEDURES

11.3.1 The Method of Steepest Descent

Let us assume a function is to be minimized by changing p_j in an iterative manner. At the i-th iteration, define the variable and gradient vectors

$$\underline{p}^i = \begin{bmatrix} p_1^i \\ p_2^i \\ \vdots \\ p_n^i \end{bmatrix} \qquad \underline{\nabla} P(\underline{p}^i) = \begin{bmatrix} \dfrac{\partial P}{\partial p_1} \\ \vdots \\ \dfrac{\partial P}{\partial p_n} \end{bmatrix}_{\underline{p} = \underline{p}^i} = \left[\frac{\partial P}{\partial \underline{p}}\right]^T_{\underline{p} = \underline{p}^i} \tag{11.25}$$

Let

$$\underline{p}^{i+1} = \underline{p}^i - \alpha^i\, \underline{\nabla} P \tag{11.26}$$

where α^i is a nonnegative scalar; $\underline{p}$ is thus being changed in the direction opposite to the gradient at each iteration. Then, from the chain rule of differentiation

$$\frac{d}{d\alpha^i} P(\underline{p}^i - \alpha^i \underline{\nabla} P)\Big|_{\alpha^i = 0} = \left[\frac{\partial P}{\partial \underline{p}}\right] \frac{d(\underline{p}^i - \alpha^i \underline{\nabla} P)}{d\alpha^i}\Big|_{\alpha^i = 0} \tag{11.27}$$

$$= - \left[\frac{\partial P}{\partial \underline{p}}\right] (\underline{\nabla} P) \tag{11.28}$$

$$= - \left[\frac{\partial P}{\partial \underline{p}}\right] \left[\frac{\partial P}{\partial \underline{p}}\right]^T < 0 \tag{11.29}$$

since the latter is the negative of the sum of squares of the components of $\partial P/\partial \underline{p}$. Equation (11.29) then, simply formalizes the obvious statement that a small motion in the direction opposite to the gradient always decreases the function. Thus, this "method of steepest descent" always converges to a minimum of P eventually.

To estimate the rate of convergence, we note the optimum occurs when $\underline{\nabla} P(\underline{p}) = \underline{0}$, in contrast to $\underline{f} = \underline{0}$ for Newton-Raphson methods. We now expand the gradient in a Taylor's series with remainder

$$\underline{\nabla} P(\underline{p} + \underline{\Delta} p) = \underline{\nabla} P(\underline{p}) + \left[\frac{\partial^2 P(\underline{\xi})}{\partial p_i \partial p_j}\right] \underline{\Delta} p \tag{11.30}$$

where $\underline{\xi}$ is contained in the interval $\underline{p} + t\underline{\Delta}p$ $(0 \le t \le 1)$. Therefore, $\underline{\nabla} P$ at the minimum $\hat{\underline{p}}$ is related to $\underline{\nabla} P(\underline{p}^i)$ by

$$\underline{\nabla} P(\hat{\underline{p}}) = \underline{\nabla} P(\underline{p}^i) + \left[\frac{\partial^2 P(\underline{\xi})}{\partial p_i \partial p_j}\right] (\hat{\underline{p}} - \underline{p}^i) \tag{11.31}$$

$$= \underline{0}$$

or

$$\underline{\nabla} P(\underline{p}^i) = - \left[\frac{\partial^2 P(\underline{\xi})}{\partial p_i \partial p_j}\right] (\hat{\underline{p}} - \underline{p}^i) \tag{11.32}$$

Therefore, the error in $\underline{p}^{i+1}$ is related to the error in $\underline{p}^i$ by

$$\hat{\underline{p}} - \underline{p}^{i+1} = \hat{\underline{p}} - \underline{p}^i - \alpha^i \underline{\nabla} P(\underline{p}^i) \tag{11.33}$$

$$= \left\{ \underset{\sim}{I} + \alpha^i \left[\frac{\partial^2 P(\underline{\xi})}{\partial p_i \partial p_j}\right] \right\} (\hat{\underline{p}} - \underline{p}^i) \tag{11.34}$$

Hence the error in $\underline{p}^{i+1}$ tends to vary <u>linearly</u> with the error in $\underline{p}^i$ so that the method is a first order minimization method.

The usual method of choosing α^i is to proceed in the direction opposite the gradient until $P(\underline{p})$ is minimum and then to recalculate $\underline{\nabla} P(\underline{p})$. This method is demonstrated in Figure 11.2a where it is evident the convergence is most rapid in regions where radius of curvature is relatively constant.

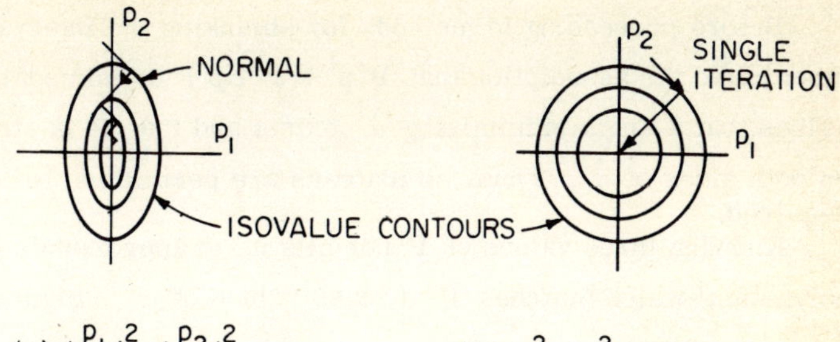

FIG. 11.2 Comparison of Gradient Minimization for Ellipsoidal, Circular Functions

This behavior is characteristic of regions far removed from the minimum, where the method is most commonly used.

Note also the shape of $P(\underline{p})$ has a considerable effect on the number of iterations to be taken (Figure 11.2a-b). A partial solution to this "ill-conditioning" is to make the logarithmic parameter transformation of Chapter VI, which shrinks elongated isovalue contours.

11.3.2 Calculation of $\hat{\alpha}^i$

The most time-consuming portion of slope-following methods such as the gradient procedure is the minimization of $P(\underline{p}^i + \alpha^i \Delta \underline{p}^i)$, i.e., the calculation of α^i. This is clearly a simpler computational task than finding $\underline{p}$ since only a single parameter is involved. Also, there is far less likelihood of multiple minima occurring with a single parameter variation. α^i_j be the j-th calculated value of α^i, so that $\alpha^i_0 = 0$. Then a value α^i_k must be found such that $0 < \hat{\alpha}^i \leq \alpha^i_k$. The most direct attack is to choose a conservative (small) value of α^i_k and then calculate

$$\alpha^i_k = \alpha^i_1 (1 + r + r^2 + . . + r^{k-1}) \qquad k = 1, . . . \tag{11.35}$$

until $P(\underline{p}^i + \alpha^i_k \Delta\underline{p}^i) > P(\underline{p}^i + \alpha^i_{k-1} \Delta\underline{p}^i)$. (The value of r chosen depends on the class of problems considered, a typical range being $1.5 \leq r \leq 2$.) In any case, the result is a set of three values of $\alpha^i (\alpha^i_{k-2}, \alpha^i_{k-1}, \alpha^i_k)$ which bound the minimum as in Figure 11.3a.

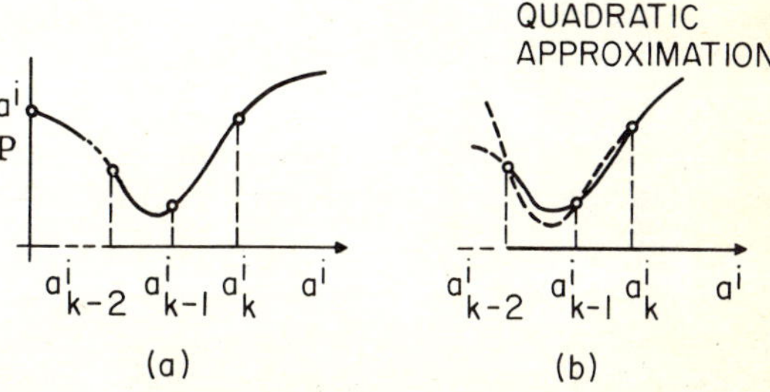

FIG. 11.3 Bounding and Shrinking Interval

Before proceeding to methods for shrinking the interval $\alpha_{k-2}^{i}, \alpha_{k}^{i}$, we must make the assumption that $P(\underline{p}^{i} + \alpha^{i} \Delta p^{i})$ is unimodal in α^{i}. This requires that a single minimizing $\hat{\alpha}^{i}$ exists and that P is strictly increasing on both sides of $\hat{\alpha}^{i}$. Thus, no plateaus are permitted, but convexity is not required.

Knowing three values of P permits us to approximate P by a quadratic polynomial which matches P at these values of α_{j}^{i} (Figure 11.3b). In detail if we let

$$a = \alpha_{k-2}^{i} \qquad b = \alpha_{k-1}^{i} \qquad c = \alpha_{k}^{i} \tag{11.36}$$

and

$$P_a = P(\underline{p}^{i} + \alpha_{k-2}^{i} \underline{\Delta p}), \text{ etc.} \tag{11.37}$$

then the turning point of the approximating polynomial is

$$d = \frac{1}{2} \frac{(b^2 - c^2)P_a + (c^2 - a^2)P_b + (a^2 - b^2)P_c}{(b - c)P_a + (c - a)P_b + (a - b)P_c} \tag{11.38}$$

It is clear from Figure 11.3b that since $P_b < P_a < P_c$, the turning point d will lie within the bounded interval [a, c]. By comparing P_b and P_d, we can shrink the bound on $\hat{\alpha}^{i}$, since

$$\begin{array}{lll} b > d & \text{and } P_b > P_d & a \le \hat{\alpha}^{i} \le b \\ & \text{and } P_d > P_b & d \le \hat{\alpha}^{i} \le c \\ d > b & \text{and } P_b > P_d & b \le \hat{\alpha}^{i} \le c \\ & \text{and } P_d > P_b & a \le \hat{\alpha}^{i} \le d \end{array} \tag{11.39}$$

Therefore, if the interval bounds are successively chosen in accordance with (11.39) the method must converge. The rate of convergence of such multipoint formulae is studied in [2.1]; this rate is necessarily less than quadratic since no derivative calculation is involved.

11.3.3 The Fibonacci Search

For certain types of criteria functions P which contain large higher derivative terms, and to avoid numerical problems in the subtraction involved in (11.38), a direct search method is sometimes used to locate $\hat{\alpha}^{i}$. The values of b and d are chosen independently to divide the interval [a, c] symmetrically and then the strategy of (11.39) is applied for interval reduction.

At first thought it might appear most logical to divide the interval into thirds; however, two calculations are then required for each reduction of the

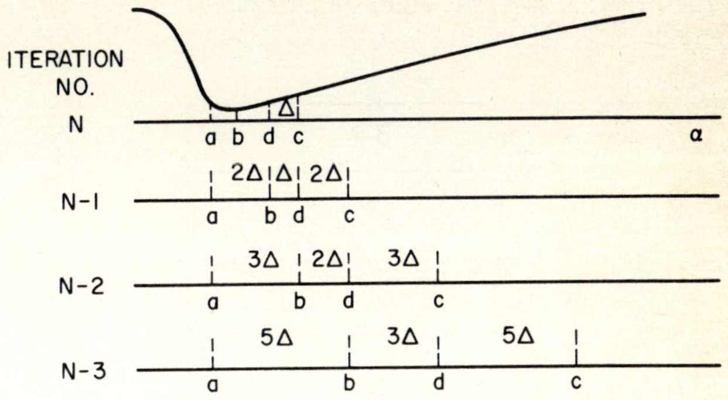

FIG. 11.4 Final Steps of a Fibonacci Search

interval. We will show that, by choosing the interior points b and d carefully, either P_b or P_d will be common to any two successive iterations.

The efficient scheme can be evolved by assuming, on the final iteration, the interval is divided into thirds. This is clearly an optimal strategy for that iteration. Proceeding backward to the (N-1)th iteration under the assumptions that

(1) inequalities of the form of Equation (11.39) were applied at the (N-1)st iteration,
(2) the Nth and (N-1)st iterations share a common interior point,
(3) b and d are symmetric about the center of the interval [a, c].

then, the intervals between the abcissas at the (N-1) iteration are [2Δ, Δ, 2Δ] respectively. A sample backward iteration is shown in Figure 11.4. From a tabular display of the interval sizes the center interval is γ_{N-j+1}, which satisfied the difference equation

$$\gamma_{M-j+1} = \gamma_{N-j} + \gamma_{N-j-1} \qquad j = 1, 2, \ldots, N-2 \tag{11.40}$$

where $\gamma_1 = \gamma_2 = 1$. The solution generates the Fibonacci sequence 1, 1, 2, 3, 5, 8, 13, etc. The end intervals satisfy a similar difference equation, but shifted one iteration number (or, alternately, satisfying the conditions $\gamma_1 = 1$, $\gamma_2 = 2$).

The search procedure begins by choosing a value for N, a number which can be determined by the total interval reduction desired. In terms of the γ_i, this reduction factor is $\gamma_{N-i+3}/\gamma_{N-i+4}$. The latter quickly asymptotes to 0.618, so the total reduction for N iterations becomes $(0.618)^N$. For example, ten iterations, requiring eleven evaluations of P, reduces the bound on $\hat{\alpha}^i$ by a factor of 100.

Another form of search, known as the Golden Section, is almost as efficient as the above and does not require prior knowledge of N.

11.3.4 The Fletcher Powell Minimization Procedure

If P is a well-behaved but general function of $\underline{p}$ and if the only knowledge of P at each iteration is its gradient, there is no better procedure than a motion in the direction of the gradient. Two methods of possible improvement are

TABLE 11.1 Interval Sizes in Fibonacci Search

Iteration	b-d	d-b	c-b
N	Δ	Δ	Δ
N-1	2Δ	Δ	2Δ
N-2	3Δ	2Δ	3Δ
N-3	5Δ	3Δ	5Δ
.	.	.	.
.	.	.	.

(1) to calculate higher order derivatives, a time consuming procedure if the derivatives are found by perturbing the elements in turn and recalculating P each time;
(2) to use knowledge of P and $\underline{\nabla}P$ at previous iterations to improve the current iteration.

The latter principle has been developed by Fletcher and Powell into the following power ful iteration scheme [11.4].

Before we examine the details of their method, first consider their overall strategy. They propose to generate a vector $\underline{s}^r$ at the r-th iteration (1) which is orthogonal to all previous vectors ($\underline{s}^i$, $i = 1, 2, \ldots r-1$), rather than just the (r-1) as in gradient minimization, and (2) which defines a downhill direction for P. The purpose of (1) is to avoid the zig-zag path of Figure 11.2a as the minimum is approached.

Let $\underset{\sim}{H}^i$ be any positive definite matrix. At the i-th iteration define the vector

$$\underline{s}^i = -\underset{\sim}{H}^i \underline{\nabla}P(\underline{p}^i) \tag{11.41}$$

If, as before,

$$\underline{p}^{i+1} = \underline{p}^i + \alpha^i \underline{s}^i \tag{11.42}$$

then choose α^i so as to minimize $P(\underline{p}^i + \alpha^i \underline{s}^i)$. Note P is a function of only one variable here. Now, denoting the minimizing value of α^i as $\hat{\alpha}^i$, define

$$\underline{\sigma}^i = \hat{\alpha}^i \underline{s}^i$$

$$\underline{d}^i = \underline{\nabla}P(\underline{p}^{i+1}) - \underline{\nabla}P(\underline{p}^i) \tag{11.43}$$

$$\underset{\sim}{A}^i = \frac{\underline{\sigma}^i \underline{\sigma}^{i^T}}{\underline{\sigma}^{i^T} \underline{d}^i} \tag{11.44}$$

$$\underset{\sim}{B}^i = \frac{\underset{\sim}{H}^i \underline{d}^i \underline{d}^{i^T} \underset{\sim}{H}^i}{\underline{d}^{i^T} \underset{\sim}{H}^i \underline{d}^i} \tag{11.45}$$

(Note the denominators of $\underset{\sim}{A}^i$ and $\underset{\sim}{B}^i$ are scalars.) Then let

$$\underset{\sim}{H}^{i+1} = \underset{\sim}{H}^i + \underset{\sim}{A}^i - \underset{\sim}{B}^i \tag{11.46}$$

It remains to be shown that selection of $\underline{p}^{i+1}$ and $\underset{\sim}{H}^{i+1}$ according to Equations (11.42) and (11.46) decreases P.

In preparation for showing the downhill behavior of P, we note that,

$$(1) \quad \frac{d}{d\alpha^i} P(\underline{p}^i + \alpha^i \underline{s}^i)\bigg|_{\alpha^i = 0} = \frac{\partial P(\underline{p})}{\partial \underline{p}}\bigg|_{\underline{p} = \underline{p}^i} \underline{s}^i \tag{11.47}$$

$$= -\underline{\nabla} P(\underline{p}^i)^T \underset{\sim}{H}^i \underline{\nabla} P(\underline{p}^i)$$

$$(2) \quad \frac{d}{d\alpha^i} P(\underline{p}^i + \alpha^i \underline{s}^i)\bigg|_{\alpha^i = \hat{\alpha}^i} = \frac{\partial P(\underline{p})}{\partial \underline{p}}\bigg|_{\underline{p} = p^{i+1}} \underline{s}^i \tag{11.48}$$

$$= 0$$

since $\hat{\alpha}^i$ is the minimizing value of α^i. The derivative represented by Equation 11.47 will assuredly be negative if $\underset{\sim}{H}^i$ is positive definite, an inductive proof will be used to show $\underset{\sim}{H}^i$ being positive definite implies $\underset{\sim}{H}^{i+1}$ is also. One method of demonstrating $\underset{\sim}{H}^{i+1}$ is positive definite is to show the quadratic form $\underline{y} H^i \underline{y}^T > 0$ for all vectors $\underline{y}$. From (11.44) through (11.46),

$$\underline{y} \underset{\sim}{H}^{i+1} \underline{y}^T = \underline{y} \underset{\sim}{H}^i \underline{y}^T + \frac{\underline{y}\, \underline{\sigma}^i \underline{\sigma}^{i^T} \underline{y}^T}{\underline{\sigma}^{i^T} \underline{d}^i} - \frac{\underline{y} \underset{\sim}{H}^i \underline{d}^i \underline{d}^{i^T} \underset{\sim}{H}^i \underline{y}^T}{\underline{d}^{i^T} \underset{\sim}{H}^i \underline{d}^i}$$

$$= \underline{y} \underset{\sim}{H} \underline{y}^T + \frac{(\underline{y}\, \underline{\sigma}^i)(\underline{y}\, \underline{\sigma}^i)^T}{\underline{\sigma}^{i^T} \underline{d}^i} - \frac{(\underline{y} \underset{\sim}{H}^i \underline{d}^i)(\underline{y} \underset{\sim}{H}^i \underline{d}^i)^T}{\left(\underline{d}^{i^T} \underset{\sim}{H}^{i^{1/2}}\right)\left(\underline{d}^{i^T} \underset{\sim}{H}^{i^{1/2}}\right)^T}$$

$$= \underline{y} \underset{\sim}{H}^i \underline{y}^T + \frac{|\underline{y}\, \underline{\sigma}^i|}{\underline{\sigma}^{i^T} \underline{d}^i} - \frac{|\underline{y} \underset{\sim}{H}^i \underline{d}^i|}{|\underline{d}^{i^T} (\underset{\sim}{H}^i)^{1/2}|}$$

$$\geq \underline{y} \underset{\sim}{H}^i \underline{y}^T + \frac{|\underline{y}\, \underline{\sigma}^i|}{\underline{\sigma}^{i^T} \underline{d}^i}$$

where $\underset{\sim}{H}^{i^{1/2}}$ exists if $\underset{\sim}{H}^i$ is positive definite. If $\underset{\sim}{H}^i$ is positive definite, then obviously $\underset{\sim}{H}^{i+1}$ is also if $\underline{\sigma}^{i^T} \underline{d}^i > 0$. But

$$\underline{\sigma}^{i^T} \underline{d}^i = \underline{\sigma}^{i^T} \left(\underline{\nabla} P(\underline{p}^{i+1}) - \underline{\nabla} P(\underline{p}^i) \right) \tag{11.49}$$

$$= -\underline{\sigma}^{i^T} \underline{\nabla} P(\underline{p}^i) \tag{11.50}$$

$$= -\hat{\alpha}^i \underline{s}^{i^T} \underline{\nabla} P(\underline{p}^i) \tag{11.51}$$

$$= \hat{\alpha}^i \left(\underline{\nabla} P(\underline{p}^i) \right)^T \underset{\sim}{H}^i \underline{\nabla} P(\underline{p}^i) \tag{11.52}$$

$$> 0 \tag{11.53}$$

since $\underset{\sim}{H}^i$ is positive definite. Therefore $\underset{\sim}{H}^{i+1}$ is positive definite, and the method does converge to a minimum.

Fletcher and Powell show that if the function to be minimized is a quadratic as in (11.13), then H^n is the inverse of the matrix of second partial derivatives, i.e., $\underset{\sim}{H}^n = \underset{\sim}{A}^{-1}$. Hence, from (11.41), (11.42), and (11.15).

$$\underline{p}^{n+1} = \underline{p}^n - \hat{\alpha}^n \underset{\sim}{A}^{-1} (\underline{C} + \underset{\sim}{A} \underline{p}^n) \tag{11.54}$$

$$= \underline{p}^n - \underline{p}^n \underset{\sim}{A}^{-1} \underline{C} - \hat{\alpha}^n \underline{p}^n \tag{11.55}$$

$$= (1-\hat{\alpha}^n) \underline{p}^n + \hat{\alpha}^n \hat{\underline{p}} \tag{11.56}$$

Choosing $\hat{\alpha}^n = 1$ yields $\underline{p}^{n+1} = \hat{\underline{p}}$ and the method converges in exactly n+1 iterations!

Let us now review the overall behavior of the method. If $\underset{\sim}{H}^o$ is chosen as the unit matrix, the first iteration is precisely the same as the method of steepest descent, from (11.42). If This is far removed from the minimum, the convergence tends to be rapid, as previously noted. As the minimum is approached, the function behaves quadratically, for which the above method converges in a finite number of iterations. Therefore, the method combines the best features of linear and quadratic convergence methods. One caution must be noted: $\underset{\sim}{H}^{i+1}$ is guaranteed positive definite only if $\hat{\alpha}^i$ is exact. In practice, it may be necessary to reset $\underset{\sim}{H}$ to the unit matrix if it is found $\underline{s}^i$ defines an increasing direction for P.

Example 11.2. Beginning with the function

$$T(s) = \frac{s^2 + 1.2s + 0.2}{s^2 + 5s + 0.2} \tag{11.57}$$

the gain response data of Table 11.2 is to be approximated in the squared error sense of Chapter VI.

Three methods of minimization are used:

(1) method of steepest descent, with $\hat{\alpha}^i$ calculated by a Fibonacci search;
(2) Fletcher-Powell minimization, with $\hat{\alpha}^i$ calculated by a Fibonacci search;
(3) Fletcher-Powell minimization, with $\hat{\alpha}^i$ calculated by successive quadratic polynomial fitting.

TABLE 11.2 Data Summary for Example 11.2

Frequency (rps)	Desired Gain (db)	Initial Gain	Optimized Gain
0.01	0.	-0.25	-0.22
0.04	-3.	-2.78	-2.56
0.1	-7.	-7.50	-7.09
0.2	-10.	-10.86	-10.35
0.5	-12.	-12.31	-11.65
1.	-10.	-10.86	-10.04
2.	-6.5	-7.50	-6.71
10.	-2.	-0.91	-1.23
100.	-0	-0.01	-0.63

The initial squared error at the nine frequencies shown is 4.15; the minimum value is found to be 1.53, with the initial and optimized gains shown in Table 11.2.

Of more interest for our purposes is the relative efficiency of each method. The number of evaluations of P and the value of P are shown in Table 11.3 for each of the three cases.

The table emphatically shows the advantages of using higher order convergence methods in both the major iteration and in the search for $\hat{\alpha}^i$.

The progress of the third method is illustrated in Table 11.4, where the G's and Q's indicate the evaluation of the gradient and the initialization of the quadratic search (respectively).

11.3.5 Search Avoidance

In some cases, there is no replacement for a careful scan to avoid overshooting α such that $P(\underline{p}^i + \alpha \underline{s}^i) > P(\underline{p}^i)$. However, for problems with a smooth error function $P(\underline{p})$, an alternative has been proposed by several authors [11.19] [11.20]. The basic notion is that rather than employ a single set of updating formulae as (11.44)-(11.46), either of the two sets of formulae can be adopted. The decision on which to use depends on the outcome of simple tests made on vectors involved in the updating process [11.20].

TABLE 11. 3 Comparison of Minimization Methods

Major Iteration No. (i+1)	1	2	3
1	(4. 15. 0)	(4. 15, 0)	(4. 15, 0)
2	(3. 92, 21)	(3. 92, 21)	(3. 92, 17)
3	(3. 70, 42)	(2. 51, 40)	(2. 48, 32)
4	(3. 53, 62)	(1. 87, 55)	(1. 88, 43)
5	(3. 24, 85)	(1. 79, 71)	(1. 79, 55)
6	(3. 02, 105)	(1. 54, 87)	(1. 54. 67)
7	(2. 95, 126)	(1. 53, 95)	(1. 53, 71)
8	(2. 88, 147)		
9	(2. 82, 168)		
10	(2. 82, 189)		
11	(2. 77, 210)		
12	(2. 71, 231)		
13	(2. 66, 252)		
14	(2. 61, 737)		
.	.		
.	.		
.	.		
25	(2. 22, 435)		

TABLE 11. 4 Output for Example 11. 2

THE NUMERATOR COEFFICIENTS ARE

0. 1200000E 01 0. 1000000E 01

THE DENOMINATOR COEFFICIENTS ARE

0. 2000000E 00 0. 5000000E 01 0. 1000000E 01

FREQ (RPS)	GAIN (DB)	PHASE (DEG)	DELAY (SEC)
0. 0100000	-0. 2468953	-10. 605	17. 5758972
0. 0400000	-2. 7866783	-31. 630	6. 8951092
0. 1000000	-7. 5010335	-36. 918	-1. 3200541
0. 2000000	-10. 8633242	-24. 600	-2. 2914934
0. 5000000	-12. 3161840	3. 618	-1. 1297989
1. 0000000	-10. 8633289	24. 600	-0. 4588988
2. 0000000	-7. 5010376	36. 918	-0. 0660028
10. 0000000	-0. 9064721	19. 755	0. 0283084
100. 0000000	-0. 0101773	2. 175	0. 0003788

TABLE 11.4 (Continued)

THE MEAN SQUARE ERRORS ARE

	4.1461163				
	4.1417637				
	4.1331043				
	4.1163092			G	
	4.0849180				1.7865934
	4.0306721				1.6504335
	3.9560871				1.5391378
	3.9400225				1.9319420
	4.4209604	G		Q	
Q			2.4717216		1.5390730
	3.9253845		1.9643869		1.5390568
	3.9253826		2.4061584	G	
G		Q			1.5390568
	3.9253826		1.8783207		1.5382118
	3.8208008		1.8777294		1.5366430
	3.6243076	G			1.5340071
	3.2810402		1.8777294		1.5308142
	2.7840261		1.8276186		1.5330343
	2.4809875		1.7866526	Q	
	4.2021179		1.9212980		1.5302582
Q		Q			1.5302639
	2.4723997		1.7866268	G	
	2.4717216		1.7865934		1.5302582

FREQ (RPS)	GAIN (DB)	PHASE (DEG)	DELAY (SEC)
0.0100000	-0.2140130	-9.903	16.4703217
0.0400000	-2.5658197	-29.986	6.7788162
0.1000000	-7.0968161	35.409	-1.2612505
0.2000000	-10.3591232	-23.294	-2.2694016
0.5000000	-11.6573191	4.588	-1.1047640
1.0000000	-10.0508308	24.469	-0.4109488
2.0000000	-6.7105942	34.214	-0.0289403
10.0000000	-1.2253532	15.051	0.0237898
100.0000000	-0.6270195	1.674	0.0002918

THE MEAN SQUARE ERRORS ARE

1.5302505

THE NUMERATOR COEFFICIENTS ARE

0.2032062E 00 0.1250307E 01 0.1113719E 01

THE DENOMINATOR COEFFICIENTS ARE

0.2029965E 00 0.4842712E 01 0.1196544E 01

11.4 MINIMIZATION WITH CONSTRAINTS

11.4.1 Statement and Interpretation of Constrained Minimization

It is not uncommon in circuit design to be presented with side conditions which constrain an otherwise straightforward optimization problem. These constraints are most generally represented by inequalities; the problem of minimizing P(p) is replaced by:

minimize

$$P(\underline{p}) \tag{11.58}$$

subject to

$$g_i(\underline{p}) \geq 0 \qquad i = 1, 2, \cdots m \tag{11.59}$$

To assure that an unconstrained $P(\underline{p})$ possessed a single minimum, we previously imposed the condition that it be convex. The presence of constraints now requires $P(\underline{p})$ be convex only within the region bounded by the constraints (termed the feasible region). On the other hand, $g_i(\underline{p})$ is required to be concave to assure a single minimum. For example, if

$$P(\underline{p}) = p_1^2 + p_2^2 \tag{11.60}$$

$$g_1(\underline{p}) = p_1^2 - p_2 - 1 \geq 0 \tag{11.61}$$

then, from Figure 11.5 it is clear that P(p) has a minimum at both (1/2, -1/2) and (-1/2, 1/2). The Hessian matrix $[\partial^2 g_1/\partial p_1 \partial p_2]$ is positive semidefinite, so that $g_1(\underline{p})$ is convex— evidently "more convex" than $P(\underline{p})$ - causing multiple minima.

Note that it is possible that one or more of the $g_i(p)$ may not be zero at the solution of (11.59). As an example, consider the similar problems

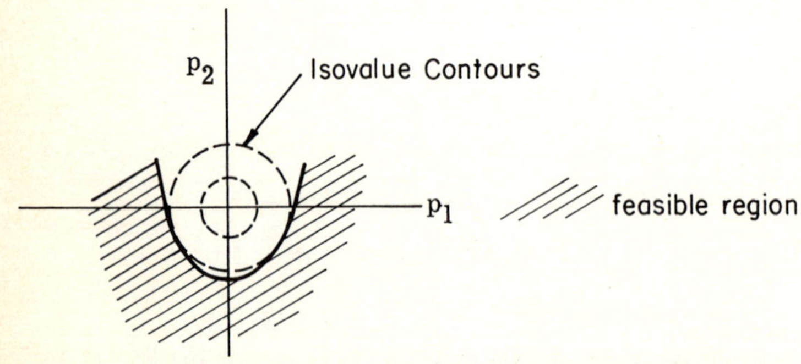

FIG. 11.5 Example of Multiple Minima Occurring When g_i Is Convex

minimize

$$P_1(\underline{p}) = (p_1 - 2)^2 + (p_2 - 1)^2, \quad P_2(\underline{p}) = p_1^2 + (p_2 - 2.5)^2 \tag{11.62}$$

subject to

$$g_1(\underline{p}) = p_2 - p_1^2 \geq 0 \tag{11.63}$$

$$g_2(\underline{p}) = -p_1 - p_2 + 2 \geq 0 \tag{11.64}$$

The problems are displayed in Figure 11.6a and 11.6b (respectively). It is clear that $g_1(\underline{p})$ could have been ignored in the second problem, since it is nonzero at the minimum.

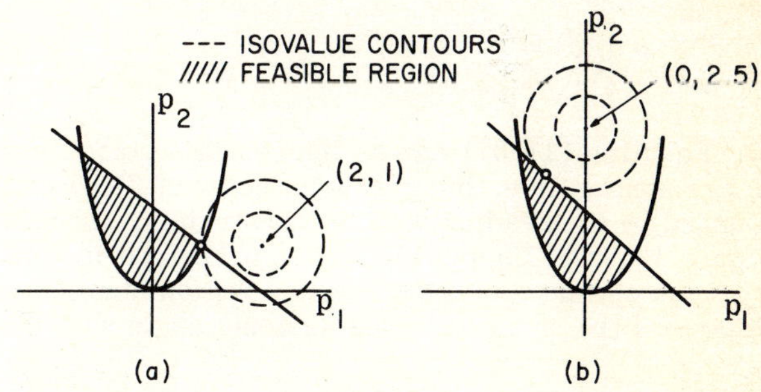

FIG. 11.6 Example Showing When Constraint Can Be Ignored

From the above observations, we can deduce the equation for $\underline{\Delta}P(p)$ at the minimum, i.e., the equivalent of $\underline{\nabla}P(\underline{p}) = 0$ for the unconstrained case. The solution must have the property that any movement $\Delta\underline{p}$ from the minimum $\underline{p}$ which points into the feasible region must increase $P(\underline{p})$. Such a (small) motion into the feasible region can be approximated by

$$g_i(\underline{p}+\underline{\Delta}p) \approx g_i(\hat{\underline{p}}) + \underline{\nabla}g_i(\underline{p})^T\underline{\Delta}p = \nabla g_i(\underline{p})^T\underline{\Delta}p = \epsilon_i \tag{11.65}$$

for all g_i which are zero at the minimum. Hence, $\epsilon_i > 0$. For the same small $\underline{\Delta}p$ $P(p)$ must increase, so by a similar approximation

$$\underline{\nabla}P(p)^T\underline{\Delta}p > 0 \tag{11.66}$$

Equation (11.66) must be satisfied at the minimum for every $\underline{\Delta}p$ satisfying (11.65). Now assume we can find an $\hat{\underline{p}}$ satisfying

$$\underline{\nabla}P(\hat{\underline{p}}) = \sum_{i=1}^{m} u_i \underline{\nabla}g_i(\hat{\underline{p}}) \tag{11.67}$$

where $u_i > 0$ for $g_i(\hat{\underline{p}}) = 0$ and $u_i = 0$ for $g_i(\hat{\underline{p}}) \neq 0$. Then (11.66) becomes

$$\underline{\nabla P}(\hat{\underline{p}})^T \, \underline{\Delta p} = \sum_{i=1}^{m} u_i \, \underline{\nabla g}_i(\hat{\underline{p}})^T \, \underline{\Delta p} \tag{11.68}$$

$$= \sum_{i=1}^{m} u_i \, \epsilon_i$$

$$> 0 \tag{11.69}$$

Equation (11.67) thus defines a sufficient condition for the minimum. The process of selecting $u_i > 0$ or $u_i = 0$ may be succinctly stated as

$$\sum_{i=1}^{m} u_i g_i(\hat{\underline{p}}) = 0 \tag{11.70}$$

Equation (11.67) has an interesting geometric interpretation. Since all u_i are nonnegative, the gradient vector of $P(\underline{p})$ must lie within the cone bounded by the gradients of the constraints, as shown for the example of Figure 11.6a in Figure 11.7. From 11.7, it is clear that any movement from the optimum into the feasible region must of necessity increase the value of $P(\underline{p})$ since it forms an acute angle with the gradient of $P(\underline{p})$ at $\hat{\underline{p}}$.

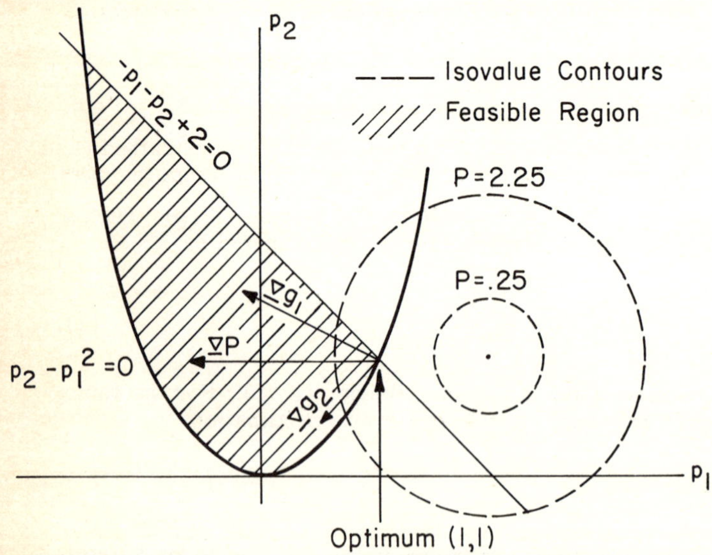

FIG. 11.7 Relation Between $\underline{\nabla P}$ and $\underline{\nabla g}_i$

To prove the necessity of (11. 66) and (11. 70) at a minimum is considerably more involved and beyond the scope of this text [11. 1].

11. 4. 2 Computational Algorithms for Constrained Minimization

A computational procedure for minimizing $P(\underline{p})$ under the constraints $g_i(\underline{p}) \geq 0$ may be developed by converting the constrained problem into an unconstrained one. This approach requires that we combine $P(\underline{p})$ and the $g_i(\underline{p})$ into a single scalar function P and then minimize P.

Any proposed combination P of a convex P and concave g_i must have the property that a vector $\underline{p}$ minimizing P must solve the original constrained problem and vice versa. Therefore, P may contain no new minima or, equivalently P must be convex in $\underline{p}$. Two common methods of combination which yield a convex P are

$$\text{(1)} \qquad \tilde{P}(\underline{p}, \underline{u}) = P(\underline{p}) - \sum_{i=1}^{m} u_i g_i(\underline{p}) \qquad (11.71)$$

where $u_i > 0$ as before and $\underline{u} = [u_1, u_2, \ldots u_m]^T$; clearly $\tilde{P}(\underline{p}, \underline{u})$ is convex, since if $g_i(\underline{p})$ is concave then $-g_i(\underline{p})$ is convex;

$$\text{(2)} \qquad \tilde{P}(\underline{p}, \underline{u}) = P(\underline{p}) + \sum_{i=1}^{m} u_i / g_i(\underline{p}) \qquad (11.72)$$

where $u_i > 0$; it is left as an exercise to show that if $g_i(\underline{p})$ is concave then $1/g_i(\underline{p})$ is convex.

The first formulation may be recognized as based on the classical method of Lagrange multipliers. The most useful computational algorithm based on this formulation converts the problem to a "saddle-point" problem and solves for the u_i by gradient means. The chief disadvantage of the method is that iteration may drift outside the constraint region, where $P(\underline{p})$ may behave quite erratically.

In contrast, it is easy to see, if $\tilde{P}(\underline{p}, \underline{u})$ of (11. 72) is minimized using any monotonically decreasing scheme (such as the gradient) and beginning in the feasible region $g_i(\underline{p}) > 0$, then the iterations will remain in the feasible region. If a somewhat less general form of P is defined as [11. 5],

$$\hat{P}(\underline{p}, r_k) = P(\underline{p}) + r_k \sum_{i=1}^{m} 1/g_i(\underline{p}) \qquad (11.73)$$

then in addition to remaining within the constraint boundary, the iteration may be shown to converge to the minimum of $P(\underline{p})$ as $r_k \to 0$ (see Problem 11.10). The procedure is then the following:

(1) choose r_1 constant and minimize $\hat{P}(\underline{p}, r_1)$;
(2) choose $r_2 < r_1$ and minimize $\hat{P}(\underline{p}, r_2)$;
(3) repeat (2) with $r_{i+1} < r_i$.

The solution is thus allowed to approach the boundary $g_i(\underline{p}) = 0$ as r_k decreases.

Problems

11.1 Show that the i-th direction defined in the method of steepest descent is orthogonal to the (i-1) direction, provided $\hat{\alpha}^{i-1}$ is found precisely.

11.2 Fletcher and Power show their method converges to $\underset{\sim}{A}^{-1}$ in the case of a quadratic by showing each vector Δp_j is linearly independent of the j-1 previous. How does this prove the result?

11.3 The n-dimensional Taylor's series with remainder is

$$P(\underline{p}) = P(\hat{\underline{p}}) + (\underline{\Delta p})^T \underline{\nabla} P(\hat{\underline{p}}) + \tfrac{1}{2}(\underline{\Delta p})^T \left[\frac{\partial^2 P(\underline{\xi})}{\partial p_i \partial p_j}\right] \underline{\Delta p}$$

where $\underline{\Delta p} = \underline{p} - \hat{\underline{p}}$ is a global minimum, and where $\underline{\xi}$ lies on the line segment $\hat{\underline{p}} = t\underline{\Delta p}$ for $0 \le t \le 1$. Show that if $\underline{p}$ is a local minimum, then the Hessian matrix cannot be positive definite on the entire line segment.

11.4 Extend the unimodal condition in mathematical terms to n-dimensions. Sketch a cup-shaped surface which is unimodal but not convex.

11.5 It may be shown the initial Fibonacci intervals approach $[(3 - \sqrt{5})/2, \sqrt{5} - 2, (3 - \sqrt{5})/2]$. If these intervals are initially chosen but with a finite number of iterations, how could the subsequent intervals be characterized? Why is this method less efficient than the Fibonacci search?

11.6 A function $P(\underline{p})$ is to be minimized subject to $G_j(\underline{p}) \le 0$ (i=1, 2, . . . , n). The procedure of Section 11.4.2 is used by defining

$$P(\underline{p}, r_k) = P(\underline{p}) + r_k \sum_{i=1}^{m} 1/G_j(\underline{p})$$

where $r_k < r_{k-1}$. If $\hat{p}^k$ represents the value of $\underline{p}$ minimizing $P(\underline{p}, r_k)$, show $P(\hat{p}^{k+1}) < P(\hat{\underline{p}}^k)$ and $\sum_{j=1}^{m} 1/G_j(\hat{\underline{p}}^{k+1}) \le \sum_{j=1}^{m} 1/G_j(\hat{p}^k)$. [P and G_j need not be convex and cocave (respectively).] If the iteration begins near a boundary $G_j = 0$, but the minimum occurs always from the boundary, how can $\sum 1/G_j$ decrease as required?
(Credit to G. C. Temes.)

REFERENCES

11.1 Kuhn, H. W., and Tucker, A. W., "Non-Linear Programming," Proc. Second Symposium on Mathematical Statistics and Probability, University of California Press; 1951.

11.2 Carrol, C. W., "The Created Response Surface Technique for Optimizing Non-Linear Constrained Systems," Opns. Res., vol. 9, pp. 169-184; 1961.

11.3 Spang, H. A., "A Review of Minimization Techniques for Non-Linear Functions," SIAM Review, vol. 4, pp. 343-365; 1962.

11.4 Fletcher, R., and Powell, M. J. D., "A Rapidly Convergent Descent Method for Minimization," The British Computer Journal, vol. 6, pp. 163-168; 1963.

11.5 Fiacco, A. V., and McCormick, G. P., "The Sequential Unconstrained Technique for Non-Linear Programming. Algorithm II, Optimum Gradients by Fibonacci Search," Research Analysis Corp. Report RAC-TP-123; June, 1964.

11.6 Fiacco, A. V., and McCormick, G. P., "The Sequential Unconstrained Minimization for Non-Linear Programming: A Primal-Dual Method," Management Service, vol. 10, pp. 360-366; 1964.

11.7 Murata, T., "The Use of Adaptive Constrained Descent in Systems Design," IEEE International Conv. Rec. Part 1, pp. 296-306; 1964.

11.8 Powell, M. J. D., "An Efficient Method for Finding the Minimum of a Function of Several Variables Without Calculating Derivatives," The British Computer Journal, pp. 155-162; 1964.

11.9 Fiacco, A. V., and McCormick, G. P., "The Sequential Unconstrained Minimization Techniques for Convex Programming and Equality Constraints," Research Analysis Corp. Report RAC-TP-155; April, 1965.

11.10 Calahan, D. A., "Computer Design of Linear Frequency Selective Networks," Proc. IEEE, vol. 53, pp. 1701-1706; November, 1965.

11.11 Calahan, D. A., "A Numerical Algorithm for the Minimization of Sensitivity," Proc. Third Allerton Conference, pp. 394-406; 1965.

11.12 Powell, M. J. D., "A Method for Minimizing a Sum of Squares of Nonlinear Functions Without Calculating Derivatives," British Computer Journal, vol. 7, pp. 394-406; 1965.

11.13 Lovi, A., and Vogl, T. P., Recent Advances in Optimization Techniques, Wiley; 1966.

11.14 Wilde, D. J., and Beightler, C. S., Foundations of Optimization, Prentice-Hall; 1967.

11.15 Broyden, C. G., "Quasi-Newton Methods and Their Application of Function Minimization," Mathematics of Computation, vol. 21, pp. 368; 1967.

11.16 Davidon, W. C., "Variance Algorithm for Minimization," The Computer Journal, vol. 10, 11., pp. 406; 1968.

11.17 Murtagh, B. A., and Sargent, R. W. H., "Computational Experience with Quadratically Convergent Minimization Methods," The Computer Journal, vol. 13, pp. 185; 1970.

11.18 Jacobson, D. H., and Oksman, W., "An Algorithm That Minimizes Homogeneous Functions of N Variables in N+2 Iterations and Rapidly Minimizes General Functions," Tech Report No. 618, Div. of Engineering and Applied Physics, Harvard, Univ., Cambridge, Mass., 1970.

11.19 Broyden, C. G., "The Convergence of a Class of Double-rank Minimization Algorithms Parts I and II," Jour. Inst. Maths. and its Applics.; 1970.

11.20 Fletcher, R., "A New Approach to Variable Metric Algorithms," Report TP. 383, Theoretical Physics Division, U. K. A. E. E. Res. Group, Atomic Energy Res. Establishment, Harwell, England.

12

Time-Domain Sensitivity Calculation

12.1 INTRODUCTION

It is probably a fitting climax to the circuit design portion of this text to discuss a topic with an apparent design potential that has been largely unexploited. Until very recently, the transient analysis of nonlinear circuits was so time consuming that one was delighted to obtain one or two analyses. With the advent of sophisticated matrix handling and numerical integration algorithms, however, gate-sized circuits can be analyzed in seconds of computer time. This development in analysis techniques was occurring coincident with that of efficient time domain gradient generation techniques based on the calculus of variations [12.1].

It remained for Director and Rohrer [12.2] to extend the notion of an adjoint network and its utility for gradient computation to include nonlinear dynamic circuits. This freed time domain sensitivity calculations from their previous attachment to the state variable formulation and gave them a topological interpretation well beyond their previous system interpretation [12.3] [12.4].

The reader of this chapter should be familiar with Chapter V. In particular, we will have need for the Tellegen summation

$$\sum_{k=1}^{n_b} dv_k \hat{i}_k - di_k \hat{v}_k = 0 \qquad (12.1)$$

which relates changes in branch currents and voltages in the original and adjoint networks.

12.2 TIME-DOMAIN SENSITIVITY CALCULATION

12.2.1 Adjoint Network Elements

If we aspire to computing time dependent sensitivities from the adjoint network, two extensions of resistive network procedures are necessary: (1) a time dependent replacement for the constant "unit" source at the output; and (2) adjoint network replacements for the capacitor.* As we will shortly discover, other considerations will arise in evolution of the adjoint network.

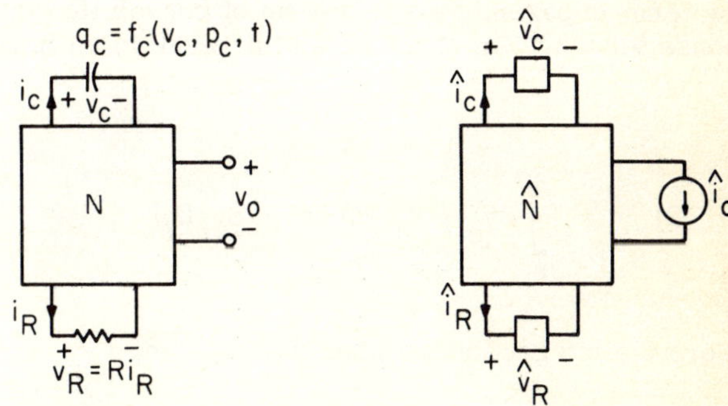

FIG. 12.1 Example Model of Sensitivity Calculation

The development will be illustrated by the simple example of Figure 12.1. The capacitor is represented by a charge-voltage relationship.

$$q_c = f_c\,(v_c,\; p_c,\; t)$$

which of course includes the linear, time invariant case $f_C = Cv_C$. We will initially derive methods for calculating $dv_O(t_f)/dp_C$ and $dv_O(t_f)/dR$, that is, the sensitivity of the output voltage at time $t = t_f$.

Let the current source $\hat{i}_o$ be of the form

$$\hat{i}_o(t) = \delta\,(t_f - t)$$

i.e., an impulse applied at $t = t_f$. The summation of Eq. (12.1) takes the form

$$dv_R\hat{i}_R - di_R\hat{v}_R + dv_o\,\delta\,(t_f - t) + dv_c\hat{i}_c - \frac{d}{dt}\,(df_c)\hat{v}_c$$

$$= i_R\hat{i}_R dR + dv_o\delta\,(t_f - t) + dv_c\hat{i}_c - \frac{d}{dt}\left(\frac{\partial f_c}{\partial v_c}\,dv_c + \frac{\partial f_c}{\partial p_c}\,dp_c\right)\hat{v}_c \qquad (12.2)$$

---

*Derivation of formulae for inductors will be left to the reader.

where the adjoint network replacement for a resistance is a resistance, as before. The terms involving dv_c and dv_o do not appear in a form convenient for sensitivity calculations. However, the presence of the impulse and the time derivative d/dt suggests integration of the expression to expose dv_c and dv_o. This operation is performed, term by term, as follows.

(1) $$\int_0^{t_f} dR\, i_R \hat{i}_R\, dt = \left(\int_0^{t_f} i_R \hat{i}_R\, dt\right) dR$$

The term in parentheses is a form of convolution integral, since i_R is the response for $0 \leq t \leq t_f$ of the adjoint network to an impulse applied at $t = t_f$.

(2) $$\int_0^{t_f} dv_o\, \delta(t_f - t)\, dt = dv_o(t_f)$$

where the integration includes the impulse.

(3) $$\int_0^{t_f} \hat{i}_c dv_c dt - \int_0^{t_f} \hat{v}_c \frac{d}{dt}\left(\frac{\partial f_c}{\partial v_c} dv_c\right) dt$$

The right hand side integral is integrated by parts by letting

$$x = \hat{v}_c$$

$$dy = \frac{d}{dt}\left(\frac{\partial f_c}{\partial v_c} dv_c\right) dt$$

$$= d\left(\frac{\partial f_c}{\partial v_c} dv_c\right)$$

Then (12.2) becomes

$$\int_0^{t_f} \hat{i}_c dv_c dt - \hat{v}_c \frac{\partial f_c}{\partial v_c} dv_c \Bigg|_0^{t_f} + \int_0^{t_f} \left(\frac{\partial f_c}{\partial v_c} dv_c\right)\left(\frac{d\hat{v}_c}{dt}\right) dt$$

$$= -\hat{v}_c \frac{\partial f_c}{\partial v_c}\Bigg|_{t=t_f} dv_c(t_f) + \hat{v}_c \frac{\partial f_c}{\partial v_c}\Bigg|_{t=0} dv_c(0) + \int_0^{t_f} \left(\left(\hat{i}_c + \frac{\partial f_c}{\partial v_c} \frac{d\hat{v}_c}{dt}\right) dt\right. \tag{12.3}$$

We recall from Chapter V that as many terms as possible from this expression should be forced identically to zero. The first term can be eliminated by setting $\hat{v}_c(t_f) = 0$, which provides an initial condition for the backward time solution of the adjoint network. The integral may be rendered zero for any dv_c if and only if the integrand is zero, i. e.,

$$\hat{i}_c = -\frac{\partial f_c}{\partial v_c} \frac{d\hat{v}_c}{dt}$$

$$= -\hat{C} \frac{d\hat{v}_c}{dt}$$

for a linear capacitor. This defines the adjoint network capacitor replacement to be a negative-valued capacitor! The second term will be temporarily preserved.

$$(4) \qquad \left(- \int_0^{t_f} \hat{v}_c \frac{d}{dt}\left(\frac{\partial f_c}{\partial p_c}\right) dt \right) dp_c$$

This expression is again a form of convolution integral, a result made obvious by letting $f_c = Cv_c$ and $p_c = C$, viz,

$$\left(-\int_0^{t_f} \hat{v}_c \left(\frac{dv_c}{dt}\right. dt\right) dC = \left(- \frac{1}{C} \int_0^{t_f} \hat{v}_c\, i_c\, dt \right) dC$$

Again, a circuit variable from N (i_c) is convolved with a variable ($\hat{v}_c$) from $\hat{N}$. Returning to the reduction at hand, collecting terms yields

$$\left(\int_0^{t_f} i_R \hat{i}_R\, dt\right) dR + dv_o(t_f) + \hat{v}_c \frac{\partial f_c}{\partial v_c}\Bigg|_{t=0} dv_c(0)$$

$$- \left(\int_0^{t_f} \hat{v}_c \frac{d}{dt}\left(\frac{\partial f_c}{\partial p_c}\right) dt\right) dp_c = 0$$

The calculation of the required sensitivities follows immediately as

$$\frac{dv_o(t_f)}{dR} = -\hat{v}_c \frac{\partial f_c}{\partial v_c}\Bigg|_{t=0} \frac{dv_c(0)}{dR} - \int_0^{t_f} i_R \hat{i}_R \, dt \qquad (12.4)$$

and

$$\frac{dv_o(t_f)}{dp_c} = \hat{v}_c \frac{\partial f_c}{\partial v_c}\Bigg|_{t=0} \frac{dv_c(0)}{dp_c} + \int_0^{t_f} \hat{v}_c \frac{d}{dt}\left(\frac{\partial f_c}{\partial p_c}\right) dt \qquad (12.5)$$

Example 12.1. The response of the network of Figure 12.2 is

$$v_o = v_c(0)\, e^{-t/RC}$$

and the sensitivities to R and C are

$$\frac{dv_o(t_f)}{dR} = -v_c(0)\frac{t_f}{R^2C}\, e^{-t_f/RC} \qquad (12.6)$$

$$\frac{dv_o(t_f)}{dC} = e^{-t/RC}\frac{dv_c(0)}{dC} - v_c(0)\frac{t_f}{RC^2}\, e^{-t_f/RC} \qquad (12.7)$$

Calculation of sensitivities from the adjoint network is preceded by calculation of

$$\hat{v}_c(t) = \frac{1}{C} e^{-(t - t_f)/(-RC)} = \frac{1}{C} e^{(t - t_f)/RC} \qquad t < t_f$$

$$\hat{i}_R(t) = \frac{1}{RC} e(t - t_f)/RC \qquad t < t_f$$

$$i_c(t) = C\frac{dv_c}{dt} = -\frac{v_c(0)}{R} e^{-t/RC}$$

$$i_R(t) = \frac{v_c(0)}{R} e^{-t/RC}$$

where it should be noted that $\hat{v}_c(t_f^-) \neq 0$ because the impulse is applied at $t = t_f$. The sensitivities of interest are then

$$\frac{dv_o(t_f)}{dR} = -C\hat{v}_c(0)\ \frac{dv_c(0)}{dR} - \int_0^{t_f} \left(\frac{v_c(0)}{R}\ e^{-t/RC}\right)\left(\frac{1}{RC}\,e^{(t-t_f)/RC}\right) dt$$

$$= -\frac{v_c(0)}{R^2C} \int_0^{t_f} e^{-t_f/RC}\ dt$$

$$= \frac{-v_c(0)\,t_f}{R^2C}\ e^{-t_f/RC}$$

$$\frac{dv_o(t_f)}{dC} = C\hat{v}_c(0)\ \frac{dv_c(0)}{dC} + \int_0^{t_f} \left(\frac{1}{C}\,e^{(t-t_f)/RC}\right)\left(\frac{-v_c(0)}{RC}\ e^{-t/RC}\right) dt$$

$$= C\left(\frac{1}{C}\,e^{-t_f/RC}\right)\frac{dv_c(0)}{dC} - \frac{v_c(0)}{RC^2}\int_0^{t_f} e^{-t_f/RC}\ dt$$

$$= e^{-t/RC}\ \frac{dv_c(0)}{dC} - \frac{v_c(0)t_f}{RC^2}\ e^{-t_f/RC}$$

which agree with Equations (12.6) - (12.7).

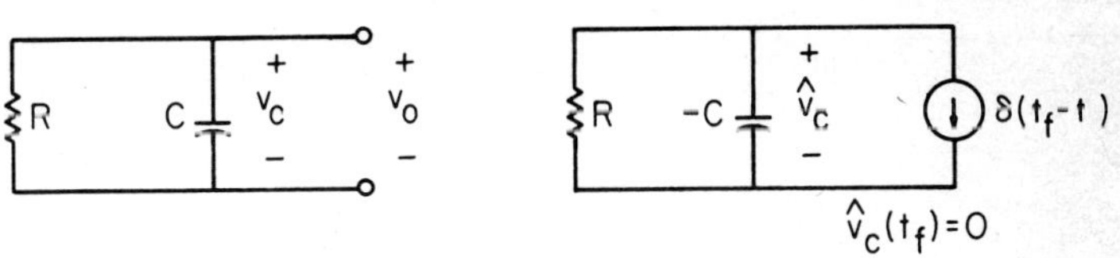

FIG. 12.2 Network Example

12.2.2 DC Sensitivity Calculation

The term $dv_c(0)/dp_j$ requires interpretation since its value depends on the assumptions made in computing $d(v_o, i_o)/dp_j$. For the circuit of Figure 12.2, conservation of stored energy $E_c(t)$ at $t = 0$ would require $dv_c(0)/dR = 0$; however, $E_c(0) = \frac{1}{2}Cv_c(0)^2$, so the sensitivity to C is nonzero, viz,

$$dE_c = \tfrac{1}{2}\,v_c^{\,2}(0)\ dC + Cv_c(0)\ dv_c(0) = 0$$

$$\frac{dv_c(0)}{dC} = - \frac{1}{2C} v_c(0)$$

In contrast, the switching circuit of Figure 12.3 is assumed to always begin in the quiescent state; i.e., if the network state equations are

$$\dot{\underline{v}}_c = \underline{f}(\underline{v}_c, t)$$

then a change dp_j is made under the assumption that $\dot{\underline{v}}_c(0) = \underline{0}$. This condition may be simulated in a network model by removing all capacitors and calculating sensitivities by methods similar to those already given for resistive networks.

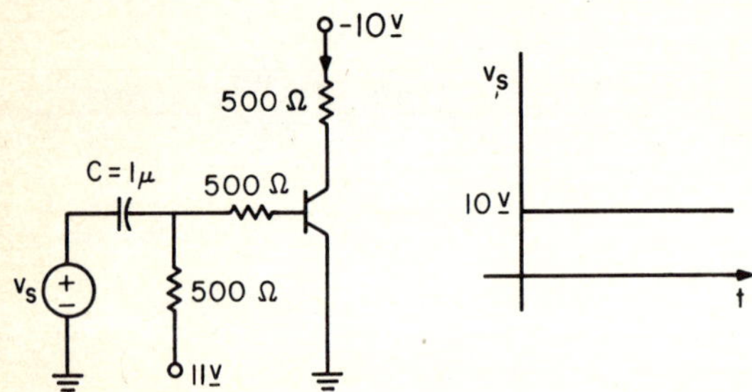

FIG. 12.3 Switching Circuit

This dc sensitivity calculation may be made quite efficient, in the following way. Assume that the sensitivities

$$\frac{dv_o}{dR_j} = \sum_{i=1}^{n} \hat{v}_{c_i}(0) \left. \frac{\partial f_{c_i}}{\partial v_{c_i}} \right|_{t=0} \frac{dv_{c_i}(0)}{dR_j} - \int_0^{t_f} i_{R_j}(t)\, \hat{i}_{R_j}(t)\, dt \qquad (12.8)$$

are to be calculated ($j = 1, 2, \ldots r$). The integral is calculated from the resistor currents in Figure 12.4a-b, similarly to Example 12.1, as a convolution integral. We also save from these analyses the initial capacitor voltages $\hat{v}_{c_i}(0)$ as well as $(\partial f_{c_i}/\partial v_{c_i})_{t=0}$. For example, the latter term for a linear capacitor is simply C_i. To determine the summation of (12.8), we view the terms $[\hat{v}_{c_i}(0)\, \partial f_{c_i}/\partial v_{c_i})_{t=0}]\, dv_{c_i}(0)/dR_j$ as weighted sensitivities to be calculated, i.e., $w_i dv_{c_i}(0)/dR_j$. The term $dv_{c_i}(0)/dR_j$ can be found by regarding v_{c_i} as an "output": we can place a unit current source across C_i and calculate $\hat{i}_{R_j}$ in the dc version of $\hat{N}(\hat{N}_{dc})$. If instead we place a current source <u>of value w_i</u> across C_i, then we obtain the complete term representing C_i in the summation of (12.8) (see Figure 12.4c). Recalling that the adjoint network $\hat{N}_{dc}$ is a linear network, we can exploit the superposition principle by

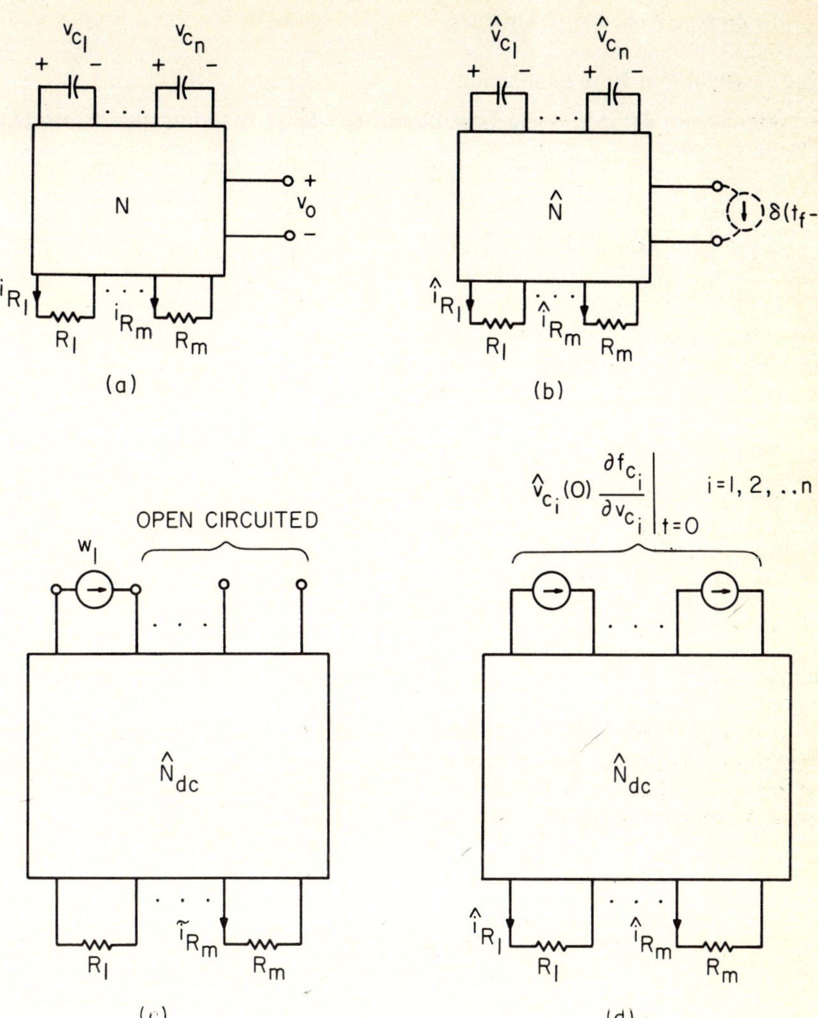

FIG. 12.4 One-step Calculation of Initial Condition Sensitivity (Equation (12.8))

applying all weighted current sources simultaneously as shown in Figure 12.4d. The complete summation term is then calculated as

$$\sum_{i=i}^{n} v_{c_i}(0) \left. \frac{\partial f_{c_i}}{\partial v_{c_i}} \right|_{t=0} \frac{dv_{c_i}(0)}{dR_j} = -i_{R_j} \hat{i}_{R_j}$$

Thus only one additional analysis is necessary, or, more precisely, one forward and one back substitution.

12.3 DESIGN CONSIDERATIONS

12.3.1 Problem Representation

Most time domain design problems can be phrased as the minimization of a functional

$$P = \int_0^{t_f} h(\underline{z}, \underline{p})\, dt \tag{12.9}$$

where $\underline{z}$ is a vector of output voltages and currents. Iteration of $\underline{p}$ with the methods of Chapter XI would then require the calculation of $\partial P/\partial \underline{p}$. For example, if $h = z(t)\delta(t_f - t)$, then $P = z(t_f)$ and $\partial P/\partial p$ could be found as described in Section 12.2.

More generally, h often takes on one of the two following forms.

(1) Matching problem:

$$P(p) = \int_0^{t_f} \sum_i \lambda_i(t)\, e_i^2(t)\, dt \tag{12.10}$$

where $e_i(t) = z_i - \hat{z}_i$, i.e., the error between a calculated and desired output, and where $\lambda_i(t)$ is a nonnegative weighting function. To illustrate, consider the problem of Figure 12.5, where a circuit response is found to have an undesirable overshoot. An idealized response is defined with an accompanying weighting function to emphasize the mismatch between the specified and desired responses for $t_2 \leq t \leq t_3$.

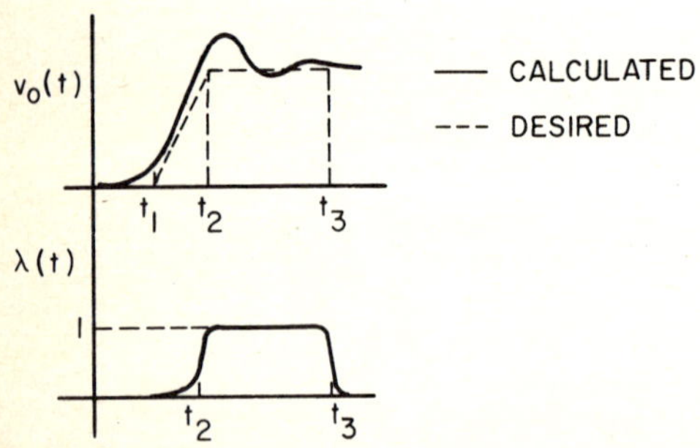

FIG. 12.5 Matching Problem Example

(2) Minimum-delay problem:

$$P = \int_0^{t_f} dt \tag{12.11}$$

$$= t_f$$

Here, the final time itself is variable; the design objective is to minimize the time t_f required to achieve a prescribed output level. For example, in Figure 12. 6 the design which achieved t_{f_1} would be considered an improvement over that yielding t_{f_2}.

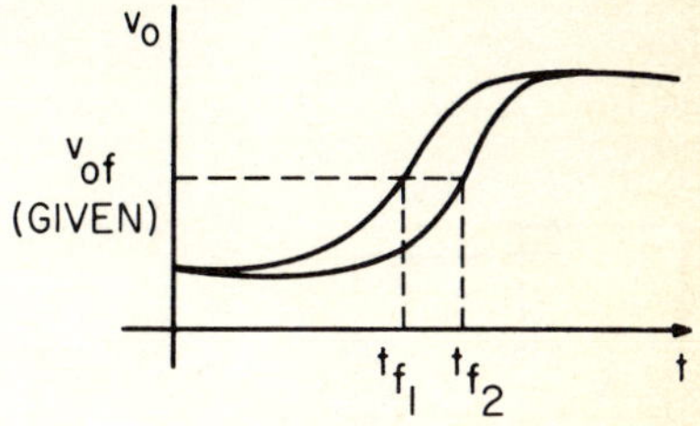

FIG. 12. 6 Minimum Delay Problem

To distinguish between these problems mathematically, we define a target set function $\theta(\underline{z}, t)$ that is zero at the problem termination [12. 11]. By definition, then

$$d\theta\Big|_{\substack{t=t_f\\ \underline{z}=\underline{z}_f}} = \left(\frac{\partial\theta}{\partial\underline{z}}\, d\underline{z} + \frac{\partial\theta}{\partial t}\, dt\right)\Bigg|_{\substack{t=t_f\\ \underline{z}=\underline{z}_f}} = 0 \tag{12. 12}$$

For the matching problem,

$$\theta = t - t_f \qquad\qquad d\theta = dt_f = 0$$

and for the minimum-delay problem,

$$\theta = z - z_f \qquad\qquad d\theta = dz_f = 0$$

12. 3. 2 Sensitivity Calculation (t_f fixed)

The key to calculation of the sensitivity $dP/d\underline{p}$ is to include dP in the Tellegen sum of (12. 1). This is done by augmenting the network N with an "objector" network [10. 13]. This consists of a controlled source-capacitor combination connected as shown in Figure 12. 7a where the controlled source has a value $h(\underline{z}, \underline{p})$ and the capacitor has zero initial conditions. From the branch equation of a capacitor, the capacitor voltage of Figure 12. 7a at $t = t_f$ is precisely the performance function of (12. 9). We can therefore determine the sensitivity $dP/d\underline{p}$ by (1) constructing the adjoint of the augmented network and (2) determining the associated Tellegen sum.

The adjoint representation of the capacitor already derived, only the nonlinear controlled source requires study. This is a nonlinear two-port which may be dependent on $\underline{p}$. From Equation (5. 29), which involves a nonlinear port adjoint determination, we conclude that the augmented network contains a reversed, linearized controlled source driving the output terminals

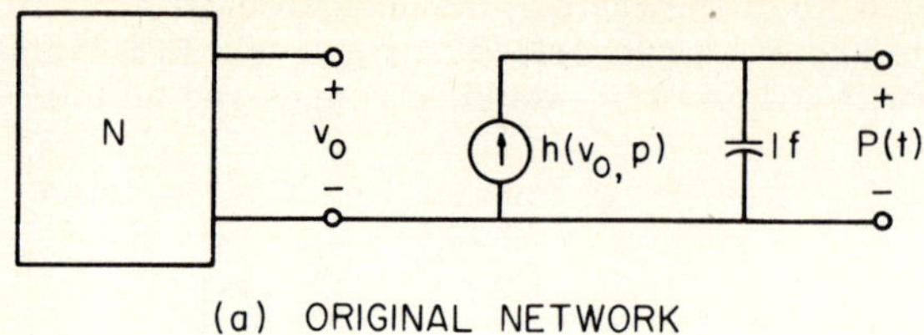

(a) ORIGINAL NETWORK

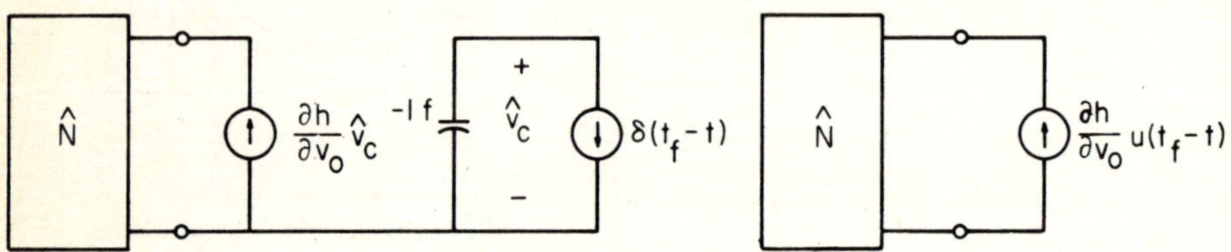

(b) EQUIVALENT ADJOINT NETWORK REPRESENTATION

FIG. 12.7 Network Interpretation of Integral Performance Function

of $\hat{N}$ (Figure 12.7b). Further, this source is controlled by the capacitor voltage that results from integration of a unit current impulse applied at $t = t_f$, i.e., a unit step beginning at $t = t_f$ and extending backward in time to $t = 0$. For example, if $h = \frac{1}{2} e^2 = \frac{1}{2} (v_o - \hat{v}_o)^2$ as in (12.10), then the error $e = v_o - \hat{v}_o)^2$ is "played back" to the network $\hat{N}$ beginning at $t = t_f$ after the solution of N itself.

The Tellegen sum associated with the augmented network has several distinguishing characteristics.

(1) Since P is represented as the output voltage, a term $(dP)\hat{i}_o = (dP)\,\delta\,(t_f - t)$ will appear in the sum in place of $dv_o\delta(t_f - t)$. Upon integration of the Tellegen sum, this will yield simply $dP(t_f)$.

(2) If h is a function of $\underline{p}$, then the controlled source must be regarded as parameter-dependent nonlinear two port. Using the methods of Chapter V, it is easily shown that a term of the form $(\partial h/\delta \underline{p})\,d\underline{p}$ will appear in the Tellegen sum. Integration of this yields

$$\left[\int_0^{t_f} \left(\frac{\partial h}{\partial \underline{p}}\right) dt\right] d\underline{p} = \sum_{j=1}^{r} \left[\int_0^{t_f} \frac{\partial h}{\partial p_j}\, dt\right] dp_j$$

<u>Example 12.2</u> We return to the problem of Example 12.1, but we replace the calculation of $dv_o(t_f)/d\underline{p}$ with the calculation of $dP/d\underline{p}$ where

$$P = \int_0^{t_f} (\tfrac{1}{2}Cv_c^2(t) - E_c(t))^2 dt$$

$$= \int_0^{t_f} h(v_c, C)dt$$

This integral has a minimum value of zero when the energy stored in capacitor is equal to a prescribed value $E_c(t)$ for $0 \le t \le t_f$.

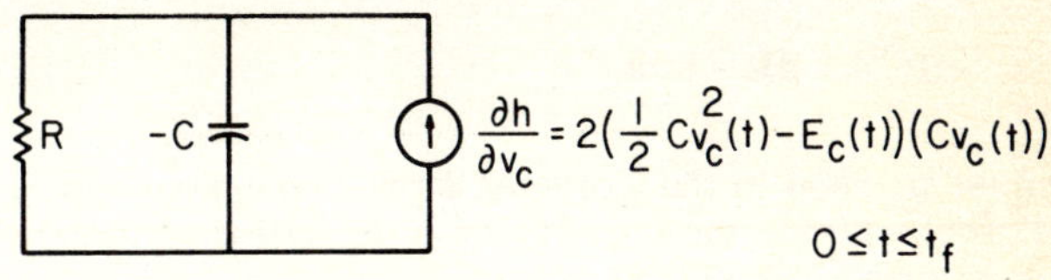

FIG. 12. 8 Adjoint Excited by Error Signal

The adjoint network has the form of Figure 12. 8; it is solved backward in time from t_f after the original network has been analyzed to generate $v_c(t)$. The Tellegen sum of (12. 2) now has the form

$$dv_R \hat{i}_R - di_R \hat{v}_R + dP\delta(t_f - t) + dv_c \hat{i}_c - \frac{d}{dt}(df_c)\hat{v}_c + \frac{\partial h}{\partial c} dC$$

where

$$\frac{\partial h}{\partial C} = 2(\tfrac{1}{2} C v_c^2 - E_c)(\tfrac{1}{2} v_c^2)$$

Then Equations (12. 4) - (12. 5) become

$$\frac{dP}{dR} = -\hat{v}_c \left. \frac{\partial f_c}{\partial v_c} \right|_{t=0} \frac{dv_c(0)}{dR} - \int_0^{t_f} i_R \hat{i}_R \, dt \tag{12. 13}$$

$$\frac{dP}{dC} = \hat{v}_c \left. \frac{\partial f_c}{\partial v_c} \right|_{t=0} \frac{dv_c(0)}{dC} + \int_0^{t_f} \hat{v}_c \dot{v}_c \, dt - \int_0^{t_f} \frac{\partial h}{\partial C} \, dt \tag{12. 14}$$

The calculation of the derivative dP/dR remains the same. Of course, the currents and voltages of $\hat{N}$ are different because it is excited as in Figure 12. 8 rather than with an impulse.

12. 3. 3 Sensitivity Calculation (General)

When t_f is variable as in Figure 12. 6, the differential dt_f enters the Tellegen sum in the following way. The total change in P has two components

$$dP = \frac{\partial P}{\partial t_f} dt_f + dP^{tel}$$

$$= h(\underline{z}_f, \underline{p})dt_f + dP^{tel}$$

$$= h_f dt_f + dP^{tel}$$

Here dP^{tel} is the change that would occur in P if t_f were constant. Since the Tellegen sum relates changes only in network parameters, the additional change in P due to the nonnetwork parameter t_f is not represented. Thus,

$$dP^{tel} = dP - h_f dt_f \tag{12.15}$$

must be used in the Tellegen sum, thus introducing dt_f into the summation. Unfortunately, the expression $h_f dt_f$ will then survive setting $dp_j = 0$ $(j \neq k)$, thus apparently foiling our previous procedures for computing dP/dp_k.

We overcome this problem by letting $\hat{v}_{c_i}(t_f) \neq 0$ in (12.3). This exposes the expression

$$-\hat{v}_{c_i}(\partial f_{c_i}/\partial v_{c_i})dv_{c_i}^{tel}\Big|_{t=t_f} \tag{12.16}$$

for all capacitive (or inductive) elements. The term $dv_{c_i}^{tel}$ again does not include the variation in v_{c_i} due to dt_f. The expression of (12.16) therefore becomes

$$-\hat{v}_{c_i}(\partial f_{c_i}/\partial v_{c_i})\left(dv_{c_i} - \frac{\partial v_{c_i}}{\partial t}dt_f\right)\Big|_{t=t_f}$$

$$= -\hat{v}_{c_i}(\partial f_{c_i}/\partial v_{c_i})(dv_{c_i} - \dot{v}_{c_i}dt_f)\Big|_{t=t_f} \tag{12.17}$$

Next we define

$$\underline{x} = \begin{bmatrix} v_{c_i} \\ \cdot \\ \cdot \\ \cdot \\ v_{c_n} \end{bmatrix} \quad \underline{\lambda} = \begin{bmatrix} \hat{v}_{c_i} \\ \cdot \\ \cdot \\ \hat{v}_{c_n} \end{bmatrix} \qquad \underset{\sim}{S} = \text{diag}\,[\,\partial f_{c_i}/\partial v_{c_i} \cdot \cdot\, \partial f_{c_n}/\partial v_{c_n}\,]$$

Now we collect all the new terms in the Tellegen sum, which we force to zero to eliminate the effect of dt_f, viz,

$$[\,(-h + \underline{\lambda}^T \underset{\sim}{S}\dot{\underline{x}})\,dt_f - \underline{\lambda}^T \underset{\sim}{S}\,d\underline{x}\,]_{t=t_f} = 0 \tag{12.18}$$

This expression may be combined with the target function differential relationship

$$d\theta\Big|_{t=t_f} = \left(\frac{\partial \theta}{\partial \underline{z}} \frac{\partial \underline{z}}{\partial \underline{x}} d\underline{x} + \frac{\partial \theta}{\partial t} dt_f\right)_{t=t_f} = 0$$

to yield

$$\underline{\lambda}(t_f) = \left\{ \frac{\underset{\sim}{S}^{-1}\left(\frac{\partial \theta}{\partial \underline{z}} \frac{\partial \underline{z}}{\partial \underline{x}}\right)^T h}{\left(\frac{\partial \theta}{\partial \underline{z}} \frac{\partial \underline{z}}{\partial \underline{x}}\right)\dot{\underline{x}} + \frac{\partial \theta}{\partial t_f}} \right\}_{t = t_f} \tag{12.19}$$

This relationship gives the final state of the adjoint network. For example, the matching problem, with $\theta = t - t_f$, yields $\partial\theta/\partial\underline{z} = 0$ and $\underline{\lambda}(t_f) = \underline{0}$. This agrees with our previous results. The minimum delay problem, with $\theta = z_o - z_{of}$, gives

$$\underline{\lambda}(t_f) = \left\{ \frac{\underset{\sim}{S}^{-1}\left(\frac{\partial z_o}{\partial \underline{x}}\right)^T}{\left(\frac{\partial z_o}{\partial \underline{x}}\right)\dot{\underline{x}}} \right\}_{t = t_f} \tag{12.20}$$

Both $\underset{\sim}{S}^{-1}$ and $\dot{\underline{x}}$ are readily determined at $t = t_f$. The calculation of the partial derivative $(\partial z_o/\partial\underline{x})_{t = t_f}$ is considered in Problem 12. 1.

12. 4 Example

An outstanding feature of the adjoint network model is that the manner of formulation and numerical solution of its descriptive equation is not specified. The adoption of a particular solution method can be based on numerical, programming, or, in the case of large networks, available computer hardware considerations. In the following example, node analysis is used, together with Gear's multistep implicit integration (Chapter IX) and sparse matrix solution methods (Chapter X). Also, for reasons explained in Chapter IX, capacitor charge rather than voltage is used as a state variable set. See references [12. 6] [1. 1] for other examples.

Example 12. 3 (Modeling [12. 8] Consider the circuit of Figure 12. 9a. It is proposed to adjust the parameters of the Ebers-Moll model until the response of the collector current i_L to the voltage step shown matches a given measured or idealized response. In particular, the model parameters are α_f, α_r, $\omega_{\alpha f}$,

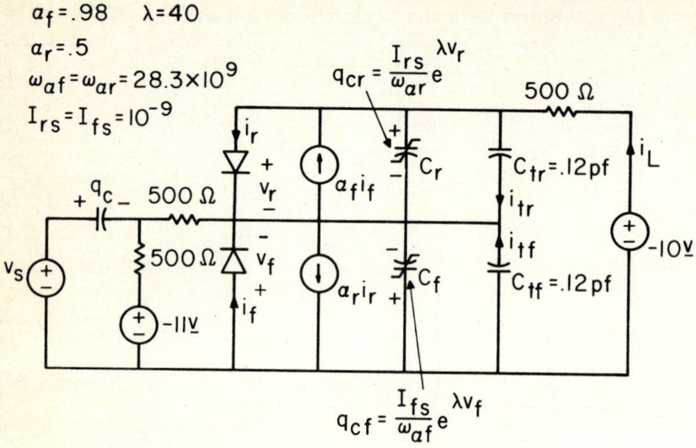

(a) SWITCHING CIRCUIT NETWORK MODEL

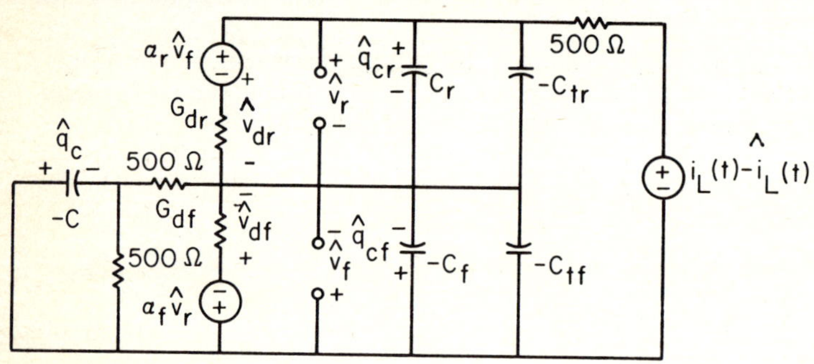

(b) ADJOINT NETWORK MODEL

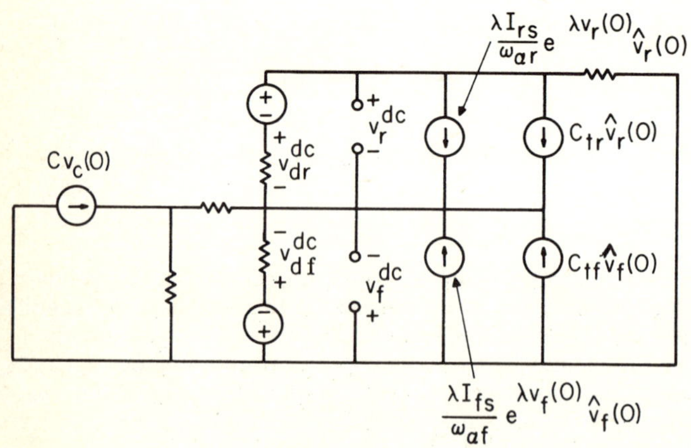

(c) DC SENSITIVITY NETWORK MODEL

FIG. 12.9 Network Models for Sensitivity Calculation

$\omega_{\alpha r}$, I_{fs}, C_{tr}, and C_{tf}. The design criterion is defined as

$$P = \int_0^{2000} \tfrac{1}{2}(i_L(t) - \hat{i}_L(t))^2 dt$$

where $\hat{i}_L$ is prescribed.

The adjoint network is obtained from the original network by following the branch replacement rules previously outlined. Of particular interest are the nonlinear capacitors, which have the charge representation

$$q_{cf} = \frac{I_{fs}}{\omega_{\alpha f}} e^{\lambda v_f} = f_c(v_f)$$

The representation of this element in the adjoint network is given by

$$\hat{i}_c = -\frac{\partial f_c}{\partial v_f} \frac{d\hat{v}_f}{dt}$$

$$= -\frac{\lambda I_{fs}}{\omega_{\alpha f}} e^{qv_f/kT} \frac{d\hat{v}_f}{dt}$$

The normalized sensitivities are calculated from network variables in N and $\hat{N}$ as follows:

$$I_{fs}\frac{\partial P}{\partial I_{fs}} = I_{fs}\hat{v}^{dc}_{df} \frac{\partial i_f(0)}{\partial I_{fs}} + I_{fs}\int_0^{t_f} \hat{v}_{df}(t) \frac{\partial i_f}{\partial I_{fs}} dt + I_{fs}\int_0^{t_f} \hat{v}_f(t) \frac{d}{dt}\left(\frac{\partial q_{cf}}{\partial I_{fs}}\right) dt$$

$$= \hat{v}^{dc}_{df} i_f(0) + \int_0^{t_f} \hat{v}_{df}(t) i_f(t)dt + \int_0^{t_f} \hat{v}_f(t) i_{cf}(t) dt$$

$$I_{rs}\frac{\partial P}{\partial I_{rs}} = \hat{v}^{dc}_{dr} i_r(0) + \int_0^{t_f} \hat{v}_{dr}(t)i_r(t)dt + \int_0^{t_f} \hat{v}_r(t)i_{cr}(t)dt$$

$$\alpha_f \frac{\partial P}{\partial \alpha_f} = \alpha_f \hat{v}^{dc}_r i_f(0) + A_f \int_0^{t_f} i_f(t)\hat{v}_r(t)dt$$

$$\alpha_r \frac{\partial P}{\partial \alpha_r} = \alpha_r \hat{v}^{dc}_f i_r(0) + \alpha_e \int_0^{t_f} i_r(t)\hat{v}_f(t)dt$$

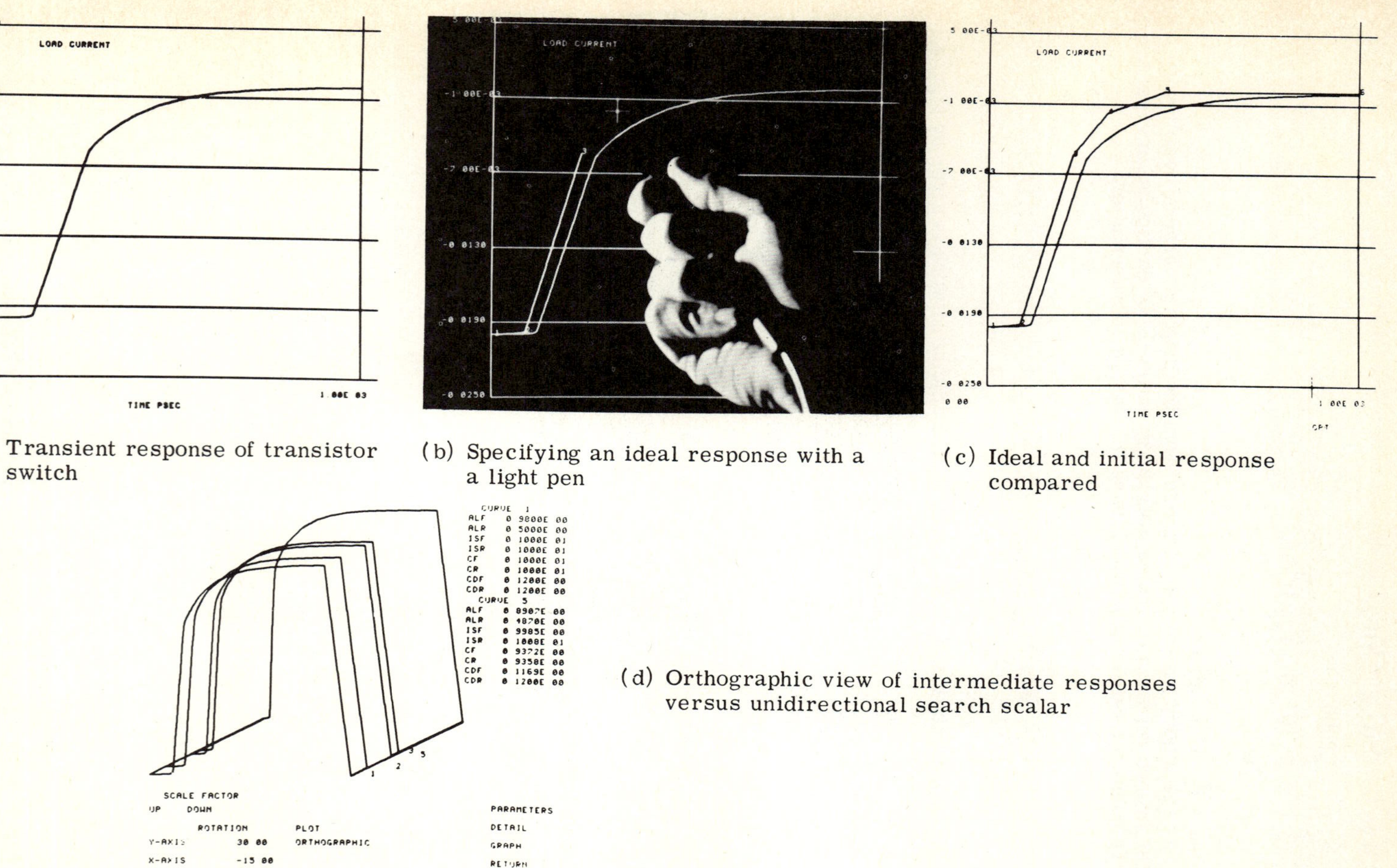

(a) Transient response of transistor switch

(b) Specifying an ideal response with a a light pen

(c) Ideal and initial response compared

(d) Orthographic view of intermediate responses versus unidirectional search scalar

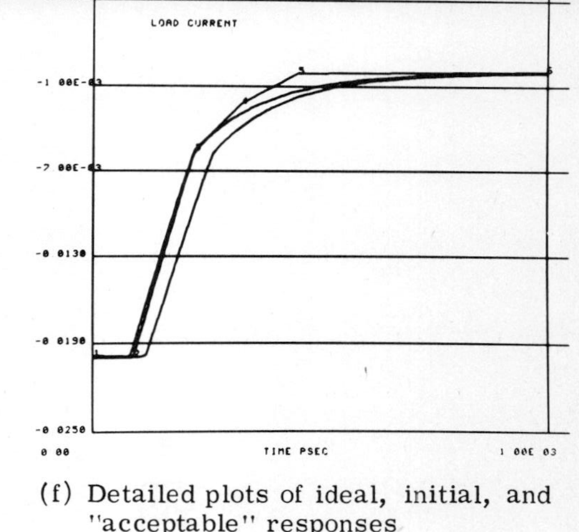

(e) Light pen selection of "acceptable" response for detailed plotting

(f) Detailed plots of ideal, initial, and "acceptable" responses

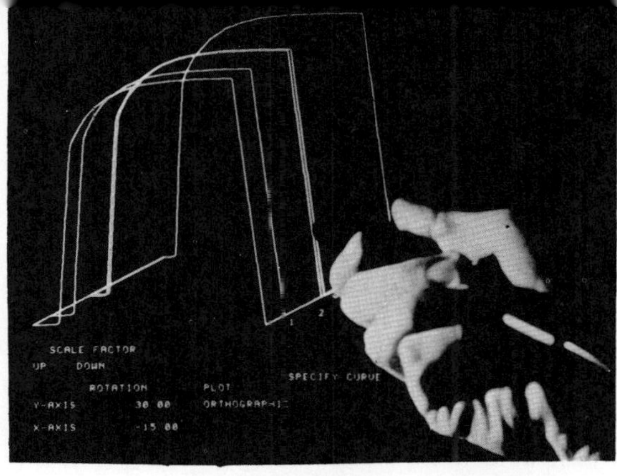

(g) Results of defining a steeper ideal response

FIG. 12.10. Steps in Solution of Modeling Problem

$$C_{tf}\,\frac{\partial P}{\partial C_{tf}} = \int_0^{t_f} \hat{v}_f(t)\,i_{tf}(t)\,dt$$

$$C_{tr}\,\frac{\partial P}{\partial C_{tr}} = \int_0^{t_f} \hat{v}_r(t)\,i_{tr}(t)\,dt$$

$$\omega_{\alpha f}\,\frac{\partial P}{\partial \omega_{\alpha f}} = \omega_{\alpha f}\int_0^{t_f} \hat{v}_f(t)\,\frac{d}{df}\left(\frac{\partial q_{cf}}{\partial \omega_{\alpha f}}\right) df$$

$$= -\int_0^{t_f} \hat{v}_f(t)\,i_{cf}(t)\,dt$$

$$\omega_{\alpha r}\,\frac{\partial P}{\partial \omega_{\alpha r}} = -\int_0^{t_f} \hat{v}_r(t)\,i_{cr}(t)\,dt$$

The solution of the problem has been programmed so that (1) the desired (or measured) output $i_L(t)$ can be entered through a graphical display with a light pen, and (2) the intermediate steps in the iteration procedure can be displayed. In this way, the designer can prematurely terminate the optimization process if he judges any intermediate response acceptable for his design purpose.

In Figure 12.10a the load current $i_L(t)$ is shown as the circuit switches from the saturated to the cutoff state. The designer then enters a piece-wise linear approximation to an ideal characteristic by light pen (Figure 12.10b). After the appropriate gradients are calculated, a search is made opposite to the gradient direction. The load current i_L is plotted in perspective* at each step in the search (Figure 12.10b). The designer can stop the iteration by designating an acceptable response with the light pen (Figure 12.10b). The result of choosing one of the curves in Figure 12.10a is shown in Figure 12.10f, and the initial and final parameter values are given in Table 12.1. The transistor α_f is seen to be the most influential parameter, due to the desired and initial responses having approximately the same slope. If the slope of the desired i_L is now changed and the optimization repeated, the characteristic of Figure 12.10g is obtained. The new parameter values (Table 12.1) show a dramatic change in the transition capacitance C_{tr}, a quite reasonable result.

*The third dimension in the search scalar α of Chapter XI.

TABLE 12.1 Parameter Values in Transistor Modeling Problem

	Initial Value	After First Optimization	After Second Optimization
α_f	.98	.898	.925
α_r	.5	.487	.451
$\omega_{\alpha f}$	28.3×10^9 rps	30.3×10^9 rps	31.2×10^9 rps
$\omega_{\alpha r}$	28.3×10^9 rps	30.3×10^9 rps	31.2×10^9 rps
I_{fs}	10^{-9} amp	10^{-9} amp	10^{-9} amp
I_{rs}	10^{-9} amp	10^{-9} amp	10^{-9} amp
C_{tf}	.12 pf	.117 pf	.089 pf
C_{tr}	.12 pf	.12 pf	.115 pf

12.5 Implementation on the Backward Integration

Most of the numerical and programming considerations relating to the backward integration are discussed in [1.1]. We will discuss only the numerical aspects of the convolution procedure.

The convolution of network variables required for the evaluation of sensitivities can proceed simultaneously with the backward integration itself by interpreting the integrand as the derivative of a variable ξ [12.7]. However, this process can be remarkably ill-conditioned. For example, as the optimum is approached, the gradient will approach zero. However, the error signal that is exciting $\hat{N}$ in the backward integration is in general nonzero at the optimum. This implies that the value of the integral will approach zero even

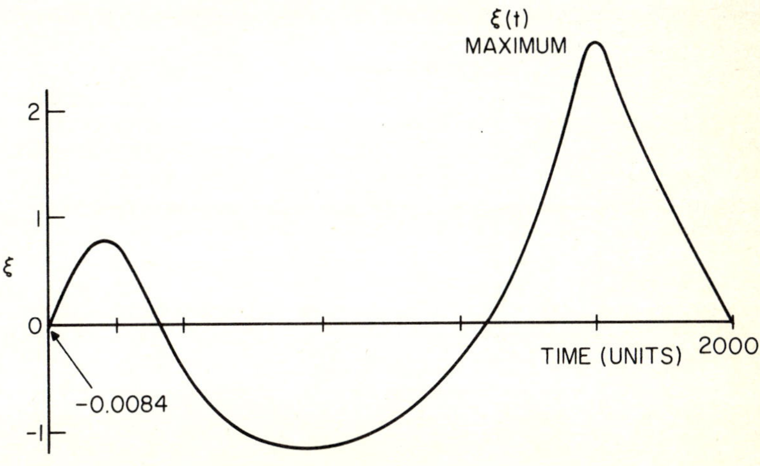

FIG. 12.11 Backward Integration Step

though the integrand will not. This in turn requires that the integrand change sign and that cancellation take place during the integration (convolution). Figure 12. 11 shows a typical form for $\xi(t)$ during the backward integration. $\xi(0)$, the value of the convolution integral, is more than two orders of magnitude smaller than $\xi(t)_{max}$.

Problems

12. 1 The problem of determining the partial derivative $(\partial v_o/\partial \underline{x})_{t=t_f}$ in (12. 20) can be regarded as the calculation of the sensitivity of an output v_o to a sequence of voltage sources of value v_{c_i}. Describe a calculation procedure consistent with the Gear multistep procedure, assuming that the node equations N and $\hat{N}$ are available.

12. 2 Construct the adjoint of the lumped model of a lossless transmission line. Suggest from this adjoint model the adjoint model of a continuous line (see Chapter IX).

REFERENCES

12. 1 Hachtel, G. D., and R. A. Rohrer, "Design and Synthesis of Switching Circuits," Proc. IEEE, vol. 55, no. 11, pp. 1864-1876; November 1967.

12. 2 Director, S. W., and R. A. Rohrer, "A Generalized Adjoint Network and Network Sensitivities," Trans. IEEE, vol. CT-16, pp. 300-336; August 1969.

12. 3 Tomovic, Rajho, Sensitivity Analysis of Dynamic Systems, McGraw-Hill; 1963.

12. 4 Kokotovic, P., and J. E. Heller, "Direct and Adjoint Sensitivity Equations for Parameter Optimization," Trans. IEEE, vol. AC-12, pp. 609-610; October 1967.

12. 5 Ho, C. W., "Time-Domain Sensitivity Computation for Networks Containing Transmission Lines," Trans. IEEE, vol. CT-18, no. 1, pp. 197-199; January 1971.

12. 6 Russo, P. M., and R. A. Rohrer, "Computer Optimization of the Transient Response of an ECL Gate," Trans. IEEE, vol. CT-18, no. 1, pp. 197-199; January 1971.

12. 7 Calahan, D. A., "Optimization of Switching Circuits," Proc. Cornell Conf. on Computerized Electronics, (Ithaca, N. Y.), pp. 282-292; August 1969.

12. 8 Calahan, D. A., "Display-Based Linear and Nonlinear Circuit Design: Interaction versus Automation," Proc. of the Thirteenth Midwest Symposium on Circuit Theory, University of Minnesota; May 1970.

Appendix A

Proofs of Tellegen's Theorems

Consider a network graph having n_b branches and n nodes. Let (v_k, i_k) be the complete set of branch voltages and currents ($k = 1, 2, \ldots n_b$) with the reference directions of Figure A1a, but otherwise arbitrary. Then Tellegen's (simple) theorem requires that

$$\sum_{k=1}^{n_b} v_k i_k = 0 \tag{A1}$$

Proof A. The proof begins by representing branch connections between <u>all</u> nodes (including self-incident branches at all nodes), where appropriate currents are in fact zero in Eq. (A1) (Figure A1c). Let the total number of branches be $n_{b'}$, so that

$$\sum_{k=1}^{n_b} v_k i_k = \sum_{k=1}^{n_{b'}} v_k i_k$$

Since one and only one branch exists between every pair of nodes, this summation over all branches can be replaced by a summation over all nodes. For this purpose, the branch currents and voltages are resubscripted as shown in Figure A1e, where the node voltages are measured with respect to

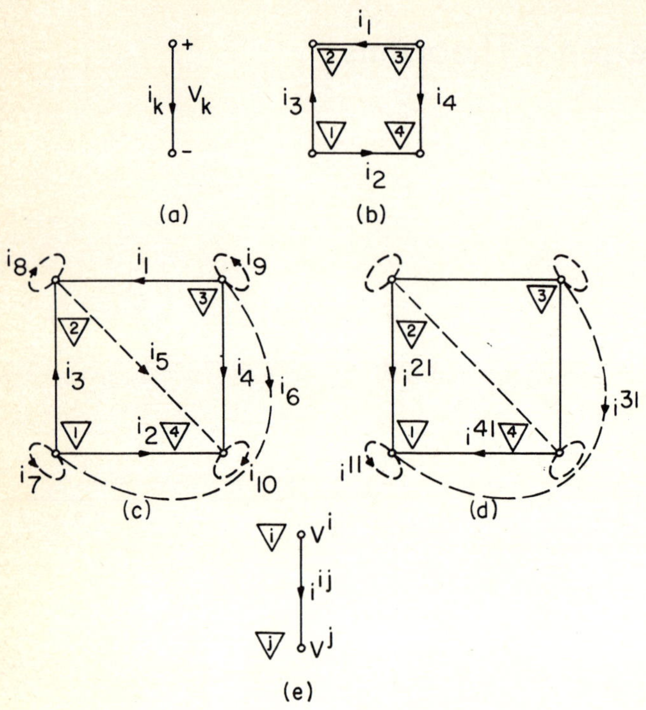

FIG. A1

an arbitrary reference. The v_k i_k summation of branches incident at the j-th node is then

$$\sum_{i=1}^{n} (v^i - v^j)\, i^{ij}$$

For example, at node 1 in Figure A1d, this sum is

$$\begin{aligned}(v^1 - v^1)i^{11} + (v^3 - v^1)i^{21} + (v^3 - v^1)i^{31} + (v^4 - v^1)i^{41} &= (v_7)(i_7) + (-v_3)(-i_3) \\ &\quad + (v_6)(i_6) + (-v_2)(-i_2) \\ &= v_7 i_7 + v_3 i_3 + v_6 i_6 + v_2 i_2 \end{aligned} \tag{A2}$$

Note that the signs of the v_k i_k terms of Eq. (A2) are always positive. If we sum over all nodes (j = 1, 2,.. n), then it is clear that each v_k i_k product will be represented twice, so that

$$\sum_{k=1}^{n_b'} v_k\, i_k = \frac{1}{2} \sum_{j=1}^{n} \sum_{i=1}^{n} (v^i - v^j)\, i^{ij}$$

$$= \frac{1}{2} \sum_{i=1}^{n} v^i \left(\sum_{j=1}^{n} i^{ij} \right) + \frac{1}{2} \sum_{j=1}^{n} v^j \left(\sum_{i=1}^{n} i^{ji} \right) \tag{A3}$$

where the indentity $i^{ij} = -i^{ji}$ has been used in the second term. But the two current summations in (A3) are both zero from KCL, since they represent the total currents flowing into nodes j and i respectively. Therefore,

$$\sum_{k=1}^{n_b} v_k i_k = 0 \tag{A4}$$

Equation (A4) is not surprising on physical grounds, since it represents the conservation of instantaneous power in a circuit. However, inspection of Eq. (A4) and its derivation reveals a rather remarkable fact: there is no requirement that the branch current i^{ij} and the voltages v^i and v^j be in the same graph! The voltage can be defined in one graph and the current in another, so long as both graphs have the same directed structure. In fact, if we permit representation of open and short circuited branches, then the two graphs need only have the same number of nodes. This gives rise to Tellegen's (extended) theorem of Eq. (5.4).

Proof B. An alternative proof of Tellegen's extended theorem follows from the fundamental cut set and fundamental loop matrices (Chapter VIII), which satisfy

$$[\underset{\sim}{I} \ \underset{\sim}{F}] \begin{bmatrix} \underline{i}_t \\ \underline{i}_\ell \end{bmatrix} = \underline{0} \qquad [-\underset{\sim}{F}^T \ \underset{\sim}{I}] \begin{bmatrix} \hat{\underline{v}}_t \\ \hat{\underline{v}}_\ell \end{bmatrix} = \underline{0}$$

Here we represent voltages with a "$\wedge$" to indicate that the loop equations apply to any graph of the same directed structure as the graph yielding the cut set equations.

Collecting all tree and link currents and voltages into vectors $\underline{i}_b$ and $\hat{\underline{v}}_b$, we have

$$\underline{i}_b \triangleq \begin{bmatrix} \underline{i}_t \\ \underline{i}_\ell \end{bmatrix} = \begin{bmatrix} -\underset{\sim}{F} \\ \underset{\sim}{I} \end{bmatrix} \underline{i}_\ell \qquad \hat{\underline{v}}_b \triangleq \begin{bmatrix} \hat{\underline{v}}_t \\ \hat{\underline{v}}_\ell \end{bmatrix} = \begin{bmatrix} \underset{\sim}{I} \\ \underset{\sim}{F}^T \end{bmatrix} \hat{\underline{v}}_t$$

Then

$$\sum_{k=1}^{n_b} \hat{v}_k i_k = \hat{\underline{v}}_b^T \underline{i}_b = \hat{\underline{v}}_t^T [\underset{\sim}{I} \ \underset{\sim}{F}] \begin{bmatrix} -\underset{\sim}{F} \\ \underset{\sim}{I} \end{bmatrix} \underline{i}_t = 0$$

Appendix B

Input Data Format for Circuit Analysis Programs (RCAP, DCAP, TCAP)

(a) A name card: cols. 1-60 will be printed on output sheet.

(b) Cards describing branches of the network. Cards may be in any order.

cols. 1-4: Branch name (left justified begining with
- V - voltage source (DCAP, TCAP)
- I - current source (RCAP, DCAP, TCAP)
- C - capacitance (TCAP)
- R - resistance (RCAP, DCAP, TCAP)
- L - inductance (TCAP)
- D - exponential diode (DCAP, TCAP)

cols. 5: Type of control (V or I) if controlled source; otherwise blank. Only VCCS is allowed in RCAP; only VCCS and CCS are allowed in general.

cols. 6-9: Name of controlling branch (left justified) if controlled source; otherwise blank.

cols. 11-12 and 14-15: Node numbers, establishing the following polarity (for all branches, including sources):

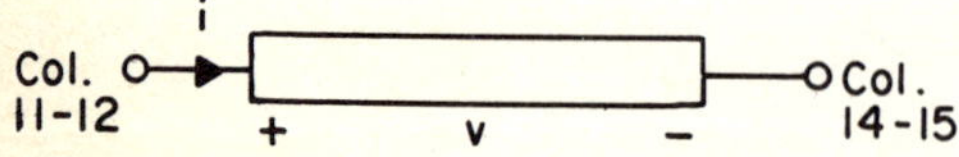

The numbering must be consecutive, beginning with 00, but ordering is arbitrary.

col. 17: Polarity of output; blank for plus, - for minus.
col. 18: If branch voltage is to ce considered an output quantity, use symbol V.
cols. 20-30: Value of element or strength of sources, or initial diode voltage.
cols. 31-40, 41-50,... 71-80: Values of stepped elements or stepped strengths of stepped controlled sources; all stepped elements must be stepped the same number of times; stepped element values may not be continued on succeeding cards. The last stepped value of an element may not be zero.

(c) A card denoting transient response information (TCAP). If a step response is desired.
cols. 1-4: TIME
cols. 6-9: Name of input source branch
cols. 10-12: Total number of time data points number is right justified with no decimal point*
cols. 31-40: Maximum time

(d) Last card
cols. 1-4: AMEN

---

*This number includes only one end point.

Appendix C

Input Data Format for DSAP, DCOP

(a) A name card; cols. 1-60 will printed on output sheet.

(b) Cards describing branches of the network. (Cards may be in any order.)

cols. 1-4: Branch name (left justified) beginning with

V-voltage source
I-current source
R-resistor
D-exponential diode
O-(in DCOP) output branch; allows specifications to be entered beyond col. 20.

col. 5: Type of control (V or I) if controlled source; blank. Only VCCS and CCCS are allowed.

cols. 6-9: Name of controlling branch (left justified) if controlled source; otherwise blank.

cols. 11-12 and 14-15: Node numbers, establishing the polarity illustrated in Appendix B. The numbering must be consecutive, beginning with 00, but ordering is arbitrary.

col. 17: Polarity of output + or -); blank is considered +.

col. 18: If branch voltage is an output, use V symbol.

col. 20: Any alphanumeric character will cause sensitivity to be calculated w. r. t. the element parameter in DSAP; in DCOP, a character identifies a parameter.

cols. 21-30: Value of element or strength of source, or initial diode voltage.

cols. 31-40, 41-50,... 71-80: Values of stepped elements or stepped strengths of stepped controlled sources; all stepped elements must be stepped the same number of times; the last stepped value of an element may not be zero.

(c) Temperature card.

This card can be used to redefine temperature-dependent diode saturation currents and λ's. If this card is absent a temperature of 290°K (17°C) is used; this corresponds to a λ of 40 and an I_S of 10^{-9}.

cols. 1-4: TEMP

cols. 21-30, 31 40, etc.: stepped temperature values

(d) Last card

cols. 1-4: AMEN.

Programs

Listings of the following programs are given in the text.

Name	Appendix Table	Page Reference
DRNET	3.1	63
DNET	3.2	68
TNET	4.1	88
DSNET	5.2	129

Copies of the following programs can be obtained by sending a 9-track tape with at least 800 BPI rating to the author.

Name	Appendix Table	Page Reference
RNET	2.1	22
MCAP	2.2	40
PEEL	2.3	47
RCAP	2.4	50
DCAP	3.3	72
SPLINE	3.4	77

Name	Appendix Table	Page Reference
TCAP	4.2	94
MSEN	5.1	123
DSAP	5.3	130
DCOP	6.1	155

```
C****  PROGRAM OF APPENDIX TABLE 3.1
C****  RESISTOR-DIODE EXAMPLE PROGRAM (DRNET)
C****  READ SATURATION CURRENT IN NANOAMPS,RESISTANCE IN OHM
       READ(5,15)CSAT,R
       CSAT=CSAT*1.E-9
       WRITE(6,8)CSAT,R
C****  INITIALIZE ITERATION PARAMETERS
       M=0
       TMAX=6.2832
       T=TMAX/20.
       TIME=0.
       WRITE(6,9)
C****  GUESS INITIAL DIODE VOLTAGE
       VD=.6
 7     DVD=1.E60
C****  UPDATE VS
       VS=1.7*(1.+.2*SIN(TIME))
C****  ENTER NEWTON RAPHSON ITERATION
       GO TO 1
C****  MODIFY THE FOLLOWING STATEMENT
3      DVD=(-R*CSAT*(EXP1-1)-VD+VS)/(1+40.*R*CSAT*EXP1)
       VD=VD+DVD
       M=M+1
 1     EXP1=EXP(40.*VD)
C****  CONVERGENCE CRITERION
       IF(M.LE.15)GO TO 2
       WRITE(6,11)M
       GO TO 12
 2     IF(ABS(DVD).GT.1.E-6)GO TO 3
 12    CUR=CSAT*(EXP1-1.)
       WRITE(6,4)TIME,VS,CUR
       IF(TIME.GT.TMAX)GO TO 5
       M=0
       TIME=TIME+T
       GO TO 7
 5     STOP
 9     FORMAT(2X,'TIME         SOURCE       CURRENT')
 4     FORMAT(3E12.4)
 8     FORMAT(2X,'CSAT CURRENT ='E12.4,2X,'RESISTANCE ='E12.4)
 11    FORMAT(2X,'NO CONVERGENCE AFTER ',I2,'ITERATIONS')
 15    FORMAT(2F10.0)
       END
1.         50.
```

```
C****  PROGRAM OF APPENDIX TABLE 3.2
C****  DC ANALYSIS OF TRANSISTOR AMPLIFIER (DNET)
      IMPLICIT REAL*8(A-H,O-Z)
      REAL*8 G(10,10),I(10),V(10)
      REAL*8 LAMDA
      DATA CSAT/1.D-6/,LAMDA/40.D0/
      NDIM=10
C****  READ INPUT DATA
    1 M=0
      CALL READ(R1,R2,RE,RL,E,ALF,VDI,NN)
      DO 4 J=1,NN
    4 V(J)=0.D0
C****  SET UP NODE EQUATIONS
    3 CALL NODEQ(R1,R2,RE,RL,E,ALF,VDI,G,V,I,NN,NDIM)
C****  SOLVE NODE EQUATIONS
      CALL FCTR(G,NN,NDIM)
      CALL BKSB(G,NN,I,NDIM)
C****  V(J)-I(J) IS THE INCREMENT IN THE JTH NODE VOLTAGE
      ERR=0.D0
      DO 2 J=1,NN
      ERR=ERR+(V(J)-I(J))**2
    2 V(J)=I(J)
C****  CALCULATE AND PRINT OUTPUT VOLTAGE
      CUR=CSAT*(DEXP(LAMDA*(V(1)-V(2)))-1)
      VCE=V(3)-V(2)
      VBE=V(1)-V(2)
      WRITE(6,5) M,VBE,VCE,CUR
      M=M+1
C****  TEST CONVERGENCE CRITERION
      IF(ERR.GT.1.D-6.AND.M.LT.15)GO TO 3
      STOP
5     FORMAT(' ITER NO.=',I3,' VBE =',F7.5,' VCE=',F7.5,
     1' IC=',F7.5)
      END
      SUBROUTINE READ(R1,R2,RE,RL,E,ALF,VDI,NN)
      IMPLICIT REAL*8(A-H,O-Z)
      DIMENSION NODOUT(2,1)
      READ(5,1)R1,R2,RE,RL,E,ALF,VDI
      READ(5,3)NN
      WRITE(6,2)R1,R2,RE,RL,E,ALF
      RETURN
    1 FORMAT(6F10.0)
    2 FORMAT(' R1 =',F10.3,' R2 =',F10.3,' RE =',F10.3/
     1' RL =',F10.3,' E =',F10.3,' ALF ='.F10.3)
3     FORMAT(I3)
      END
      SUBROUTINE NODEQ(R1,R2,RE,RL,E,ALF,VDI,G,V,I,NN,NDIM)
      IMPLICIT REAL*8 (A-H,O-Z)
      REAL*8 G(NDIM,1),I(1),V(1),LAMDA
      DATA INIT/0/
```

```
      DATA CSAT/1.D-6/,LAMDA/40.D0/
C****  ZERO NODE MATRIX AND CURRENT VECTOR
      DO 1 K=1,NN
      I(K)=0.D0
      DO 1 J=1,NN
    1 G(J,K)=0.D0
C****  CALCULATE DIODE PARAMETERS
      VDIODE=V(1)-V(2)
C****  APPLY INITIAL DIODE VOLTAGE IF ON FIRST ITERATION
      IF(INIT.EQ.0)VDIODE=VDI
      DEX=DEXP(LAMDA*VDIODE)
      GD=LAMDA*CSAT*DEX
      CUR=CSAT*(DEX-1.D0)-GD*VDIODE
C****  LOAD NODE MATRIX
      G(1,2)=-GD*(1.D0-ALF)
      G(1,3)=-1.D0/R1
      G(1,1)=-G(1,2)-G(1,3)+1.D0/R2
      G(2,1)=-GD
      G(3,1)=ALF*GD-1.D0/R1
      G(2,2)=GD+1.D0/RE
      G(3,2)=-ALF*GD
      G(3,3)=1.D0/RL+1.D0/R1
C****  LOAD CURRENT VECTOR
      I(1)=(ALF-1.D0)*CUR
      I(2)=CUR
      I(3)=-ALF*CUR+E/RL
      INIT=1
      RETURN
      END
      SUBROUTINE FCTR(A,N,NDIM)
      IMPLICIT REAL*8(A-H,O-Z)
      DIMENSION A(NDIM,1)
      I=0
    1 I=I+1
      IP1=I+1
      IM1=I-1
C****  FORM ROW OF U
      DO 2 L=IP1,N
      IF(IM1.EQ.0)GO TO 2
      DO 3 LL=1,IM1
    3 A(I,L)=A(I,L)-A(I,LL)*A(LL,L)
    2 A(I,L)=A(I,L)/A(I,I)
      L=IP1
      LM1=L-1
C****  FORM COLUMN OF L
      DO 4 II=L,N
      DO 4 LL=1,LM1
    4 A(II,L)=A(II,L)-A(II,LL)*A(LL,L)
      IF(L.EQ.N)RETURN
      GO TO 1
```

```
      END
      SUBROUTINE BKSB(A,N,B,NDIM)
      IMPLICIT REAL*8(A-H,O-Z)
      DIMENSION A(NDIM,1),B(1)
C****  SOLVE L*Y=B
      DO 2 K=1,N
      KM1=K-1
      IF(KM1.EQ.0)GO TO 2
      DO 3 I=1,KM1
 3    B(K)=B(K)-A(K,I)*B(I)
 2    B(K)=B(K)/A(K,K)
      NM1=N-1
C****  SOLVE U*X=Y
      DO 30 KK=1,NM1
      K=N-KK
      KP1=K+1
      DO 30 JJ=KP1,N
   30 B(K)=B(K)-A(K,JJ)*B(JJ)
      RETURN
      END
   20.       3.        .3          5.      10.        .98
.4
  3
```

```
C****  PROGRAM OF APPENDIX TABLE 4.1
C****  TRANSIENT ANALYSIS OF TRANSISTOR AMPLIFIER (TNET)
       IMPLICIT REAL*8(A-H,O-Z)
       REAL*8 G(10,10),I(10),V(10),VSAV(10),INPUT
       DIMENSION NODOUT(2,5)
       COMMON TMAX,NUMPTS
       NDIM=10
     1 M=0
       N=0
C****  READ INPUT DATA
       CALL READ(R1,R2,RE,RL,E,ALF,VDI,CE,INPUT,
      1NN,NODOUT,NOUT)
       T=TMAX/NUMPTS
       DO 6 J=1,NN
     6 V(J)=0.D0
     7 TIME=N*T
C****  SET UP NODE EQUATIONS
     3 CALL NODEQ(R1,R2,RE,RL,E,ALF,VDI,CE,INPUT,
      1G,V,I,NN,NDIM,VSAV,TIME,T)
C****  SOLVE NODE EQUATIONS
       CALL FCTR(G,NN,NDIM)
       CALL BKSB(G,NN,I,NDIM)
C****  V(J)-I(J) IS THE INCREMENT IN THE JTH NODE VOLTAGE
       ERR=0.D0
       DO 2 J=1,NN
       ERR=ERR+(V(J)-I(J))**2
     2 V(J)=I(J)
C****  TEST CONVERGENCE CRITERION
       IF(ERR.GT.1.D-6.AND.M.LT.15)GO TO 3
C****  CALCULATE AND PRINT OUTPUT VOLTAGES
       WRITE(6,4)TIME
       CALL OUTPUT(NODOUT,NOUT,V)
       IF(TIME.GE.TMAX)STOP
C****  STORE NODE VOLTAGES FOR UPDATING CAPACITORS
       DO 5 J=1,NN
     5 VSAV(J)=V(J)
       M=0
       N=N+1
       GO TO 7
     4 FORMAT(' TIME =',F10.6)
       END
       SUBROUTINE READ(R1,R2,RE,RL,E,ALF,VDI,CE,INPUT,
      1NN,NODOUT,NOUT)
       IMPLICIT REAL*8(A-H,O-Z)
       REAL*8 INPUT
       DIMENSION NODOUT(2,1)
       COMMON TMAX,NUMPTS
       READ(5,1)R1,R2,RE,RL,E,ALF,VDI,CE,INPUT,TMAX,NUMPTS
       READ(5,3)NN,NOUT,(NODOUT(1,J),NODOUT(2,J),J=1,NOUT)
       WRITE(6,2)R1,R2,RE,RL,E,ALF,CE,INPUT
```

```
      RETURN
    1 FORMAT(6F10.0/(4F10.0,I2))
    2 FORMAT(' R1 =',F10.4,' R2 =',F10.4,' RE =',F10.4/
     1' RL =',F10.4,' E =',F10.4,' ALF =',F10.4/
     2' CE =',F10.8,' INPUT =',F10.4)
    3 FORMAT(10I3)
      END
      SUBROUTINE NODEQ(R1,R2,RE,RL,E,ALF,VDI,CE,INPUT,
     1G,V,I,NN,NDIM,VSAV,TIME,T)
      IMPLICIT REAL*8 (A-H,O-Z)
      REAL*8 G(NDIM,1),I(1),V(1),LAMDA,VSAV(1),INPUT
      DATA INIT/0/
      DATA CSAT/1.D-9/,LAMDA/40.D0/
C****  ZERO NODE MATRIX AND CURRENT VECTOR
      DO 1 K=1,NN
      I(K)=0.D0
      DO 1 J=1,NN
    1 G(J,K)=0.D0
C****  CALCULATE DIODE PARAMETERS
      VDIODE=V(1)-V(2)
C****  APPLY INITIAL DIODE VOLTAGE IF ON FIRST ITERATION
      IF(INIT.EQ.0)VDIODE=VDI
      DEX=DEXP(LAMDA*VDIODE)
      GD=LAMDA*CSAT*DEX
      CUR=CSAT*(DEX-1.D0)-GD*VDIODE
C****  LOAD NODE MATRIX
      G(1,2)=-GD*(1.D0-ALF)
      G(1,3)=-1.D0/R1
      G(1,1)=-G(1,2)-G(1,3)+1.D0/R2
      G(2,1)=-GD
      G(3,1)=ALF*GD-1.D0/R1
      G(2,2)=GD+1.D0/RE
      G(3,2)=-ALF*GD
      G(3,3)=1.D0/RL+1.D0/R1
      IF(TIME.NE.0.D0)G(2,2)=G(2,2)+CE/T
C****  LOAD CURRENT VECTOR
      I(1)=(ALF-1.D0)*CUR
      I(2)=CUR
      I(3)=-ALF*CUR+E/RL
      IF(TIME.EQ.0.D0)GO TO 2
      I(1)=I(1)+INPUT*DSIN(1.D4*TIME)
      I(2)=I(2)+CE*VSAV(2)/T
    2 INIT=1
      RETURN
      END
      SUBROUTINE OUTPUT(NODOUT,NOUT,V)
      IMPLICIT REAL*8 (A-H,O-Z)
      DIMENSION NODOUT(2,1),V(1)
      DO 2 I=1,NOUT
      VOUT=0.D0
```

```
      N1=NODOUT(1,I)
      N2=NODOUT(2,I)
      IF(N1.NE.0)VOUT=VOUT+V(N1)
      IF(N2.NE.0)VOUT=VOUT-V(N2)
      WRITE(6,3)N1,N2,VOUT
    2 CONTINUE
      RETURN
    3 FORMAT('OUTPUT NODES =',2I3,'  OUTPUT VOLTAGE =',F15.7)
      END
      SUBROUTINE FCTR(A,N,NDIM)
      IMPLICIT REAL*8(A-H,O-Z)
      DIMENSION A(NDIM,1)
      I=0
    1 I=I+1
      IP1=I+1
      IM1=I-1
C****  FORM ROW OF U
      DO 2 L=IP1,N
      IF(IM1.EQ.0)GO TO 2
      DO 3 LL=1,IM1
    3 A(I,L)=A(I,L)-A(I,LL)*A(LL,L)
    2 A(I,L)=A(I,L)/A(I,I)
      L=IP1
      LM1=L-1
C****  FORM COLUMN OF L
      DO 4 II=L,N
      DO 4 LL=1,LM1
    4 A(II,L)=A(II,L)-A(II,LL)*A(LL,L)
      IF(L.EQ.N)RETURN
      GO TO 1
      END
      SUBROUTINE BKSB(A,N,B,NDIM)
      IMPLICIT REAL*8(A-H,O-Z)
      DIMENSION A(NDIM,1),B(1)
C****  SOLVE L*Y=B
      DO 2 K=1,N
      KM1=K-1
      IF(KM1.EQ.0)GO TO 2
      DO 3 I=1,KM1
 3    B(K)=B(K)-A(K,I)*B(I)
 2    B(K)=B(K)/A(K,K)
      NM1=N-1
C****  SOLVE U*X=Y
      DO 30 KK=1,NM1
      K=N-KK
      KP1=K+1
      DO 30 JJ=KP1,N
   30 B(K)=B(K)-A(K,JJ)*B(JJ)
      RETURN
      END
```

```
20000.     3000.      300.      5000.      10.        .98
.4         .000001    .0001     .00031416  16
  3  2  3  1  1  2
```

```
C****  PROGRAM OF APPENDIX TABLE 5.2
C****  SENSITIVITY ANALYSIS OF BIAS NETWORK EXAMPLE (DSNET)
      IMPLICIT REAL*8(A-H,O-Z)
      REAL*8 G(10,10),I(10),V(10)
      DIMENSION NODOUT(2,5)
      NDIM=10
C****  READ INPUT DATA
    1 M=0
      CALL READ(R1,R2,RE,RL,E,ALF,VDI,NN,NODOUT,NOUT)
      DO 4 J=1,NN
    4 V(J)=0.D0
C****  SET UP NODE EQUATIONS
    3 CALL NODEQ(R1,R2,RE,RL,E,ALF,VDI,G,V,I,NN,NDIM)
C****  SOLVE NODE EQUATIONS
      CALL FCTR(G,NN,NDIM)
      CALL BKSB(G,NN,I,NDIM)
C****  V(J)-I(J) IS THE INCREMENT IN THE JTH NODE VOLTAGE
      ERR=0.D0
      DO 2 J=1,NN
      ERR=ERR+(V(J)-I(J))**2
    2 V(J)=I(J)
C****  CALCULATE AND PRINT OUTPUT VOLTAGE
      CALL OUTPUT(NODOUT,NOUT,V)
      M=M+1
C****  TEST CONVERGENCE CRITERION
      IF(ERR.GT.1.D-6.AND.M.LT.15)GO TO 3
      CALL SEN(G,NN,V,RE,ALF,RL,NDIM)
      STOP
      END
      SUBROUTINE READ(R1,R2,RE,RL,E,ALF,VDI,NN,NODOUT,NOUT)
      IMPLICIT REAL*8(A-H,O-Z)
      DIMENSION NODOUT(2,1)
      READ(5,1)R1,R2,RE,RL,E,ALF,VDI
      READ(5,3)NN,NOUT,(NODOUT(1,J),NODOUT(2,J),J=1,NOUT)
      WRITE(6,2)R1,R2,RE,RL,E,ALF
      RETURN
    1 FORMAT(6F10.0)
    2 FORMAT(' R1 =',F10.3,' R2 =',F10.3,' RE =',F10.3/
     1' RL =',F10.3,' E =',F10.3,' ALF =',F10.3)
    3 FORMAT(10I3)
      END
      SUBROUTINE SEN(G,NN,V,RE,ALF,RL,NDIM)
      IMPLICIT REAL*8(A-H,O-Z)
      REAL*8 G(NDIM,1),V(1),VADJ(3),LAMDA
      DATA CSAT/1.D-9/,LAMDA/40.D0/
C****  SET UP UNIT CURRENT SOURCE BETWEEN C AND E
      VADJ(1)=0.D0
      VADJ(2)=1.D0
      VADJ(3)=-1.D0
C****  BACK SOLVE ADJOINT EQUATION
```

```
      CALL BKSBT(G,NN,VADJ,NDIM)
C****  CALCULATE SENSITIVITIES
      SENRE=-V(2)*VADJ(2)/RE**2
      SENALF=CSAT*(DEXP(LAMDA*(V(1)-V(2)))-1.D0)*(VADJ(3)
     1-VADJ(1))
      T=290.D0
      VDADJ=VADJ(1)-VADJ(2)+ALF*(VADJ(3)-VADJ(1))
      SENT=-CSAT*LAMDA*(V(1)-V(2))*DEXP(LAMDA*(V(1)-V(2))).*VDADJ/
      SENE=-VADJ(3)/RL
      WRITE(6,13)SENRE,SENALF,SENT,SENE
 13   FORMAT(' RE SEN =',D12.4/' ALF SEN =',D12.4/
     1' TEMP SEN =',D12.4/' E SEN =',D12.4/)
      RETURN
      END
      SUBROUTINE NODEQ(R1,R2,RE,RL,E,ALF,VDI,G,V,I,NN,NDIM)
      IMPLICIT REAL*8 (A-H,O-Z)
      REAL*8 G(NDIM,1),I(1),V(1),LAMDA
      DATA INIT/0/
      DATA CSAT/1.D-9/,LAMDA/40.D0/
C****  ZERO NODE MATRIX AND CURRENT VECTOR
      DO 1 K=1,NN
      I(K)=0.D0
      DO 1 J=1,NN
    1 G(J,K)=0.D0
C****  CALCULATE DIODE PARAMETERS
      VDIODE=V(1)-V(2)
C****  APPLY INITIAL DIODE VOLTAGE IF ON FIRST ITERATION
      IF(INIT.EQ.0)VDIODE=VDI
      DEX=DEXP(LAMDA*VDIODE)
      GD=LAMDA*CSAT*DEX
      CUR=CSAT*(DEX-1.D0)-GD*VDIODE
C****  LOAD NODE MATRIX
      G(1,2)=-GD*(1.D0-ALF)
      G(1,3)=-1.D0/R1
      G(1,1)=-G(1,2)-G(1,3)+1.D0/R2
      G(2,1)=-GD
      G(3,1)=ALF*GD-1.D0/R1
      G(2,2)=GD+1.D0/RE
      G(3,2)=-ALF*GD
      G(3,3)=1.D0/RL+1.D0/R1
C****  LOAD CURRENT VECTOR
      I(1)=(ALF-1.D0)*CUR
      I(2)=CUR
      I(3)=-ALF*CUR+E/RL
      INIT=1
      RETURN
      END
      SUBROUTINE OUTPUT(NODOUT,NOUT,V)
      IMPLICIT REAL*8 (A-H,O-Z)
      DIMENSION NODOUT(2,1),V(1)
```

```
      DO 2 I=1,NOUT
      VOUT=0.D0
      N1=NODOUT(1,I)
      N2=NODOUT(2,I)
      IF(N1.NE.0)VOUT=VOUT+V(N1)
      IF(N2.NE.0)VOUT=VOUT-V(N2)
      WRITE(6,3)N1,N2,VOUT
    2 CONTINUE
      RETURN
    3 FORMAT('OUTPUT NODES =',2I3,'  OUTPUT VOLTAGE =',F15.7)
      END
      SUBROUTINE FCTR(A,N,NDIM)
      IMPLICIT REAL*8(A-H,O-Z)
      DIMENSION A(NDIM,1)
      I=0
    1 I=I+1
      IP1=I+1
      IM1=I-1
C****  FORM ROW OF U
      DO 2 L=IP1,N
      IF(IM1.EQ.0)GO TO 2
      DO 3 LL=1,IM1
    3 A(I,L)=A(I,L)-A(I,LL)*A(LL,L)
    2 A(I,L)=A(I,L)/A(I,I)
      L=IP1
      LM1=L-1
C****  FORM COLUMN OF L
      DO 4 II=L,N
      DO 4 LL=1,LM1
    4 A(II,L)=A(II,L)-A(II,LL)*A(LL,L)
      IF(L.EQ.N)RETURN
      GO TO 1
      END
      SUBROUTINE BKSB(A,N,B,NDIM)
      IMPLICIT REAL*8(A-H,O-Z)
      DIMENSION A(NDIM,1),B(1)
C****  SOLVE L*Y=B
      DO 2 K=1,N
      KM1=K-1
      IF(KM1.EQ.0)GO TO 2
      DO 3 I=1,KM1
 3    B(K)=B(K)-A(K,I)*B(I)
 2    B(K)=B(K)/A(K,K)
      NM1=N-1
C****  SOLVE U*X=Y
      DO 30 KK=1,NM1
      K=N-KK
      KP1=K+1
      DO 30 JJ=KP1,N
   30 B(K)=B(K)-A(K,JJ)*B(JJ)
```

```
      RETURN
      END
      SUBROUTINE BKSBT(A,N,B,NDIM)
      REAL*8 A,B
      DIMENSION A(NDIM,1),B(1)
C****  SOLVE UT*Y=B
      DO 3 K=2,N
      KM1=K-1
      DO 3 I=1,KM1
 3    B(K)=B(K)-A(I,K)*B(I)
      NM1=N-1
C****  SOLVE LT*X=Y
      DO 2 KK=1,N
      K=N-KK+1
      KP1=K+1
      IF(KP1.GT.N)GO TO 2
      DO 29 JJ=KP1,N
 29   B(K)=B(K)-A(JJ,K)*B(JJ)
 2    B(K)=B(K)/A(K,K)
      RETURN
      END
20000.    3000.     300.      5000.     10.       .98
.4
  3  2  3  1  1  2
```

Index